T0313844

Seabirds of the Farallon Islands

Seabirds

Contributors
David G. Ainley, Robert J. Boekelheide,
Malcolm C. Coulter, R. Philip Henderson, Harriet R. Huber,
T. James Lewis, Stephen H. Morrell,
Teresa M. Penniman, Larry B. Spear, Craig S. Strong

STANFORD UNIVERSITY PRESS

of the Farallon Islands

Ecology, Dynamics, and Structure of an Upwelling-System Community

Edited by

David G. Ainley and Robert J. Boekelheide

STANFORD, CALIFORNIA 1990

Stanford University Press
Stanford, California

Printed and bound by CPI Group (UK) Ltd,
Croydon, CR0 4YY

*This work was published with the assistance of
the Point Reyes Bird Observatory*

Acknowledgments

The efforts of a great many people made initiation and completion of this project possible. Not the least of these efforts were those of S. Goldhaber, who, with help from E. Tuomi, prepared the many drafts of the manuscript but, unlike many others, did not have the benefit of working in the field at that incredible place, Southeast Farallon Island.

No data would ever have been collected were it not for the existence of the Farallon Island Research Station of the Point Reyes Bird Observatory (PRBO). The station was established through the efforts and foresight of C. J. Ralph, L. R. Mewaldt, and R. Stallcup. C. Merrill organized and coordinated the Farallon Patrol through its infant years, and, with O. Fisher and T. Charkins, through many of the years that followed. J. Tasley worked out some basic long-term funding that anchored the Farallon effort through the vagaries of an otherwise hand-to-mouth existence. The cooperation and assistance given to our venture by the U.S. Fish and Wildlife Service (USFWS) was coordinated initially by R. Bauer and subsequently by personnel of the San Francisco Bay National Wildlife Refuges: R. Coleman, B. Crabb, T. Harvey, R. Johnson, E. Lindeman, R. Lowe, J. G. Moss, C. Osugi, R. Personius, W. Stieglitz, and J. Takekawa. Their work has been and continues to be greatly appreciated. The USFWS Office of Migratory Bird Management also contributed to our effort. The logistic and maintenance support given by the USFWS, U.S. Coast Guard, Group San Francisco, and the Farallon Patrol of the Oceanic Society, San Francisco Chapter, and by M. Whitt, W. Holdon, and E. Harrold, has been indispensable. Use of the National Oceanic and Atmospheric Administration (NOAA) Ship *David Starr Jordan*, to census seabirds in 1985 and 1986, was a priceless asset to our efforts; we appreciate the professional expertise of the ship's officers and crew, which aided our efforts immensely.

Our long-term efforts, of course, depended heavily on the generous financial support of PRBO members and a number of agencies and foundations: Atlantic Richfield Foundation, C. E. Merrill Trust, Chevron USA, Compton Foundation, Inc., David and Lucile Packard Foundation, Dean Witter Foundation, Exxon USA Foundation, Foremost-McKesson Foundation, Inc., Giles W. and Elise G. Mead Foundation, Lurline B. Roth Charity Foundation, Marcia Brady Tucker Foundation, Marine Mammal Commission, Minerals Management Service, National Marine Fisheries Service, National Weather Service, Sanctuary Programs Office (NOAA), Standard Oil Company of California, S. H. Cowell Foundation, Texaco, The San Francisco Foundation, The Sierra Club Foundation, The Upjohn California Fund, Union Oil of California Foundation, U.S. Fish and Wildlife Service, and William R. Hewlett Foundation.

PRBO researchers and volunteers contributed many hours to the field effort, especially P. Abbott, G. Ainley, E. Akers, S. Allison, C. Annable, J. Arnold, L. Astheimer, B. Bainbridge, S. Barbour, M. Bradstreet, H. Carter, R. Clark, W. Clow, C. Coleman, K. Darling, L. Doerflinger, J. Feldman, R. Ferris, T. Fowler, M. Fry, W. Harrington-Tweit, T. Harvey, M. Heidt, J. Higbee, H. Hill, S. Johnston, J. Kelly, R. LeValley, B. Lewis, B. Manion, B. Manolis, T. Manolis, D. Manuwal, L. Meyers, D. Nelson, J. Nisbet, J. Nusbaum, D. O'Keefe, K. Opiate, T. Parmenter, L. Parrish, W. Parsons, S. Peaslee, J. Penniman, C. Peterson, S. Peterson, E. Piccolo, R. Pierotti, A. Rovetta, P. Sawyer, K. Schafer, D. Smith, N. Spear, S. Speich, E. Strauss, M. Sundove, C. Swarth, J. Swenson, W. Sydeman, B. Webb, D. Whitacre, M. Whitt, and J. Young. The PRBO administrative staff helped to coordinate logistics and funding: S. Allen, J. Bauser, A. Bruce, J. Church, D. A. Clarke, D. B. Clarke, H. Crandall, P. Daley, C. Gerstman, B. Heneman, J. Hesterly, K. Kuhl, D. McCrimmon, J. Smail, H. Strong, and O'B. Young. J. Hesterly, P. Geiss, C. Ribic, and D. Rugg helped in some of the computer analyses. K. Hamilton, I. Gaffney, and J. Penniman prepared the figures and maps. A. Bakun of the National Marine Fisheries Service (NMFS), D. Manuwal, and A. Thoresen kindly provided unpublished data. The information and discussion of fish and other seabird prey offered by NMFS personnel has been most valuable; we are particularly indebted to P. Adams, T. Wyllie Echeverria, W. Lenarz, A. MacCall, R. Parrish, S. Smith, and D. Woodbury. Helpful comments on various drafts of all or portions of the manuscript were provided by D. Boersma, H. Carter, J. Hand, K. Kuletz, W. Lenarz, A. MacCall, D. Manuwal, D. Nelson, J. Piatt, R. Pierotti, D. Siegel-Causey, K. Vermeer, and especially A. Gaston and G. Hunt. The painstaking care of W. Carver and P. Psoinos, Editor and Associate Editor, respectively, at Stanford University Press, and of V. Iorio and P. Unitt, as copy editors, was vital to the production of the finished manuscript. The efforts by V. Iorio and P. Psoinos were remarkable.

This is contribution 358 of the Point Reyes Bird Observatory.

Point Reyes, September 1988

D.G.A.
R.J.B.

Contents

Contributors

The project upon which this book is based was conceived and initiated by David G. Ainley, and was overseen in the field by him and, subsequently, Robert J. Boekelheide. All contributors were Farallon Biologists of the Point Reyes Bird Observatory, except for two, Coulter and Spear, who were Research Associates and volunteer workers. The dates of involvement as Farallon Biologist or Volunteer are given, respectively, for each person below.

DAVID G. AINLEY (1971–77) is Director of Marine Studies at Point Reyes Bird Observatory. He received his Ph.D., with a dissertation on the breeding behavior of antarctic penguins, from Johns Hopkins University. His research interests involve seabird and pinniped population dynamics and ecology, particularly with respect to food-web dynamics.

ROBERT J. BOEKELHEIDE (1973–74, 1979–84) is teaching high school science in the state of Washington. He received his M.S., with a thesis on the breeding ecology of terns in the Arctic, from the University of California, Davis.

MALCOLM C. COULTER (1970–75) heads a research program on Wood Storks at the Savannah River Ecology Laboratory, South Carolina. He received his M.S. from Oxford University and Ph.D. from the University of Pennsylvania, and in both cases his degree research dealt with the ecology of gulls on the Farallon Islands. He engages intermittently in research at the Farallones and the Galápagos Islands.

R. PHILIP HENDERSON (1974–88) is a biologist and computer specialist at Point Reyes Bird Observatory, and works part-time as a biologist on the Farallones. He received a B.S. degree from the University of California, Berkeley. His main research interest involves the phenology of migrant landbirds on the Farallon Islands.

Harriet R. Huber (1974–86) is a biologist at the National Marine Mammal Laboratory in Seattle. She received her B.S. from the University of California, Berkeley. Her research interests concern the population dynamics of seals and sea lions.

T. James Lewis (1970–76) is the operator of a nature tour company in Costa Rica. He received his B.S. from the University of Iowa, and spent several years conducting seabird research in the Central Pacific for the Smithsonian Institution before coming to the Observatory.

Stephen H. Morrell (1972–73, 1976–79) is a high-school teacher in the state of Washington. He received his M.S., with a thesis on the ecology of island Kit Foxes, from the University of California, Santa Barbara. He has conducted research on seabirds in the Antarctic and Alaska, and has been a commercial fisherman as well.

Teresa M. Penniman (1980–85) is a contract biologist living in New Mexico. She received her B.S. from Evergreen State College. Before coming to the Farallon program she spent several years in Alaska studying seabirds in the Outer Continental Shelf Environmental Assessment Program.

Larry B. Spear (1978–88) is a Research Associate and part-time biologist at Point Reyes Bird Observatory. He received his M.S., with a thesis on the population ecology of Farallon gulls, from Moss Landing Marine Laboratories (California State University). His research interests continue with gulls, as well as the ecology of tropical seabirds.

Craig S. Strong (1978–81) is completing his M.S. at Moss Landing Marine Laboratories (California State University); his thesis concerns the ecology of baleen whales in the Sea of Cortez. Besides cetaceans, his research interests concern the place of higher vertebrates in marine food-web dynamics.

Seabirds of the Farallon Islands

1

Introduction

David G. Ainley

The Farallon Islands, lying just 35 km offshore from San Francisco, California, are the home for a large and remarkably diverse assemblage of seabird species. The word "farallon" is Spanish (plural, "farallones") for "rock rising from the sea," and on these particular small islands and rocks, which together comprise somewhat more than 100 acres (46 hectares), nest two species of storm-petrels (family Oceanitidae), three species of cormorants (family Phalacrocoracidae), a gull (family Laridae), and five species of auks (family Alcidae). Such a diverse assemblage of seabirds on so small a parcel of land is not duplicated in many places in the world, even on many larger parcels, and certainly not within thousands of kilometers of central California. The birds are numerous, as well. Sir Francis Drake, after visiting the islands in 1579, wrote that they "held plentiful and great store of seals and birds" (from Doughty 1971). Today, still, on these stark rocks just outside one of the busiest ports on the North American West Coast, hundreds of thousands of seabirds occur year round or return annually to attempt breeding and to raise their young.

Our 15-year study of the 11-member Farallon seabird community, a study that began in earnest in 1971, was inspired by two Russian classics of the marine bird literature: *The Bird Bazaars of Novaya Zemlya*, by S. M. Uspenski, and *The Ecology of Sea Colony Birds of the Barents Sea*, by L. O. Belopol'skii. Though published in English as long ago as 1958 and 1961, respectively, and in spite of the awakening Golden Age of seabird research in the 1960's and 1970's, these books, particularly the latter, stand apart as detailed, long-term attempts at making order out of the seeming chaos of entire, large, and complex seabird nesting communities, or bazaars as they are known in the Arctic of the Eastern Hemisphere. To be sure, countless studies have since appeared that detail different aspects of the lives of single species or species groups, or patch together series of studies completed over time at single locations, and, as a result, much has been learned about seabirds. Nevertheless, variation among species in their response to environmental conditions, and interactions within the entire spectrum of species within a community, are best perceived from simultaneous studies of many species together, over extended periods. The following book is the result of our attempt to accomplish this at the Farallones.

From the outset, the main purpose of our project was to investigate, comparatively, how fluctuations in the marine environment and the availability of food resources affected variability in the breeding biology of the members of the Farallon nesting community. Belopol'skii noted that the incidence and success of breeding of most of the Barents Sea seabirds were keyed to the appearance of a few prey species each year. The prey base for these birds appeared to be much more diverse both before and after the breeding season (see also Bédard 1969a,b), as well as during an occasional year when breeding failed. Such insight rarely has been attained, because most studies before or since have been confined to only one or two years, and few have considered both feeding ecology and breeding biology simultaneously. One study that has come close, it too a study of arctic breeding colonies, is that by Hunt and co-workers (Hunt, Burgeson, and Sanger 1981, Hunt, Eppley, and Drury 1981, Hunt, Eppley, and Schneider 1986, Schneider and Hunt 1984). We thought we could complement and contrast the arctic studies through the study of a rookery in a rich, upwelling-based system that brings temperate sea temperatures to subtropical latitudes. As Belopol'skii had done, we would continue our comparative ecological research for a series of successive years.

As a result of the infusion of cold water to rather low latitudes by the process of upwelling (see Chapter 2), the mix of species constituting the Farallon avifauna has an unusual variety of geographic affinities (see Haley 1984 for distribution maps). Leach's Storm-Petrel, *Oceanodroma leucorhoa*, nests on islands from subtropical to subarctic waters around the northern rim of the Pacific Basin, but as the most highly migratory of the Farallon seabirds, it winters along the Equator. The Pelagic Cormorant, *Phalacrocorax pelagicus*, Common Murre, *Uria aalge*, Pigeon Guillemot, *Cepphus columba*, and Tufted Puffin, *Fratercula cirrhata*, nest from temperate to arctic waters, principally in subarctic waters, and reach the southernmost extent of their range either at the Farallones or not much farther south (i.e., within California). Despite their more northerly distributions, the Farallon populations of these species do not seem to move much at all during the nonbreeding season, except for the guillemot, which apparently migrates far northward, and the puffin, which moves out to sea. Cassin's Auklet, *Ptychoramphus aleuticus*, which occurs from northern subtropical waters to the subarctic, and Rhinoceros Auklet, *Cerorhinca monocerata* (actually a puffin), whose nesting distribution is shifted slightly northward of that of Cassin's, also move little during the nonbreeding season. The ranges of three other species, the Ashy Storm-Petrel, *Oceanodroma homochroa*, Brandt's Cormorant, *Phalacrocorax penicillatus*, and Western Gull, *Larus occidentalis*, center around the California Current system, and these birds are resident in central California waters throughout the year. Finally, although the Double-crested Cormorant, *Phalacrocorax auritus*, is a bird of coastal and interior North America, scattered marine populations of this species, such as that at the Farallones, occur along the Pacific Coast. It too remains in central California waters the year round, although in estuarine rather than in oceanic habitats. Like Leach's Storm-Petrel and the Common Murre, the Double-crested Cormorant also occurs in Atlantic waters. Thus, the Farallon (and California Current) breeding avifauna comprises a mix of migratory and sedentary species; some are endemic to the region, whereas others are widespread and have more southern or more northern zoogeographic affinities. In contrast to this heterogeneity, most arctic breeding colonies are homogeneous in zoogeographic affinities, and most species move southward for the winter; most tropical seabird communities are also homogeneous, and most species reside year round in the waters around the localities where they breed.

Not only is the Farallon nesting community diverse, but the numbers of birds are large. Regional compilations of population estimates by Varoujean (1979) and Sowls et al. (1980) demonstrate the significant if not dominant contribution of Farallon numbers to West Coast seabird populations south of Alaska. Farallon breeding populations of the Ashy Storm-Petrel, Brandt's Cormorant, and the Western Gull are by far the largest for those species in all the world, and at only a few sites do more Pigeon Guillemots and Cassin's Auklets nest. As a consequence of such avifaunal diversity and size, the Farallones made an ideal location for our studies.

INVESTIGATION OF ECOLOGICAL ORGANIZATION

The question arises, and is often asked in a more general form by ecologists, How do 11 species of seabirds manage to coexist during the nesting season on this small group of rocks? Or, how do they segregate ecologically so that their niches do not overlap and they do not compete for the same resources? Or, do they compete for resources at all? These questions are especially pertinent within the three groups of taxonomically and morphologically similar species—the two storm-petrels, the three cormorants, and three of the alcids (the guillemot and the two puffins). These groups might be construed as being three ecological "guilds," i.e., organisms that exploit the same class of environmental resources in a similar way (Root 1967). Usually, ecologists separate guilds on the basis of one type of resource, generally space or food. We can, however, merge these two niche dimensions. The two storm-petrels are cavity-nesting, sea-surface gleaners of micronekton and zooplankton; the three cormorants are surface-nesting, deep-diving piscivores (and, though taxonomically different, the Common Murre is a fourth species in this category); and the three alcids are cavity-nesting, diving piscivores. Though Connor and Simberloff (1984) argued that properly defining a guild is in practice exceedingly difficult, and has rarely been done well, a wealth of studies (e.g., Whittam and Siegel-Causey 1981) points to the existence of these three seabird guilds at least as a starting point for further discussion.

In addition to complementing the arctic community studies with one of a temperate/subtropical avifauna, we had a second main purpose in pursuing the Farallon project. Inspired by Belopol'skii, as well as by Lack (1945, 1946), Bédard (1967), Cody (1968), Wiens (1969), and others (see J. Diamond and Case 1986), we hypothesized that some mechanism of niche separation organized (or structured, as an ecologist would say) the seeming chaos of the Farallon seabird bazaar, and we sought to investigate how ecological segregation is actually managed at the Farallones. Foremost in our minds was the role of competition for resources in the structuring process. Grant (1986) pointed out a need in ecology for studies from a variety of communities, studies that demonstrate how often competition for resources occurs, between which organisms, under what circumstances, for what duration, and with what consequences. Such studies are rare at present, but through the Farallon studies we have been able to address these issues for at least one seabird community. To do this we had to investigate the nesting and feeding niches of each species. Such a task had not before been accomplished for an entire temperate seabird community, and, in fact, at the time we began the project, had not been completed for any seabird community except that of the Barents Sea and to some degree another arctic community in the Chukchi Sea (Swartz 1966). Separation of feeding niches, but not of nesting niches, had been investigated for a tropical community (Ashmole and Ashmole 1967) and for an Atlantic temperate community (Pearson 1968). After we began our study, Cody (1973, 1974) described rigid segregation by feeding habitat in a temperate Pacific seabird community, a study since disputed by many seabird researchers (e.g., Bédard 1976). Then Croxall and Prince (1980) and Croxall (1984) described ecological segregation, mainly of food resources, in the subantarctic community at South Georgia, as did Hunt, Burgeson, and Sanger (1981), Hunt, Eppley, and Drury (1981), and Schneider and Hunt (1984) for the arctic Pribilof Islands. Adding another dimension to the Pribilof studies were the investigations of nesting habitat by Whittam and Siegel-Causey (1981) and Squibb and Hunt (1983).

A niche can be defined as the means by which a species exploits significant resources. In order for two or more species to coexist in the same place at the same time, each has to exploit the available resources in a different way. If they exploit them in the same way, then competition ensues. If the resource is limited, either one species disappears or one or both change behaviorally or morphologically until effective partitioning of the resource comes about. This is the process by which natural selection and evolution occur.

Many studies indicate that animal communities are structured by competition for, or allocation of, two especially important resources, space and food (J. Diamond and Case 1986, and papers therein). This seems true especially in vertebrate communities (Schoener 1986). Furthermore, Roughgarden (1986) has proposed that through competition for these resources, communities are assembled either by invasion or by the coevolution of species. In invasion-structured communities, species are assembled sequentially, and competition between the invader and those species already present determines whether coexistence will occur. In such communities, the result of species interaction is the disappearance of either the resident or the invading species, depending on which can most effectively exploit the available resources (niche). In coevolution-structured communities, invasion and coevolution alternate, so that both the resident and the invading species adjust their niches.

These patterns in species assemblages have been most easily observed where terrestrial or marine species occur in a limited habitat, such as islands or reefs. At larger spatial scales, these patterns have been more difficult to perceive, for a number of reasons. For one, most populations are not closed (as they often are for terrestrial creatures on islands); i.e., recruits come from elsewhere. For another, the mobility of the organisms subjects them to different environmental pressures in a number of areas. Finally, the environment itself varies over different space and time scales, and the resident species (and different populations thereof) exhibit different lags in response time, owing to such factors as generation time and degree of philopatry (propensity to return repeatedly to the same place, usually the place of birth) (Wiens 1986). Besides lacking community studies pursued for an adequate number of years, seabird ecologists, especially, by the nature of their study subjects, have had to cope with these factors, and thus the organizing principles governing the assembly of seabird species into communities have been elusive. In some instances competition for space has seemed to rule, whereas in others competition for food has seemed the more important factor (Furness and Birkhead 1984, Furness and Monaghan 1987).

Coexistence of species is usually mediated by differences in prey species, in food particle size, in foraging strategy, in micro- and macrohabitat selection, and in aggressive ability (Lack 1971, Cody 1974, J. H. Brown and Bowers 1984, J. Diamond 1986). Many of these niche dimensions, as they are referred to by ecologists, are a function of body size or the size of the feeding structure, which in the case of birds, of course, is the bill. Body size becomes important in the selection of nesting habitat (especially for cavity nesters) and in aggressive ability; and bill size relates closely to species and size of prey selected. The information in Figure 1.1 provides some basis for concluding that we should not expect complete overlap in the nesting and feeding niches of Farallon seabirds. Body size (as indicated by weight) and bill size both show an approximately even progression from smallest to largest.

Competition, if it occurs, should be most intense between species most similar in these measures of size, or those species similar in size should diverge in important ways unaffected by size. Similarity, or overlap, in size is measured by the size of one species divided by that of the other species. Ratios of overlap greater than 2.2 for body size or greater than 1.3 in bill length have been found to be characteristic of coexisting avian species in terrestrial ecosystems (see discussion in J. Diamond 1986). Diamond found in the avifauna of New Guinea that

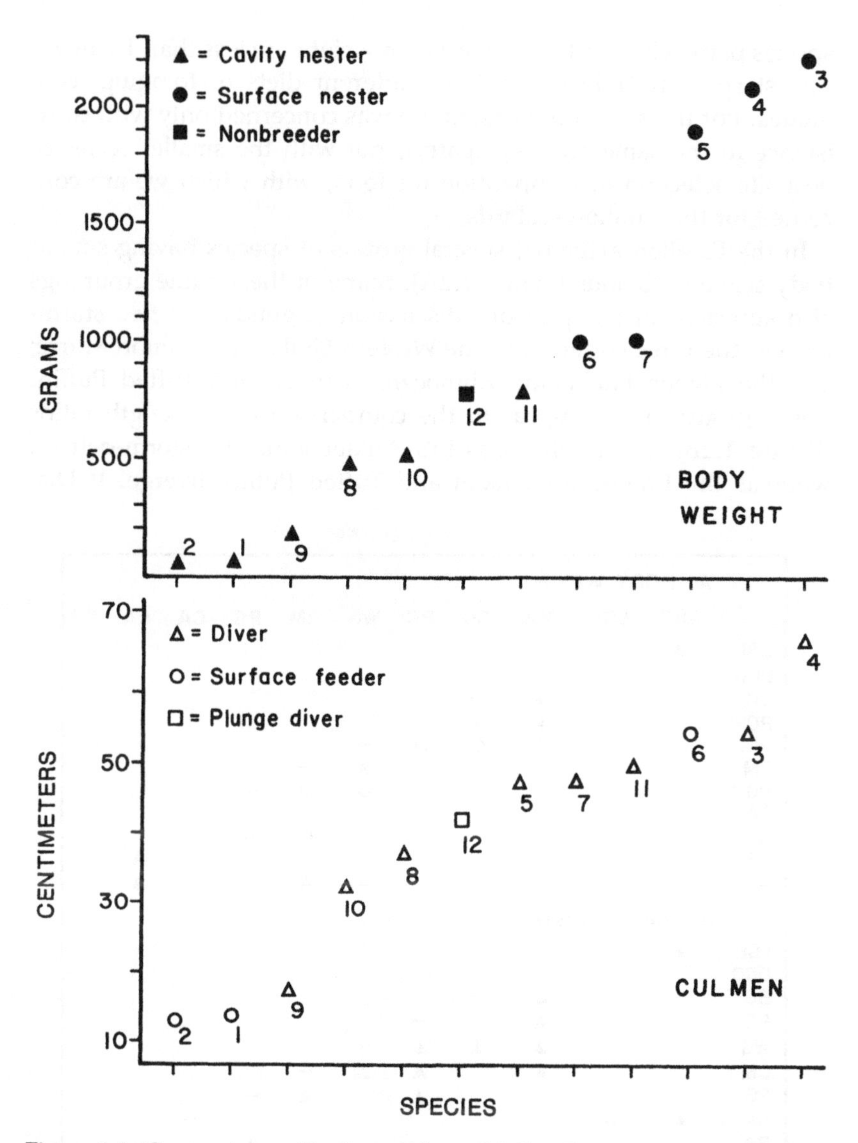

Figure 1.1. Comparison of body weight and bill (culmen) length in Farallon seabirds: 1, Leach's Storm-Petrel (LSP); 2, Ashy Storm-Petrel (ASP); 3, Double-crested Cormorant (DCC); 4, Brandt's Cormorant (BC); 5, Pelagic Cormorant (PC); 6, Western Gull (WG); 7, Common Murre (CM); 8, Pigeon Guillemot (PG); 9, Cassin's Auklet (CA); 10, Rhinoceros Auklet (RA); 11, Tufted Puffin (TP); 12, an abundant summer visitor, Sooty Shearwater (SS). Measurements are averaged from Ridgway (1919), Storer (1952), Palmer (1962), and Dunning (1984), as well as from a few recent specimens.

species pairs whose ratios of overlap in weight are less than 1.8 never coexist spatially unless they have different diets or foraging techniques. For the spatial dimension, he was concerned only with coexistence in the same area (sympatry), not with the smaller scales of nest-site selection or competition for food, with which we are concerned for the Farallon seabirds.

In the Farallon avifauna, several groups of species having similar body size are obvious (Figure 1.2A). Some of these same groupings also appeared in our previous discussion of guilds: the two storm-petrels; the three cormorants; the Western Gull and Common Murre; and the Pigeon Guillemot, Rhinoceros Auklet, and Tufted Puffin. These groups again appear in the comparison of bill-length ratios (Figure 1.2B), except that Cassin's Auklet joins the storm-petrels, whereas the Rhinoceros Auklet and Tufted Puffin diverge. If Dia-

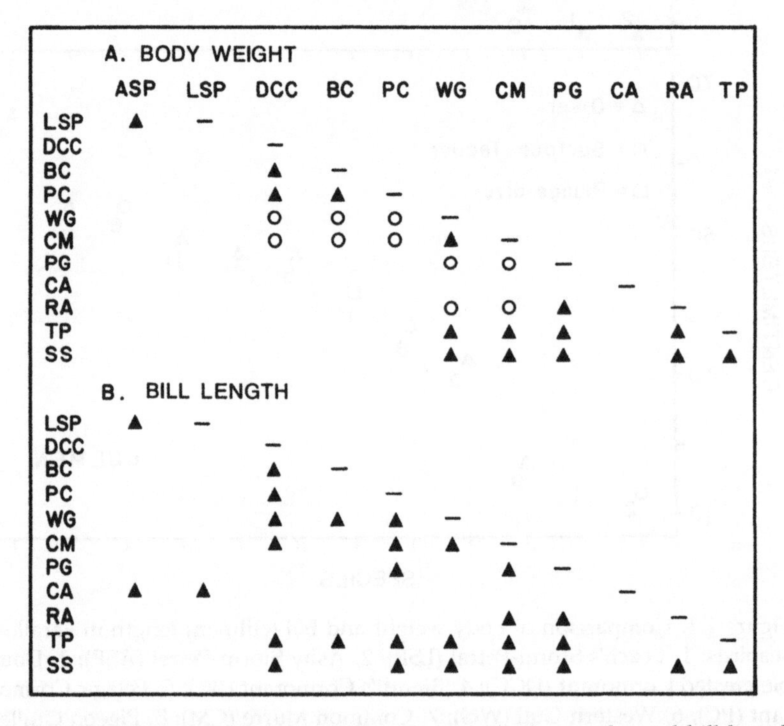

A. BODY WEIGHT

	ASP	LSP	DCC	BC	PC	WG	CM	PG	CA	RA	TP
LSP	▲	—									
DCC			—								
BC			▲	—							
PC			▲	▲	—						
WG			○	○	○	—					
CM			○	○	○	▲	—				
PG						○	○	—			
CA									—		
RA						○	○	▲		—	
TP						▲	▲	▲		▲	—
SS						▲	▲	▲		▲	▲

B. BILL LENGTH

	ASP	LSP	DCC	BC	PC	WG	CM	PG	CA	RA	TP
LSP	▲	—									
DCC			—								
BC			▲	—							
PC			▲		—						
WG			▲	▲	▲	—					
CM			▲		▲	▲	—				
PG					▲		▲	—			
CA	▲	▲							—		
RA							▲	▲		—	
TP			▲	▲	▲	▲	▲	▲			—
SS			▲		▲	▲	▲	▲		▲	▲

Figure 1.2. Matrices of Hutchinsonian ratios in body weight and bill length for all combinations of species pairs within the Gulf of the Farallones seabird community; data and species abbreviations as in Figure 1.1. Indicated, for weight, are ratios between 1.0 and 1.7 (▲) and between 1.8 and 2.2 (○) only, and, for bill length, ratios <1.4 (▲) only; all other ratios were larger.

mond's analyses apply to marine bird communities, we should find within these groups important differences in natural history patterns that overcome the size overlaps, or else we should find evidence that one species is replacing another. Accordingly, in the following chapters, working up to a synthesis in Chapter 12, we will reveal aspects of the several species' natural histories that pertain to competition or coexistence within these species groups.

Also included in the weight and bill-length ratio comparisons of Figure 1.2 is the Sooty Shearwater, *Puffinus griseus*. If competition for food during the breeding season proves to be an important factor structuring the Farallones seabird community, then surely this visitor from the Southern Hemisphere should be involved, at the least as a force reducing prey availability for breeding species. It is by far the most abundant seabird in the California Current during the late spring and early summer, its numbers probably exceeding those of all other species combined during most years (Ainley 1976, Briggs and Chu 1986, Briggs et al. 1987). The shearwater's weight is similar to that of the gull and the four large alcids, especially the Tufted Puffin, and its bill size is similar to that of seven of the Farallon species. In our discussions of segregation along the trophic dimensions of niches, then, we will also be including this species. Other summer visitors to central California waters, i.e., the Black-footed Albatross, *Diomedea nigripes*, and the Brown Pelican, *Pelecanus occidentalis*, are too few and far too large to merit inclusion in our comparisons.

The major difference between our project and others that have investigated niche overlap and separation in seabirds is that we have collected information over a period long enough that one can relate annual variability in the degree of ecological segregation with annual variability in the production of offspring. In other words, recognizing that variability in niche overlap occurs (e.g., Belopol'skii 1961, Wiens 1984, Chesson 1986, Grant 1986), one could assume that, through evolution, species reduce competition in order to allocate resources more efficiently for the production of offspring. Accordingly, we hypothesized initially that for each Farallon species we should find high offspring production during years when its niche was well defined, i.e., when its niche did not overlap much with other species' (hence not inviting competition). This hypothesis has come to seem a bit naive. Recently, the concept of the "ecological crunch," or crisis of resource scarcity, has come to the forefront in the study of community ecology (Grant 1986, Wiens 1986). The idea is that resources are not always limiting, so one cannot always observe competition. What is currently missing in the study of vertebrate communities, where generation times of component species are long, are case studies that have lasted sufficiently long to reveal the frequency and consequences of such crises. We were lucky, from a scientific perspective, that the Farallon seabird community experienced three "crunches" of varying intensity during the 15 years of our study.

OBJECTIVES AND ORGANIZATION OF THE BOOK

Therefore, we had two purposes for conducting our 15-year investigation. The first was to complement the arctic study by Belopol'skii by comparing, on an annual basis, the variability in breeding biology and feeding ecology among species in relation to variability in the marine environment in a highly productive, temperate community. Our second goal was to describe ecological segregation and structuring factors in this community, and especially to relate annual variation in niche overlaps with breeding productivity.

The available seabird literature had not prepared us for the extent of variation that we encountered in the biology of Farallon seabirds. As a consequence, throughout the book we emphasize within- and between-year variability in life-history patterns and, in fact, also include between-year differences in the extent of within-year variations. If nothing else, we hope that this book directs the attention of ecologists to the difficulty (or impossibility) of properly perceiving the typical ecological pattern of a species or community in just a few seasons of investigation. We also hope that researchers will be encouraged to address not just negative excursions from the "norm" of life-history patterns (i.e., ecological disasters, breeding failures, etc.) but the positive anomalies as well. Both extremes of variation taught us much about the Farallon seabirds.

Following this Introduction, Chapter 2 describes the marine environment in the Gulf of the Farallones and discusses seasonal and between-year variability in weather, physical oceanographic conditions, and the lower trophic levels of the seabird food web. During the 15-year study period, significant, but not unexpected, environmental variation occurred. In Chapter 3, the feeding ecology of seabirds in the Gulf of the Farallones is described, including feeding behavior and habitat as well as diet. It is this chapter that contains the evidence for annual variability in the extent of overlap among the feeding niches of various species, as well as annual variability in the feeding niches themselves. Chapters 4 through 11 present our information on annual variability in the breeding biology and productivity of the species of Farallon seabirds. Chapters 5, 7, 8, 9, and 10 each deal with a separate species; Chapters 4, 6, and 11 each deal with two closely related species. These chapters discuss the particular study methods pursued for each species, and then present information on population size (and saturation) and seasonal and interannual population fluctuations, the species' geographic range, data on recovery of Farallon-banded individuals, diurnal cycles of island occupation, nesting-habitat preference and competition for nesting space with other species, nesting phenology, and nesting success—including such aspects of natural history as clutch and brood size, hatching success, chick growth and fledging weight, and fledging success. Finally, these chapters address each species' breeding strategy and compare the strategy to those of closely related species. Most chapters compare not only interannual differences in breeding parameters but also within-year variability (early vs. late nesters), including interannual variation of within-season variability. Important to each of these chapters is a fairly complete review of published information gathered elsewhere on the breeding biology of the species. For some species, such as the Ashy Storm-Petrel and Brandt's Cormorant, we present information on breeding biology essentially for the first time; for others, such as the Pelagic Cormorant and Pacific populations of the Common Murre, we significantly augment the scant information available. In these two chapters, we review information on related species. The final chapter (Chapter 12) summarizes some of our findings through between-species comparisons and discusses niche overlap with respect to a relatively stable resource (nesting space) and an unstable resource (food). It also discusses interannual variation in niche overlap and the relationship between degree of overlap and the production of offspring. Finally, we attempt to relate our findings to the broader questions of how seabird nesting communities are structured ecologically and how the structuring processes affect population regulation.

At the end of the book are Appendix tables, which contain data that should be useful to persons attempting to compare avifaunas elsewhere to this one; Appendix tables are numbered according to the chapter to which they pertain. Finally, in addition to the Literature Cited and Index, a Glossary defines technical terms employed in the text.

METHODS AND PHILOSOPHY OF DATA COLLECTION AND ANALYSIS

Our basic field work, as just described, spanned the years 1971 through 1983, a 13-year period. We then reduced our efforts while expanding into other aspects of seabird ecology, analyzing data, and preparing initial drafts of the present work. Nevertheless, we continued to collect basic information on population size, incidence of breeding, chicks produced per breeding pair (for all species), and diet (for murres, guillemots, and auklets). We also launched a full-scale investigation into the foraging areas of Farallon seabirds. During the revision of the manuscript in 1986 we incorporated into the chapters on individual species additional data on population trends, 1984–86, and into Chapter 3 information on feeding areas in 1985 and 1986. Preliminary information on breeding productivity, 1984–86, is presented in Appendix 3.1.

In our studies at the Farallones, we attempted to cause as little disturbance as possible, either to our research subjects or to adjacent nesting birds. For one thing, we wanted our data to be little affected by our activities, thereby increasing the chances that our perceptions of breeding and feeding ecology would be as close to nature as possible. For another, because islands such as the Farallones are small and few, our disrupting the nesting activities could reduce the number of birds over large regions of ocean. As a result of these concerns, most of the data for five surface-nesting species (three cormorants, the Western Gull, and the Common Murre) were collected by observation from blinds. This required many hours of sitting patiently, or not so patiently, waiting for nest reliefs and for birds to stretch or change position on nests, allowing us to observe the nest contents. In the case of the Western Gull, our efforts evolved from the usual method of studies of larid breeding biology, i.e., to walk through plots observing the contents of nests as the owners circled screaming above, to a method in which nests in plots were observed from blinds or distant observation sites; we overlapped techniques in a few years in order to compare results. Where we collected data from blinds, the birds were minimally aware of us, and we believe that our results were not affected by our presence or activity.

The concerns for disturbance also affected our methods of studying the nests of other species, all of them cavity nesters, as well as the kinds of data we collected. We did not visit all nests daily; many we visited at two- to five-day intervals. We did not handle adults prior to egg laying or during incubation, and we seldom handled chicks if adults were present. Furthermore, we did not visit study nests from the time that clutches were completed until the eggs were due to hatch. In most cases, we present data that indicate the degrees to which our various methods disturbed the birds and affected our subsequent results.

The great effort required to collect enormous amounts of data on 11 species simultaneously, and our desire to disturb our subjects as little as possible, precluded our collecting certain kinds of data that other seabird researchers have typically included in their studies. For instance, visiting nests less than daily meant that we could not collect precise data on such things as laying intervals (for species with multiple-egg clutches), incubation periods, or hatching intervals. For these aspects of natural history we relied on seabird literature. However, because we maintained the same visitation schedules year after year, or could express the data collected by means of different schedules in terms of a standard, we could detect within- and between-year variability in such parameters. In cases where many other studies had already measured some parameter, such as incubation period, we were more interested in detecting temporal variability. The chapters include ample reference to the studies that contain the more precise measurements.

During the height of the nesting season, April through August, to minimize disturbance we constrained ourselves and others to visiting only certain parts of the island via strictly defined walkways (we were not, however, limited to only peripheral colonies or nests). This constraint limited our choice of study colonies, of observation sites, and, in some cases, even of nests. Thus, one must realize that our results are but an index to the breeding activities of the entire Farallon population of each species, albeit an index that was consistently constructed year after year. More important, perhaps, is the fact that from a statistical standpoint our between-year comparisons were not entirely "independent," since many of the same nest sites were in our samples every year. Because seabirds are long-lived and are highly faithful to the same mate and nest site, our use of the same nests probably meant we were observing the same birds, which in a few cases was confirmed by banding. From an ideal statistical standpoint, we should have randomly selected the study nests each year, but we were unable to. This is, in fact, a problem that most field biology studies have had to accept. In our situation we at least had appreciable sample sizes with which to work.

Statistical procedures

We used both parametric and nonparametric statistics to compare trends and differences in the natural history variables we measured. Our main sources of reference were Sokal and Rohlf (1969) and Steel and Torrie (1960). Theoretically, the use of many parametric statistical tests requires independent, normally distributed observations. Although the latter requirement was usually satisfied, the lack of independence was a possible problem, as stated above. In many cases, choices for nonparametric alternatives were slim. In any case, the reader should consider our concerns in the few cases where statistical trends and differences are not clearly obvious.

Throughout the text certain abbreviations represent statistical tests or terms, as follows: degrees of freedom, DF; correlation coefficient, r; Spearman rank correlation coefficient, r_S; standard deviation, SD; Student-Newman-Keuls multiple comparison test, SNK test; Student's t test, t test. Means are usually stated with plus or minus the standard deviation. In ANOVA (analysis of variance) requiring two DF, the DF's will appear without spacing as per statistical conformity (e.g., DF = 2,3 or 2,301).

Analysis of environmental patterns

We identify in Chapters 2 and 3 several years in which sea-surface temperatures were much warmer than usual. Other anomalous patterns in the physical environment were also evident, and the food web changed in ways that generally made prey less available. In subsequent chapters, we show that the biology of the Farallon seabirds was quite different in these years. These years are identified by **boldface** type in the various tables and figures accompanying each chapter. Similarly, other years are identified in Chapters 2 and 3 as ones in which conditions brought especially favorable prey availability. These years will be identified by *italic* type.

We recorded weather three times daily throughout the period of study, as part of a program for the National Weather Service, with equipment provided by the Service. Measurements recorded included wind strength and direction, sea state, swell height, air temperature, rainfall, and cloud cover. Similarly, as part of a program for Scripps Institution of Oceanography, we daily recorded sea-surface temperature (at noon), 1973 to the present, and also collected a daily water sample at East Landing to determine sea-surface salinity. During the period 1970 to 1973 we did not measure sea-surface temperature, but to estimate temperatures for that period we used a regression between 1973–83 Farallon temperatures and those calculated for

Bodega Marine Laboratory (also participating in the Scripps program): $y = 0.91x + 0.18$ ($r = 0.923$). Temperatures at the Bodega lab, located about 50 km north of the Farallones on the California coast, had been taken since the mid-1960's. Finally, we collected and filtered phytoplankton from water samples taken daily during 1972 and 1973 as part of a program for the National Aeronautics and Space Administration; these samples were then analyzed to determine the concentration of phytoplankton as indicated by chlorophyll *a*.

Analysis of activity patterns in nocturnal species

We fitted ten Cassin's Auklet and eight Ashy Storm-Petrel nest sites with treadle-activated microswitches. These nest sites were part of the samples we used to study breeding productivity. The treadles did not affect the birds' behavior; at least their reproductive success did not differ from that of individuals in other nests. When these nocturnally active birds crossed the treadle into or out of their burrows, a mark was made on an Esterline-Angus multichannel event recorder. Daily we checked the operation of the treadles and the recorder clock. This system was in operation from early 1972 until 1980. From the charts of burrow activity, we extracted for each night the time of first mark (evening; birds arrive), time of last mark (morning; birds depart), and whether or not a mark was made at all. We combined these data with information on breeding phenology, moon phase, and sky condition (cloudy, partly cloudy, clear) to determine the nightly, seasonal, and between-year visitation patterns of these species, as well as some of the factors that might affect the patterns observed.

Analysis of growth in chicks

We weighed Ashy Storm-Petrel, Western Gull, Pigeon Guillemot, and Cassin's Auklet chicks daily during a number of years. Up to 300 gm, chicks were weighed to the nearest gram, and thereafter to the nearest five grams, by means of Pesola spring scales accurate to 1 and 5 gm, respectively. These were checked periodically for accuracy with an electronic balance.

To calculate growth curves for purposes of determining intra- and interannual variation, we first screened the data to separate out those chicks that disappeared prematurely. For multiple-chick nests, we segregated nests by number of chicks, ordered the chicks in nests by date of appearance, that is, first, second, or third chick, and then did paired analyses between chicks. Nests in which chicks hatched on the same date were excluded from within-brood analyses. Within each species and chick-order category, we fitted a logistic curve for each individual and estimated hatching weight, asymptotic (i.e., peak) weight, and growth rate. These curves were then averaged and analyzed with multivariate ANOVA to determine the amount of variance explained by year and by hatching date within years. We used univariate ANOVA, including regression, to determine which differences were significant.

A BRIEF HISTORY OF THE FARALLON SEABIRD POPULATIONS

In any discussion of the occupation of niches and competition for resources among species, it is important to know where the populations reside between the states of equilibrium and disequilibrium (Wiens 1984). In some cases, species populations may never be in equilibrium; but in equilibrium, one expects resources to be limiting, competition severe, and ecological segregation readily apparent. To provide some perspective on the stability or constancy of Farallon populations, the following section summarizes the articles by Ainley and Lewis (1974) and Doughty (1971, 1974). Additional details of seabird population fluctuations and the factors behind them are supplied in the various chapters.

This book and most reports about the Farallones pertain to Southeast Farallon Island and West End, two islands that together constitute the South Farallones and are about 44 ha in area. The North Farallones lie 11 km northwest and comprise just a few hectares. Humans have occupied only the South Farallones and particularly Southeast Farallon (Figure 1.3).

American and Russian sealers occupied the Farallones for various periods from 1807 into the 1830's. They exterminated several populations of marine mammals for fur, oil, and meat. They also took birds and their eggs for food, but it is not known whether they affected avian populations. Beginning in 1848, when the human population of San Francisco started to increase rapidly, seabird eggs, especially those of the Common Murre, were taken commercially. In 1855, the Farallon Egg Company was founded; it operated until 1881, but the commercial collection of eggs continued on a smaller, less organized scale until the early 1900's. Over the 45-year period of egg harvesting, over 14 million murre eggs were taken.

In 1855, the Farallon lighthouse, the first lighthouse on the West Coast, also began operation. Until 1965 it was attended by several keepers and their families (15 to 20 people), along with their domestic animals, which included dogs, cats, hogs, and a mule. During World

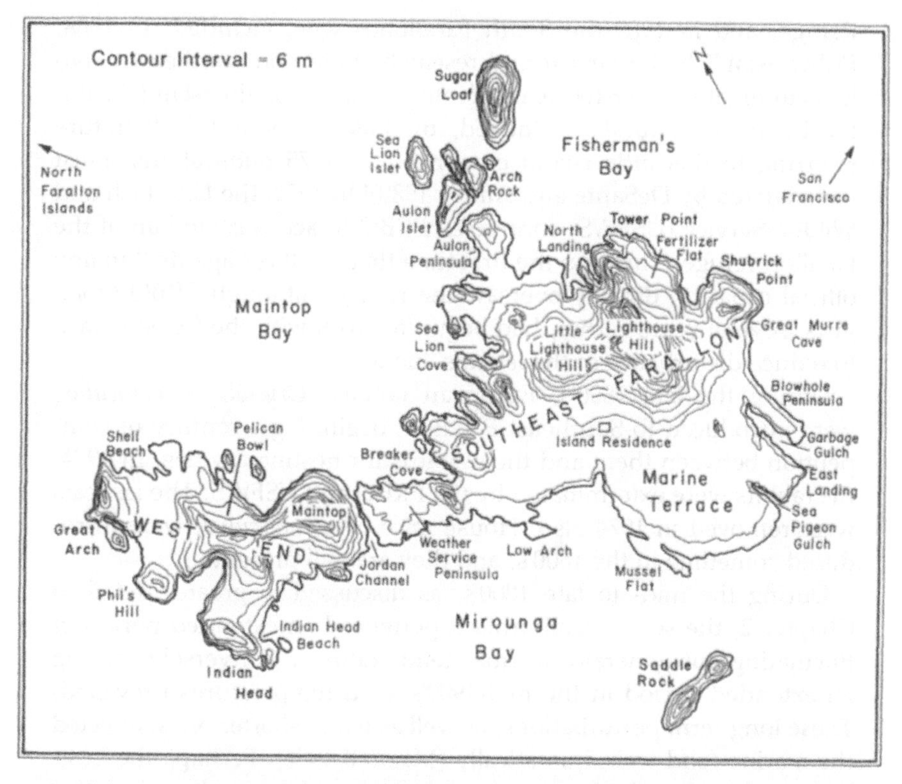

Figure 1.3. Map of Southeast Farallon, West End, and associated rocks, which collectively constitute the South Farallones; place names used in the text are indicated.

War II, a naval outpost was established to operate a gun emplacement. In 1965, the Coast Guard removed keepers' families, leaving only a complement of enlisted personnel to operate the aids to navigation. In 1972, the equipment was all automated, and Coast Guard personnel then visited only to repair and maintain it.

From the early 1900's until the 1940's, oil tankers routinely flushed tanks near the islands before entering San Francisco Bay, where several large refineries are located. Lightkeepers' logs are replete with entries concerning oiled birds on island shores. In addition, several catastrophic oil spills occurred in the area as a result of oil tankers running aground or experiencing other mishaps. Huge numbers of birds died as a result. By the 1960's, through concerted efforts and stricter controls, oil pollution had become much less of a problem in the Gulf of the Farallones.

In 1909, the North Farallones were designated a National Wildlife

Refuge, and in 1969 the South Farallones were included. In 1968, PRBO established a year-round research station on Southeast Farallon, culminating over 100 years of intense interest in the islands' wildlife by many naturalists. Indeed, by 1980 the scientific literature referring to this little island comprised over 75 publications (most summarized by DeSante and Ainley 1980)! In 1972, the U.S. Fish and Wildife Service (USFWS) contracted PRBO to act as custodian of the Farallon refuge. This was the first time that wildlife "agents," in any official capacity, directly oversaw the refuge, although PRBO biologists, several years earlier, had begun to work with the Coast Guard to reduce disturbance of the islands' biota.

During the late 1800's, European rabbits, *Oryctolagus cuniculus*, were introduced to Southeast Farallon, beginning a century of competition between them and the seabirds for nesting cavities. In 1974, the rabbits were exterminated by the PRBO and USFWS. The last cats were removed in 1974 also. House mice, *Mus musculus*, were introduced sometime in the 1800's, and they are still present.

During the mid- to late 1800's, as discussed in greater detail in Chapter 2, the sea off California experienced an extended period of fluctuating but generally warmer temperatures. Conversely, during an extended period in the mid-1900's, cold temperatures prevailed. These long-term perturbations, as well as many shorter ones, affected the marine food web dramatically (MacCall 1986). Perhaps the best-known effect was the increase in populations of the sardine *Sardinops caerulea* during the late 1800's and their demise in the mid-1900's, their disappearance very likely being helped along by overexploitation.

As a result of egging, oil pollution, and the episode of intense warm water (see Chapter 2) of 1957–59 (as well as, perhaps, the warm water of the mid-1800's), the murre population decreased from about 400,000 birds in the mid-1800's to only about 6,000 birds by the 1960's. It has been increasing since. Eggers decimated the gull population to reduce competition for murre eggs, and cormorant populations decreased significantly as well during that time as a result of disturbance. Puffin and guillemot populations declined in the early 1900's as a result of oil pollution and the introduction of rabbits. The latter, along with scientific collecting, may have exterminated the Rhinoceros Auklet by the late 1800's.

Its population recovering after the warm water of the mid-1800's, the Cassin's Auklet apparently became abundant by the turn of the century. Storm-petrels, which visit only at night and nest deep within talus slopes, may have been unaffected by any of the factors that changed the populations of other species. Storm-petrels may have benefited, however, from the additional nesting sites provided by walls, foundations, and pathways built of rocks by human occupants.

During the early to mid-1900's, the State of California killed sea lions and seals en masse along the California coast to reduce supposed competition between sea lions and fishermen. This practice was discontinued a few decades ago. In more recent decades, elephant seals and fur seals, exterminated by sealers, have been reestablishing populations on the islands, and sea lion populations have been increasing. More and more, pinnipeds have been occupying areas on Southeast Farallon that have been used recently by gulls and cormorants for nesting (Figure 1.4). This is forcing seabird colonies to move and is reducing the habitat available for seabird nesting.

In conclusion, then, Farallon seabird populations have not been

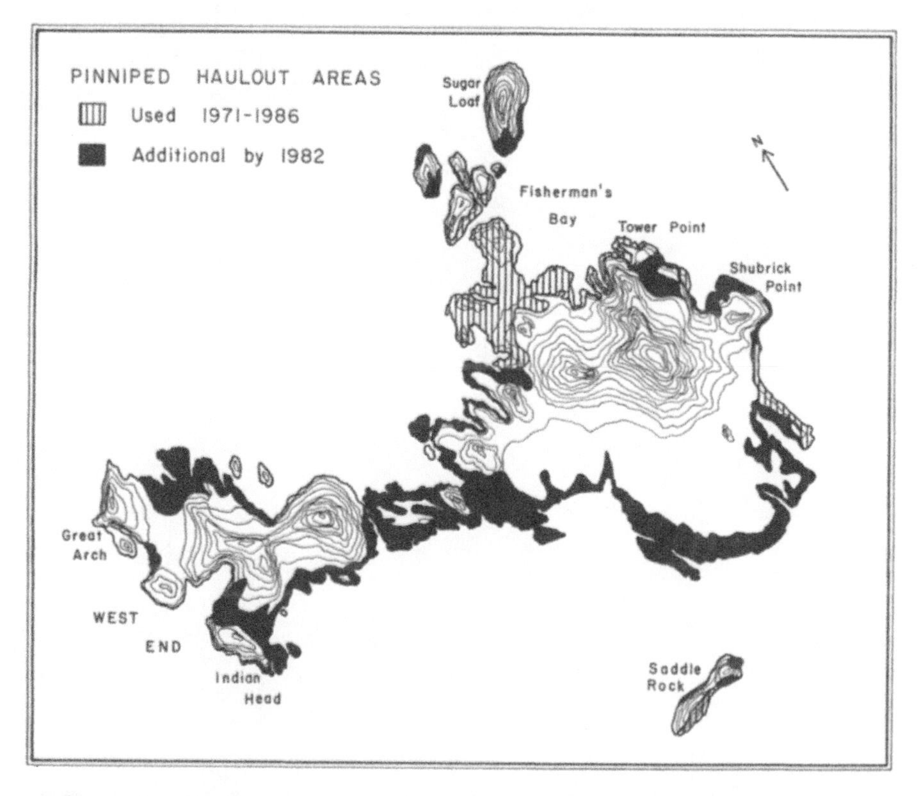

Figure 1.4. Haulout areas used by pinnipeds in the early 1970's compared with those used by increased populations in the early 1980's.

stable for at least the last 200 years, and probably not for millennia. The causes of the instability have been both natural and human-related. After reviewing population fluctuations of individual species in greater detail within respective species' chapters, we will return to the subject of population stability in Chapter 12.

2

Seasonal and Annual Patterns in the Marine Environment Near the Farallones

David G. Ainley

The Farallon Islands, at 37°42′ N, 123°00′ W, lie at the northern edge of the Californian Coastal Biogeographic Province (Udvardy 1978). The main characteristics of climate shared with the remainder of this province are mild winters and only slightly warmer summers. At the Farallones, air temperatures year round range on average only from about 9 to 13° C, reaching the higher end of the scale during the late summer and early fall. As elsewhere in the province, rain falls only during the winter and early spring (Figure 2.1A), but fog is frequent during the summer. Along the coast, about 200 km to the north, begins the Oregonian Rain Forest, a province in which rain is much heavier and falls year round.

Although the Farallones certainly share the weather patterns of the coastal province, there is little doubt that the biogeographic composition and size of the Farallon marine avifauna are influenced by marine rather than terrestrial conditions. Three major oceanographic systems or processes directly affect the ocean environment in the vicinity of the islands and, therefore, affect the avifauna as well.

First, the islands lie just east of the California Current, which is one of the more productive stretches of the world's oceans (Thompson 1981, Parrish et al. 1982). It and four other eastern boundary currents (i.e., currents on the eastern edge of large oceanic gyres and along the western edge of continents), which together constitute no more than 1% of the ocean surface, provide 40–50% of the fishery biomass harvested annually by man. The cool, productive current is no doubt responsible for the largely temperate character of the marine avifauna breeding on the islands, the large number of species, and the large size of their populations. Displaced west from the California coast by the cool current lie warm waters whose subtropical characteristics

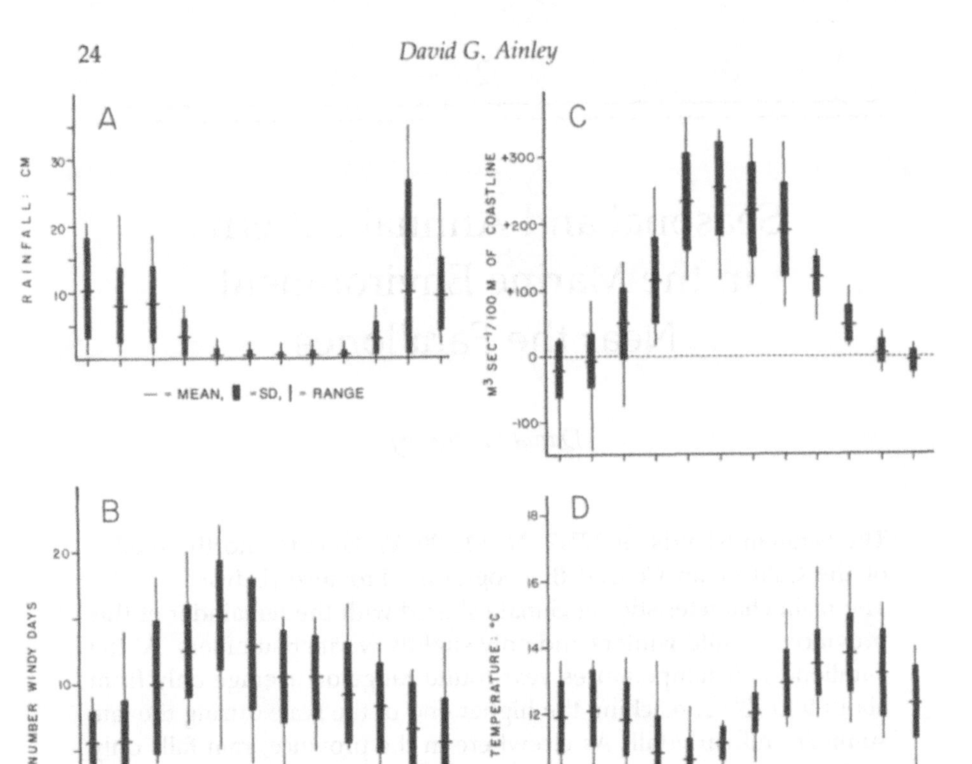

Figure 2.1. Characteristics of the physical environment at Southeast Farallon, 1970–83 (monthly means, standard deviations, and ranges are given): A, rainfall; B, days on which upwelling-favorable winds (north/northwest) blew in excess of 14 knots; C, intensity of coastal upwelling (at 39° N; data from Bakun 1975, Mason and Bakun 1986); D, sea-surface temperature.

seem more consistent with the biogeography of the mainland and of the ocean west of the California Current (see Hubbs 1960, Parrish et al. 1982).

Second, being at the edge of the California continental shelf, the Farallones lie in an area strongly influenced by coastal upwelling, which brings cool, nutrient-rich waters to the ocean surface. Along with water advected south by the California Current, upwelling helps to maintain the cool temperatures of surrounding waters and contributes to the increased productivity of the ocean in the region. The Gulf of the Farallones, in fact, lies at the southern boundary, and downstream, of the largest upwelling center of the California eastern boundary current (Parrish et al. 1982).

Third, but equally important, the islands lie at the edge of or within the Davidson Current, which is a subsurface countercurrent flowing north, opposite the California Current, and between the latter and the coast. This current brings warm water north but reaches the surface off central California only in winter (McLain and Thomas 1983). Thus it has little direct effect on the breeding ecology of the Farallon seabirds, but it has an important effect on the transition from winter to upwelling conditions and on the intensity of warm-water episodes (see below).

The waxing and waning of flow in the California Current and the Davidson Current, and of coastal upwelling, along with at least two important remote phenomena—El Niño in the eastern tropical Pacific and the relative strengths of the Aleutian Low and North Pacific High pressure systems—are the factors that cause seasonal and year-to-year variation in the weather and the ocean productivity in the Gulf of the Farallones. These processes and the way in which they interrelate are important to understanding the seasonal and interannual variation in the breeding ecology of Farallon seabirds.

One final "environmental" factor, which has a major impact on the seabirds of the Farallones and of the California Current in general, is common to all eastern boundary current regions (Parrish et al. 1982). Not only do all these regions have narrow continental shelves—thus promoting the dominance of more pelagic species of fishes—but all are associated with oceanic trenches where oceanic plates of the earth's crust are being subducted beneath the continental plates. A consequence important to seabirds is that subduction disfavors the creation of islands on which the birds can nest. When a limit to nesting sites is coupled with a spectacular abundance of food, the likelihood of competition for nesting space becomes high. The way in which Farallon seabirds adjust to both a limited but stable space resource and a copious but variable food resource strongly affects the structuring of the community. We will return to this important subject in the various species' chapters as well as in the final chapter.

SEASONAL PATTERNS IN THE PHYSICAL ENVIRONMENT

The annual pattern of currents off California was reviewed by Sverdrup, Fleming, and Johnson (1942), J. L. Reid, Roden, and Wyllie (1958), and Hickey (1979). Skogsberg (1936) and Bolin and Abbott (1963) described the marine climate of nearshore waters off central California and the seasonal patterns in climate that correspond to annual fluctuations in the various currents. The California Current is

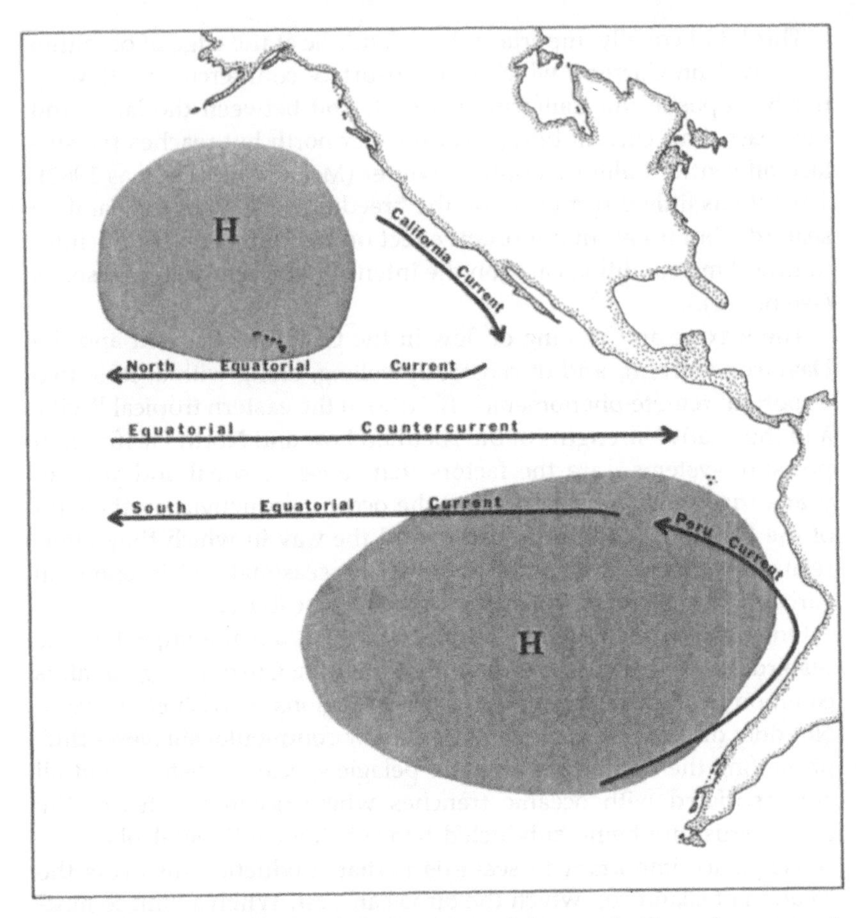

Figure 2.2. Pacific Basin and the position of the eastern boundary current known as the California Current; shaded areas represent the North and South Pacific gyres and corresponding regions of generally high (H) atmospheric pressure.

part of the North Pacific Gyre, a huge, sluggish whirlpool of water that rotates about the center of the northern half of the Pacific Basin (Figure 2.2). Over the center of the gyre sits a high-pressure area (the North Pacific High) around the north of which, from west to east, pass low-pressure systems. The winds passing from the high- to low-pressure systems are deflected to their right because of the earth's rotation (the Coriolis effect). The North Pacific Gyre, propelled by these winds, thus moves clockwise.

The water around the edge of the North Pacific Gyre moves much faster than that in the gyre's center. The rapid movement, or current, at the gyre's eastern edge is known as the California Current. When the latter reaches the latitude of Mexico it begins to swing west and eventually reaches an area where, because the water has become warmer and saltier (because of increased insolation and evaporation), it becomes known as the North Equatorial Current. There is, of course, a transitional zone between where the California Current ends and the North Equatorial Current begins. Where the North Equatorial Current reaches Asia and begins to swing northward it becomes known as the Kuroshio Current, and eventually this current swings east to become the West Wind Drift. After the latter moves through the Gulf of Alaska and begins to turn south at about 45° N, it becomes the California Current. This connection from the north brings cool water south to central California (and beyond), and the process of upwelling, which we mentioned above and will discuss more below, helps to maintain the cool temperatures as well. The water in the center of the gyre does not move much. It is subtropical in temperature and of relatively high salinity; it is also rather depleted of nutrients. These central waters thus contrast in many ways with those of the California Current.

The main flow of the California Current off central California lies within a corridor about 200 to 500 km off the coast. Another, much more variable zone exists near shore (Hickey 1979). The current is a response to wind patterns, which in turn result from the strength and position of atmospheric pressure systems and the air that flows from high to low pressure; the variable zone inshore results from perturbations of the flow due to benthic and coastal topographic features. Though wind patterns around the entire North Pacific contribute to the overall flow, the current gets its strongest push from West Coast alongshore winds during the boreal spring and summer. During these seasons, western North America warms dramatically, producing a persistent and rather deep, thermally induced low-pressure system that contrasts sharply with the high pressure offshore over the central Pacific. The central North Pacific high-pressure system migrates north and farther offshore (it migrates seasonally; as winter approaches it moves south and closer to the coast). This movement intensifies the flow of air, giving rise to the northwesterly winds that prevail parallel to the northern California and Oregon coast during spring and summer. On the average, from March through August these winds blow on at least one-third of all days; from April through June they blow on almost half of all days (Figure 2.1B). Velocities of

30–35 knots are typical of the latter months and sometimes persist for days on end. (To say the least, such conditions can certainly dampen one's enthusiasm for long hours in blinds studying bird behavior.) During the winter months, when the Pacific high-pressure system has migrated south and inshore, the intense summertime onshore–offshore pressure gradient is no longer as consistently in place, and thus northwesterly winds do not develop to the strength or frequency that they do during summer. In addition, the low-pressure centers that develop in the higher latitudes of the North Pacific are able during winter to move onto the northern California coast instead of being deflected farther north by the high pressure. Thus come the winter rains mentioned above (Figure 2.1A). These seasonal differences in the winds are illustrated by Parrish et al. (1982).

During spring and summer, the strong, persistent northwesterlies are responsible for three processes that contribute to the physical characteristics of surface waters and to their nutrient enrichment. First, these winds are responsible for coastal upwelling, and this part of the year has become known as the "upwelling period" in coastal waters (Bolin and Abbott 1963). It is generally recognized that coastal upwelling is a process restricted to within about 20 to 50 km of the coast (Gill and Clarke 1974), although the upwelled water can be advected farther offshore. The waters from Cape Blanco, Oregon, to Point Conception, California—a stretch that includes central California—constitute the region of maximum upwelling along the West Coast (Parrish et al. 1982). A second process related to northwesterly winds and contributing to the enrichment of central California waters, though on a larger spatial scale and a less immediate temporal scale than coastal upwelling, is the increased flow of the California Current and the resulting advection of nutrient-rich water from the north (Bernal 1981, Chelton 1981). Finally, the process of offshore upwelling is important also (Chelton 1982). It too affects productivity in the region on a broad scale.

At the local level, especially in the Gulf of the Farallones, which is coastal and just downstream from major upwelling centers (Point Reyes and Cape Mendocino), coastal upwelling probably has an appreciable influence on marine climate and productivity, and we will discuss this in more detail below. Enrichment of waters, however, is not the only factor affecting production in the lowest trophic levels of the central California food web. The intensity and persistence of coastal upwelling probably bears strongly on the survival and recruitment of young fish into the seabird food web, but not just as a positive factor of nutrient enrichment leading to higher food availability (Bakun and Parrish 1982, Parrish, Nelson, and Bakun 1981). Upwelling can also have negative effects, particularly if too strong and persistent. We will return to this subject as well, later in the present chapter.

The process of coastal upwelling occurs as follows. As surface waters off California are blown alongshore by the northerly and northwesterly winds, they are deflected to their right by the Coriolis effect. This causes an offshore transport of surface water, which is then replaced by waters that flow upward, or upwell, at the continental margin. These waters come from a depth of 100 to 200 m, where the sun's light and heat do not penetrate, and thus they are cooler and contain a rich supply of nutrients undepleted by the algae that grow in the upper layers of the ocean (the photic zone). When these nutrients are brought into the photic zone, photosynthesis takes place, and the amount of phytoplankton available to higher trophic levels increases.

Bakun (1973, 1975) has developed an index of the intensity of coastal upwelling, an index based on the strength and direction of the wind or, more properly, the amount of alongshore stress placed by the wind on surface waters. In essence, the index is equivalent to the amount of alongshore wind stress on surface waters divided by the Coriolis parameter. Of course, southerly or onshore winds cause negative upwelling, or downwelling, and the upwelling index measures this as well. These indexes, calculated daily, are available from the National Oceanic and Atmospheric Administration. We averaged them by month, and the results are shown in Figure 2.1C for the waters of central California. The curve corresponds rather closely to the frequency of northerly, upwelling-suitable winds measured at the Farallones (compare Figures 2.1B and C). As a result of this seasonal pattern of coastal upwelling, sea-surface temperatures at the Farallones reach their annual low from April to June, when the winds and upwelling are strongest (Figure 2.1D). April to June is indeed the peak of the upwelling period, although in total the period runs from about March to August. To some extent, upwelling occurs year round; all that is required is a few days of suitable winds. (Note that from December to February the upwelling index is negative, indicating downwelling.) Favorite, Dodimead, and Nasu (1976), who classified the various regions of surface waters in the North Pacific, called this narrow band of upwelling along the West Coast the Upwelling Domain.

The same northwesterly winds that cause coastal upwelling are also responsible for vertical movement of deeper waters on a much larger scale offshore in the California Current. Because the winds are, on average, strongest some distance offshore (200 km or more), the stress placed on inshore surface waters by these winds is weaker (C. S. Nelson 1977, Hickey 1979). This and other factors are responsible for the poleward, inshore flow of the California Countercurrent (see below). As discussed by Chelton (1982), because of the earth's spin, fluid dynamics, and other factors, the equatorward flow of the California Current offshore coincides with a sea surface sloping toward the coast, and the poleward flow of the countercurrent inshore coincides with a sea surface sloping away from the coast. Along the zone parallel to the coast where the two opposing slopes meet, about 100 km or so offshore, there is zero flow and actually a trough in sea level. Theoretically, the subsurface thermocline should also form a dome in response to this trough. Indeed, measurements show not only that the thermocline moves upward but that a chlorophyll peak occurs there as well, presumably in response to the increased nutrient levels resulting from the upwelling. Thus, this offshore nutrient enrichment appears to be an important process. Through the diffusion and mixing of surface waters it probably influences conditions in the waters frequented by Farallon birds.

In addition to intensifying upwelling in inshore and offshore California waters, the strong northwesterly winds during spring and early summer give a strong boost to the California Current. As a result, nutrient-rich water is advected from the north. These nutrients were incorporated into California Current waters previously, perhaps as much as a year earlier, by wind stress and upwelling in the Alaskan Gyre (Wickett 1967). As shown by Bernal (1981), Chelton (1981), and Chelton, Bernal, and McGowan (1982), this influx of nutrient-rich water is a primary cause of the large-scale seasonal and interannual variation in phytoplankton and microzooplankton abundance in California waters.

During July and August, the northwesterlies slacken. The result is a lessening of upwelling and a slowing of flow in the California Current, especially near shore (Hickey 1979). As the current and the transport of water offshore subside, the warm subtropical water lying west of the California Current begins to flow toward the coast. The result is a rather sharp rise in sea-surface temperature. Temperatures reach their annual maximum in September and October (Figure 2.1D). This period of rather passive onshore flow of warm, nutrient-depleted water is known as the "oceanic period" (Bolin and Abbott

1963). As noted above, the California Countercurrent flows poleward inshore of and, for most of the year, beneath the California Current, i.e., at depths of 200 to 500 m. During the summer, the prevailing northwesterly winds blow surface waters south, and these waters cover the countercurrent (Hickey 1979). Beginning in November and December, southerly winds associated with storm fronts come to dominate coastal wind patterns. The storms produce increased rainfall (Figure 2.1A) as well as downwelling. The southerly winds, and other factors, drive surface waters near the coast to the north and bring the countercurrent to the surface. It is then known as the Davidson Current. The other factors responsible for the countercurrent, as reviewed by McLain and Thomas (1983) and McLain, Brainard, and Norton (1985), include (1) a coastward deepening of the thermal and salinity structure (i.e., the isotherms and isopycnals, or imaginary lines connecting waters of equal temperature and salinity, curve downward), (2) the onshore movement of waters caused by winds, (3) the lessened stress of northwesterly winds on inshore waters relative to offshore waters, where the northwesterlies are stronger (as noted above; Chelton 1982), and (4) occasional aseasonal depression of the thermocline due to El Niño in the eastern tropical Pacific and onshore advection of warm offshore water. The period from November to February, because of the dominance of the countercurrent in inshore waters, is known as the "Davidson Current period" (Bolin and Abbott 1963).

SEASONAL PATTERNS IN THE BIOLOGICAL ENVIRONMENT

The strong seasonality in the physical marine environment off central California imparts a seasonality to the biology. Bolin and Abbott (1963) described the annual, cyclic nature of the standing crop of phytoplankton in Monterey Bay. Phytoplankton declines sharply during the oceanic period, reaches its lowest concentration in the midst of the Davidson Current period, and does not begin to increase appreciably until the upwelling period. The standing crop of phytoplankton usually reaches its maximum from May through July. Such a pattern also occurs in the vicinity of the Farallones, as indicated by seasonal changes in the concentration of chlorophyll and phaeophytin (the two main algal pigments; Figure 2.3A). This pattern in phytoplankton availability is probably affected greatly by the injection of nutrients into surface waters by the process of coastal upwelling.

The abundance of zooplankton in California waters is also strongly

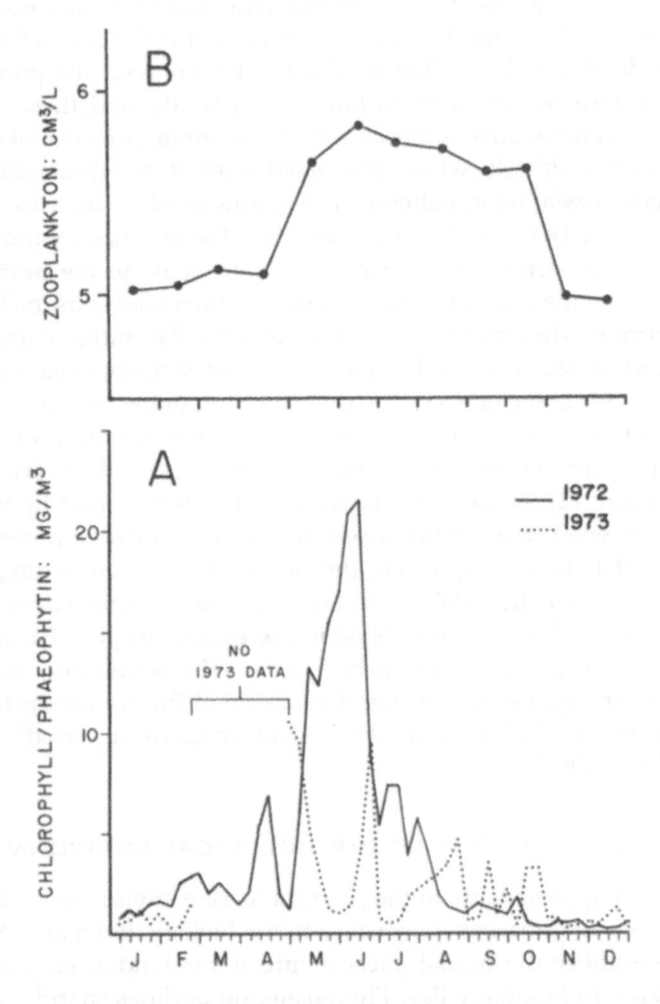

Figure 2.3. A, Seasonal change in phytoplankton abundance in waters near the Farallones, 1972 and 1973, as shown by concentrations of plant pigments; B, seasonal change in zooplankton volume in waters near the Farallon Islands, 1949–79 (after Chelton 1981, by permission).

seasonal. This has been shown for specific taxa, such as euphausiids (Brinton 1981), as well as on a general basis by Chelton (1981), who averaged microzooplankton volumes measured at a large number of oceanographic stations over a 30-year period by the California Cooperative Oceanic Fisheries Investigations (Cal-COFI). Off central

California (Farallones to Monterey), zooplankton volumes increased sharply in April and decreased sharply in October (Figure 2.3B). Although this pattern is influenced strongly by coastal upwelling, the incomplete fit of zooplankton abundance with indexes to upwelling led Chelton to look for other factors that may be involved. Following the lead of Wickett (1967), who found a correlation between wind stress in the Alaskan Gyre and zooplankton volume off California one year later, Chelton found that the strength of advection of nutrients from the north explains more of the variability in zooplankton abundance than does coastal upwelling, at least as perceived on a large regional scale. This pattern was true on a seasonal as well as an interannual basis. On the other hand, most of the Cal-COFI stations were in offshore waters rather than in waters influenced strongly by coastal upwelling (D. M. Husby pers. comm.). Thus, the weak connection between indexes of coastal upwelling and offshore zooplankton volumes is not surprising.

EL NIÑO

El Niño has a strong effect on the year-to-year variability of the California marine environment, and because it is a special circumstance that originates in an area remote from California, the eastern tropical Pacific, it is best to introduce it here before a discussion of the actual interannual variation that we observed in the marine environment near the Farallones.

El Niño has been known for a long time in the Peru Current system, because the warm water that characterizes it leads to the disappearance of pelagic fish and to high mortality of the so-called guano birds (Murphy 1936, 1981, Paulik 1971, Idyll 1973). The phenomenon of El Niño has been intensively studied off Peru because the decline in guano production, and in fish meal production during later years, was critical to the economy of Peru and other countries. Only in the last 30 years or so, however, have scientists learned that El Niño is a much more widespread phenomenon.

The term "El Niño" has been a source of misunderstanding. For hundreds of years it designated the warm-water countercurrent that appears for a short time annually in coastal surface waters off Peru. This current is analogous to the Davidson Current off California. Its appearance signaled the end of the fishing season (because fishing success declined) right around Christmastime, and thus Peruvian fishermen nicknamed the countercurrent "El Niño," The Child. Then the rest of the world rediscovered Peru and, specifically, its rich guano deposits and the immense pelagic fish populations in its ocean waters. (The Peru Current, analogous to the California Current, is the richest of all the eastern boundary currents.) European and North American economies became dependent to some extent on Peruvian guano and fish meal, but fortunes were made and lost from time to time when the conditions of El Niño persisted much longer than usual, not just for the Christmas season, but for the several months following as well. Fishing would come to an enduring standstill and the guano birds would die en masse. After scientists began to study the phenomenon, the present meaning of the term "El Niño" evolved. It now designates the unusually persistent warm-water conditions that occur every two to seven years in the Peru Current.

Thanks to programs such as Cal-COFI, which have been gathering consistent observations for over 30 years, and the Farallon seabird studies we report in this volume, it is becoming increasingly clear that El Niño has repercussions over a large part of the eastern North and South Pacific oceans (Briggs et al. 1987, Ainley et al. 1986, Duffy et al. 1986). The task of unraveling the complicated causative factors and far-reaching effects of the phenomenon, however, is difficult, because even today we lack certain data gathered over sufficiently long periods. Nevertheless, there is now little doubt that events in the equatorial Pacific are important for explaining El Niño.

The present working hypothesis about what happens in the eastern equatorial Pacific during El Niño may be summarized as follows (see the reviews by Wyrtki 1975, Thompson 1981, O'Brien, Busalacchi, and Kindle 1981, and Enfield 1981 for more details). Normally, strong southerly winds blow persistently alongshore or slightly offshore over the coastal waters, generating the same effects as described above for California, namely, upwellings and the advection of water toward the equator. When nutrient-rich water is brought into the photic zone, photosynthesis becomes possible and phytoplankton begins to increase in abundance. The continual process of nutrient enrichment off Peru means that the phytoplankton bloom can be continuous, supplying abundant food for algae-eating animals and so on higher into the food web.

The winds alongshore also impart a local boost to the Peru Current, which is part of the huge South Pacific Gyre, an analog of the North Pacific Gyre. As the Peru Current nears Ecuador and the Equator it begins to swing westward and becomes known as the South Equatorial Current (Figure 2.2). Running parallel is the North Equatorial

Current, which, as mentioned earlier, is derived partially from the California Current in the path of water circulation around the North Pacific Gyre. Because the North and South equatorial currents move along the Equator, the effect of the earth's rotation, the Coriolis effect, has minimal impact. We have thus entered the relatively simple system in which oceanographers have been interested. The picture, then, is of a lot of water heading west driven partially by the equatorial portions of the two gyres, but more importantly by the persistent trade winds that blow east to west across the equatorial region. Some equatorial water, to be sure, moves east in a compensatory flow known as the Equatorial Countercurrent.

Prior to El Niño, trade winds are stronger than normal and cause water to accumulate and the sea level to be higher in the western Pacific than in the eastern Pacific. Because of the accumulation of warm water in the western equatorial Pacific, the thermocline and pycnocline (subsurface boundaries between warm saline surface waters and the colder, less saline waters below) are much deeper there than in the eastern Pacific. (Note that this situation differs from that in the California Current region, where the more saline water is deep. The high salinity of equatorial water is due to intense evaporation at the surface.) In addition, upwelling generally is not a feature in the western equatorial Pacific; the ocean layers are quite stable and unproductive.

For unknown reasons the pressure differential between the South Pacific High and Indian Ocean Low periodically weakens. This phenomenon is known as the Southern Oscillation (Quinn et al. 1978). With a rather sudden lessening of the air pressure difference, the trade winds in the far western Pacific slacken or even reverse. Sometimes, again for reasons unknown but probably related to "teleconnections" in the atmosphere and between the atmosphere and ocean (see Namais 1976), the South Pacific and the North Pacific high-pressure systems weaken and migrate farther apart at this time. This apparently contributes to a further weakening of the trade winds in the western Pacific. Reduction in the pressure differential and slackening of the trade winds allow water in the western Pacific to "slop" back toward equatorial America. It comes in a long-period wave (Kelvin wave), much of which is internal in the sea, at the thermocline.

The slowly moving wave is detectable as a short-term (on the order of weeks) rise in sea level, as well as a deepening of the warm mixed layer above the thermocline. The wave reaches the American boundary of the equatorial region and is reflected both downward and poleward. The depression of the thermocline travels along the North and South American coasts (to the north, it is detectable as far as British Columbia). The deepening of the warm layer off Peru means that warm, nutrient-depleted water, rather than cold, nutrient-rich water, upwells. The same happens off California as the wave travels northward. Not until the wave has passed do conditions begin to return to normal.

The above, of course, is a much-simplified scenario. What I have not mentioned is how the temperature of the ocean affects the temperature of the air above it, and vice versa. Such teleconnections between ocean and atmosphere have important effects on weather and wind and ocean circulation. Not surprisingly, a huge mass of warm water in an unusual location can have profound repercussions on the usual climate of a region. Furthermore, as reviewed by Michaelson (1977), such a pool of warm water can complete a circuit around the North Pacific Gyre to affect conditions in the same region a few years later. During many (but not all) Niños, winds that normally produce coastal upwelling in Peru and California subside. Perhaps this is one of those indirect effects of El Niño by which the presence of warm water reduces the pressure differential between the oceanic high and the thermally (warm) induced continental low that was discussed earlier. In addition, the warm water weakens the effects of the high pressure as a block against low-pressure systems, and often rainfall increases.

The term "El Niño" has evolved to "ENSO," which is short for El Niño–Southern Oscillation. This new term is perhaps applicable to the wider effects of El Niño. During our Farallon bird studies, ENSO in the eastern Pacific occurred in 1972–73, 1976, and 1982–83. Within the lifetimes of adult Farallon seabirds alive when our study began in 1971, some individuals may have experienced even the strong ENSO events of 1957–59 and 1965, and 1969 as well.

ANNUAL VARIATION IN PHYSICAL CONDITIONS

So far, we have discussed the seasonal cycle of the physical marine environment in the vicinity of the Farallones, as well as the ENSO phenomenon, which can have strong effects on the cycle. The earlier discussion of zooplankton volumes, an environmental factor more directly relevant to seabirds, introduced the fact that the environment of Farallon seabirds can vary significantly from year to year. We now consider this variability in greater detail.

In essence, almost every seasonal characteristic reviewed above showed strong interannual variability. Rainfall at the Farallones was unusually high during the winters of 1969–70 (as inferred from data in Michaelson 1977, because we were not collecting rainfall data in that year), 1972–73 (as well as later in 1973), 1978–79, 1981–82, and 1982–83 (Figures 2.4, 2.5). Rainfall was abnormally low from late 1970 to 1972 and again from late 1975 to early 1977. Our study saw the wettest, 1981–83, and the driest, 1975–77, periods in the entire recorded history of California. These extremes were also reflected in rainfall anomalies measured at the Farallones. The approximately five- to six-year cycle of rainfall peaks and troughs at the islands (Figure 2.5) carried through from the 1960's and even much earlier, as noted by Michaelson (1977) for California at large. According to Michaelson, the amount of rainfall in California is a function of sea-surface temperature anomalies in the North Pacific Gyre. These anomalies persist and seem to fluctuate with a five- to six-year periodicity, which theoretically is the amount of time required for water to circle the gyre. Michaelson also noted that ENSO in the eastern tropical Pacific has a strong influence on California rainfall.

In view of the patterns noted by Michaelson (1977), it is not surprising that sea-surface temperature anomalies in central California, as measured at the Farallones, varied in a pattern that complemented the rainfall anomalies (Figure 2.4), or vice versa. The winters, or sometimes longer periods, when rainfall was high tended to be those when sea-surface temperatures were also anomalously high. This was evident during the winters of 1969–70, 1972–73, and 1977–78, throughout the latter half of 1976, and from early 1982 through 1983. Among these periods, the only one that did not correspond to ENSO in the eastern tropical Pacific was 1977–78. Rather cold temperatures prevailed during the dry periods of 1970–72 and 1974–76. McLain, Brainard, and Norton (1985) noted that warm sea temperatures off California in the winters and early springs of 1977–78, 1979–80, and 1980–81, which did not correspond to ENSO events, were the result of the onshore movement of warm, subtropical waters. This movement in turn resulted from wind patterns changed, through atmospheric teleconnections, by a weakening of the North Pacific High coinciding with a strengthening of the Aleutian Low.

The periods of high rainfall were no doubt accompanied by strong southerly and onshore winds. At the Farallones, northwesterly winds were anomalously absent early in 1970, during 1973, 1977–78, and 1979–80, and from early 1982 through mid-1983 (Figure 2.4C); they

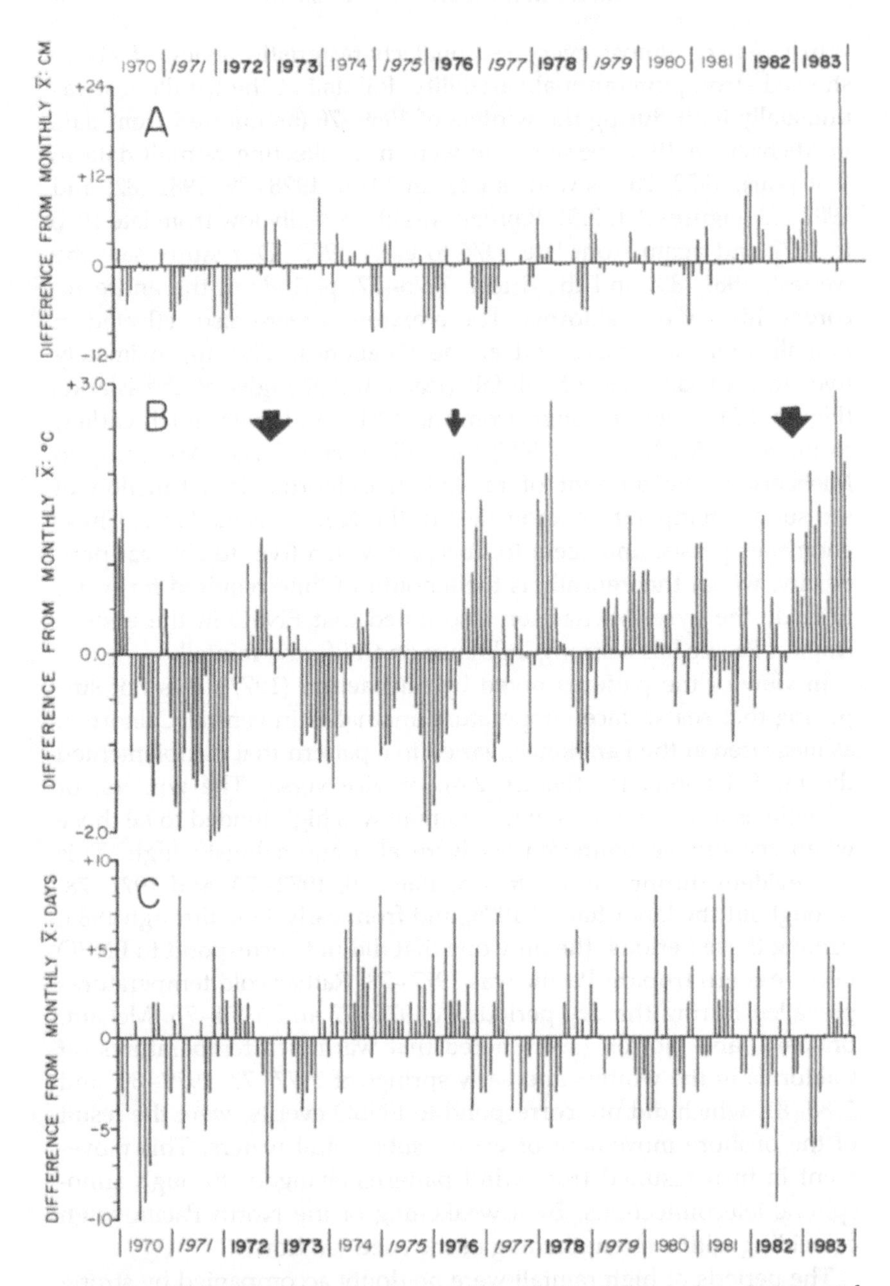

Figure 2.4. Departures from mean monthly values of physical environmental factors at Southeast Farallon, 1970–83: A, rainfall; B, sea-surface temperature; C, days per month that upwelling-favorable winds blew in excess of 14 knots. Arrows indicate onset of strong or weak ENSO's.

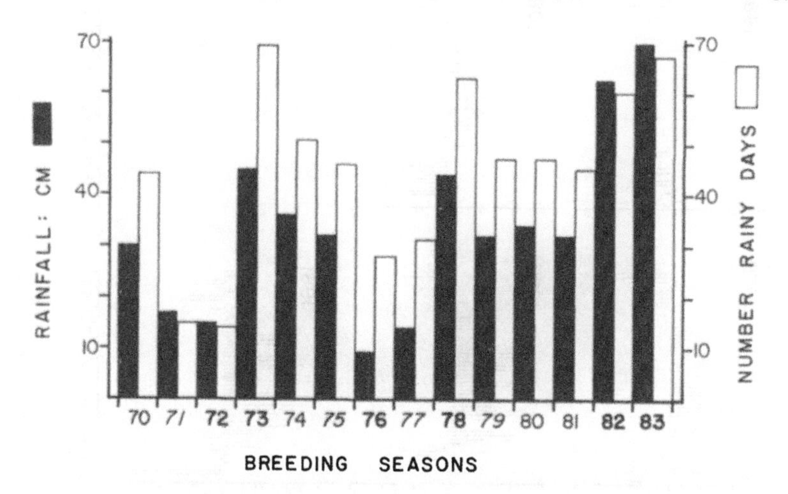

Figure 2.5. Amount of rainfall and number of days on which measurable rain fell at Southeast Farallon from January through April each year, 1970–83.

were intermittent during early 1971, 1974, 1977, 1979, 1981, and 1985, and were especially strong throughout the cold, dry 1975–77 period.

Corresponding to wind patterns, the upwelling index was strongly negative during the late winter and early spring of 1970, 1973, 1978, 1980, and 1983 (Figure 2.6, Appendix 2.1; McLain, Brainard, and Norton 1985). The cumulative annual anomalies for these periods are even more striking (Figure 2.7). These periods, as noted above, were characterized by infrequent northwesterly winds, strong southerly winds, warm temperatures, and rain. Upwelling and downwelling alternated in pulses during the late winters and early springs of 1971, 1974–75, 1977, 1979, 1981, and 1985 (Appendix 2.1). During 1976, upwelling was more or less continuous. These trends coincide fairly well with the observed patterns in the persistence of strong northwesterly winds (Figure 2.4).

The variability in the environment around the Farallones described above is consistent with other factors and phenomena. On the one hand, the years 1969, 1972, 1976, and 1982–83 were ENSO years, and thus the high sea-surface temperatures and rainfall were "expected" (in hindsight?). Interestingly, ENSO years tended to be followed by "anti–El Niño" conditions: windy, dry, and cold. Even 1978, a year of ENSO-like conditions in California, but not a real ENSO year as noted above, was followed by anti–El Niño conditions. On the other hand, the winters of 1969–70, 1972–73, 1975–76, 1977–78, and 1982–83 were those in which the Davidson Current was particularly strong (McLain and Thomas 1983, McLain, Brainard, and Norton 1985). As

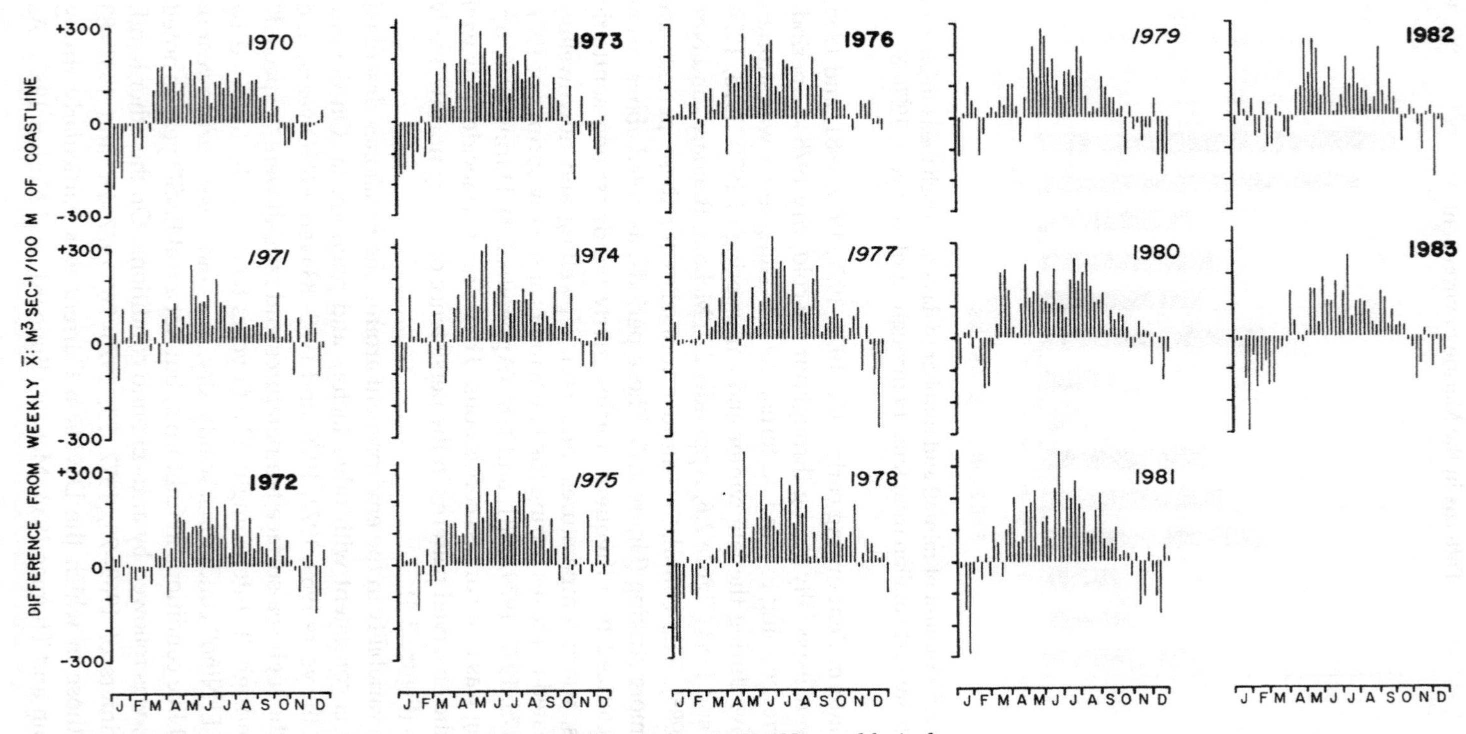

Figure 2.6. Departures from the mean weekly upwelling index, 1970–83; weekly indexes calculated by NOAA-NMFS.

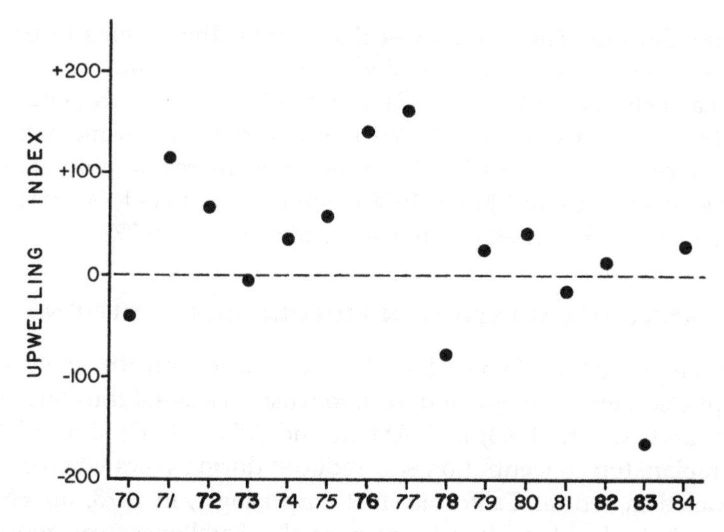

Figure 2.7. Departures from the mean seasonal upwelling index (m^3sec^{-1}/100m coast), January–April 1970–84.

noted by McLain and Thomas, the movement of Pelagic Red Crabs, *Pleuroncodes planipes*, north of southern California is evidence for a strong Davidson Current. During 1982 the crabs reached Monterey Bay, which they have only rarely done in the past, and during 1983 they reached at least as far as the Farallones, their most northerly penetration ever recorded. Sea level along the California coast was also high in 1982 and 1983 (Lynn 1983, Fiedler 1984), another indication of strong northerly flow (Hickey 1979, Chelton 1981, McLain and Thomas 1983).

Of the years from 1970 to 1985, 1976 was somewhat strange in that northwesterly winds and upwelling occurred throughout the winter and spring, but sea-surface temperatures were high, particularly from late spring onward. The warm water was the far-reaching result of ENSO, which likely deepened the thermocline, causing warm instead of cold water to be upwelled. It was the only ENSO year during this period when rainfall was not high. Another unusual year was 1978, which, as noted above, was not a year of ENSO but was characterized by ENSO-like conditions. The strange anomalies of that year have been discussed by a number of authors, including Brinton (1981), Chelton (1982), and McLain and Thomas (1983). Apparently, atypical meteorological conditions in 1978 brought a pool of warm, nutrient-depleted water of low salinity northward and onto the California coast. This may have contributed to especially strong flow in the Davidson Current. The strength of that current thus seems to be the common denominator among all years when the marine climate was anomalously warm off central California (McLain, Brainard, and Norton 1985). As a response to the warm water, rainfall was anomalously high during the winter of 1977–78, and it is interesting that because of this, even without ENSO, 1978 continued the five- to six-year periodicity in peak rainfall mentioned by Michaelson (1977).

ANNUAL VARIATION IN BIOLOGICAL CONDITIONS

Three groups of biological studies bear closely on the analysis of the physical environment and the anomalies discussed thus far. First, Bolin and Abbott (1963) and Abbott and Albee (1967) showed that phytoplankton concentration was reduced during years when ENSO prevailed in central California. Not surprisingly, in 1973, an ENSO year, phytoplankton abundance near the Farallones was reduced from the levels of 1972 (Figure 2.3A), the other year in which we measured chlorophyll (i.e., phytoplankton) concentration.

Second, and more important, perhaps, because they were concerned with factors higher in the food web, the studies by Chelton (1981) and Chelton, Bernal, and McGowan (1982) showed that microzooplankton volumes on a regional scale are largely a function of the strength of the California Current and the advection of nutrient-rich waters from the north. Furthermore, along with others (e.g., Hickey 1979, McLain and Thomas 1983, McLain, Brainard, and Norton 1985), they were able to show that sea level, as measured at coastal tide gauges, provides an extremely good indication of the strength of the current flow, either northward (countercurrent dominant; high sea level) or southward (California Current dominant; low sea level). The studies of Chelton, Bernal, and McGowan ran from 1950 through 1979, and for sea-level data they were extended by Fiedler (1984). During our bird study, zooplankton volumes were particularly high during 1971, 1974, 1975, and 1977, and particularly low during 1970 (early), 1973, 1976 (late), and 1978 (Figure 2.8). Not surprisingly, these fluctuations fit the El Niño/anti–El Niño patterns noted above. Without a doubt, zooplankton volumes must have been quite low during 1982 and 1983, years of high sea level and strong flow in the California Countercurrent (Fiedler 1984, McLain, Brainard, and Norton 1985). Sea level was also elevated somewhat in 1980 (Fiedler 1984).

In view of earlier discussions of the relative importance of advec-

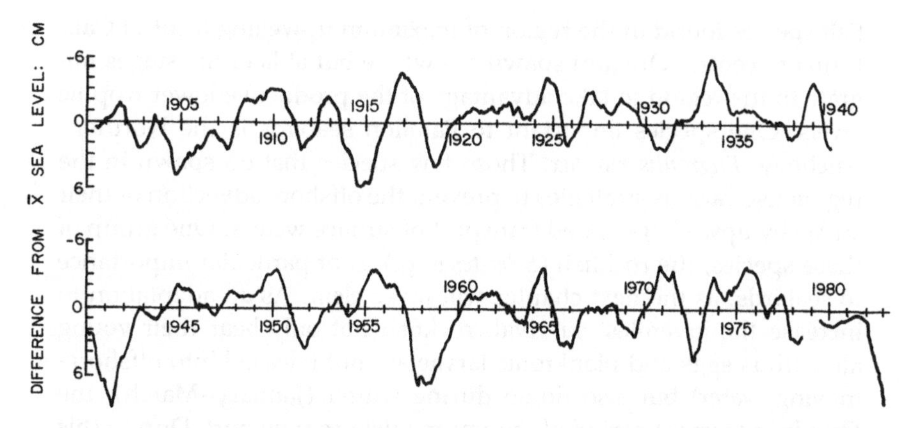

Figure 2.8. Departures from mean sea level, as an index to zooplankton volume, and ENSO/anti-ENSO events in the California Current, 1900–1983 (after Chelton, Bernal, and McGowan 1982, by permission, and Fiedler 1984; see text). High sea level indicates ENSO; no data for 1979.

tion vs. upwelling in bringing nutrients into the food web off California, a comparison among Figures 2.3, 2.6, and 2.8 shows some interesting patterns. These perhaps bring the interrelationship of those oceanographic processes into local focus for the Gulf of the Farallones. During 1972, when advection was strong (Figure 2.8), there was little correspondence between upwelling pulses (Figure 2.6) and a large seasonal phytoplankton maximum (Figure 2.3A). During 1973, however, a year of weak flow and southward advection in the California Current, each of the small phytoplankton peaks followed intense upwelling events by about two weeks. This may indicate that during years when advection is poor, coastal upwelling in the Gulf of the Farallones is more important than during years of strong advection. Strong advection may mask any effects of coastal upwelling on nutrient enhancement in the area.

The final group of studies important to understanding interannual variation in the seabird food webs includes Bakun and Parrish (1982), Parrish, Nelson, and Bakun (1981), Parrish et al. (1982), and Husby and Nelson (1982). These studies have shown that although productivity at lower trophic levels is important to levels higher in the food web, it certainly is not the whole story; indeed, seabirds do not eat phytoplankton, nor most of the kinds of zooplankton measured by Chelton, Bernal, and McGowan (1982).

Parrish and co-workers have pointed out that most major pelagic fish species found in the region of maximum upwelling (central California to central Oregon) spawn elsewhere but at later life stages migrate to the region to take advantage of the productive lower trophic levels. One species important to Farallon seabirds is the Northern Anchovy, *Engraulis mordax*. Those few species that do spawn in the region use various strategies to prevent the offshore advection of their larvae by upwelling-related transport of surface waters. One group of these species, the rockfish (*Sebastes* spp.), is of particular importance to seabirds, as the next chapter will make clear. As an adaptation to increase the juveniles' survival, rockfish not only bear their young alive (thus eggs and planktonic larvae are not released into offshore-moving water) but also do so during winter (January–March), the Davidson Current period of minimum offshore transport. During this period of the year, however, they must cope with winter storms that, if frequent and intense, increase turbulence and degrade feeding conditions. The situation is compounded by the fact that winter storms are most intense during warm-water years when the lower trophic levels are also depleted. The best situations are winters with intermittent weak upwelling (slight offshore transport) interspersed with weak pulses of downwelling (onshore transport, minimum of turbulence). These conditions occurred in 1971, 1974, 1975, 1977, 1979, 1981–82, and 1985 (Appendix 2.1), especially in 1977, 1979, and 1985. Other years exhibited either strong downwelling or, in 1976, persistent upwelling during winter.

ANNUAL VARIATION IN THE AVAILABILITY OF PREY

The major prey of central California seabirds are juvenile rockfish, *Sebastes* spp., Northern Anchovy, *Engraulis mordax*, Market Squid, *Loligo opalescens*, and the euphausiid *Thysanoessa spinifera* (Manuwal 1974b, Follett and Ainley 1976, Pierotti 1976, Baltz and Morejohn 1977, Ainley, Anderson, and Kelly 1981, Chu 1984). During the winter and early spring, Pacific Herring, *Clupea harengus*, can also be important to some seabird species, especially gulls and probably cormorants (Spratt 1981a). In addition to these abundant schooling organisms, California seabirds also feed heavily on a variety of other more dispersed or more highly localized prey (e.g., Ainley, Anderson, and Kelly 1981). Unfortunately, for most of the latter species little is known of their biology or abundance because few are of direct commercial importance.

Reviewed here is information on the natural history of the major prey listed above, particularly aspects of their biology that affect availability to seabirds. Included is information on between-year differences in abundance from 1971 to 1985.

Northern Anchovy

The abundance of anchovies off California has been increasing dramatically since the early 1950's. There are three subpopulations; the central subpopulation occurs from northern Baja California, Mexico, to about Point Reyes, California (Vrooman, Palona, and Zweifel 1981). Anchovies occur from the shoreline out to waters overlying depths of more than 1,000 m. Anchovies through one year of age predominate in waters out to 300 m deep; one- and two-year-olds predominate over depths between 300 and 600 m; older fish predominate in deeper waters. In the central subpopulation, individuals enter their second year measuring about 125 mm (Parrish, Mallicoate, and Mais 1985). Almost all anchovies mature by the end of their second year.

Anchovies move offshore and south during winter, their major period of spawning (Mais 1974). Spawning takes place in a number of locations along the coast, including the Gulf of the Farallones, but the major spawning ground is off southern California. Beginning in late summer, anchovies tend to form large, dense schools that remain at depths of 180 to 300 m during the day. They move to the surface and disperse during the night. In the spring and early summer, schools occur near the surface during the day (Frey 1971). During some years, anchovies are found more inshore; in others, more offshore (Mais 1974). Larger anchovies move north during years of warmer sea-surface temperatures (R. H. Parrish pers. comm.).

The Northern Anchovy is one of the most intensively studied creatures in the California Current, if not in the whole world. Much information is available on its relative abundance from year to year, and is expressed in terms of spawning biomass (Hewitt 1985, MacCall 1986). The abundance estimates, of course, are rather broad in scale, and may not necessarily be interpretable on a level as local, for instance, the Gulf of the Farallones. This point aside, the spawning biomass of the central subpopulation decreased from 1970 to 1972 but recovered to about 3 million metric tons by 1975 and 1976 (Figure 2.9). It then entered a period of decline, reaching especially low levels in 1978, 1980, and 1984. Part of the decline in biomass resulted from a reduction in fish size (MacCall 1986). The population recovered somewhat in 1986 (A. D. MacCall pers. comm.).

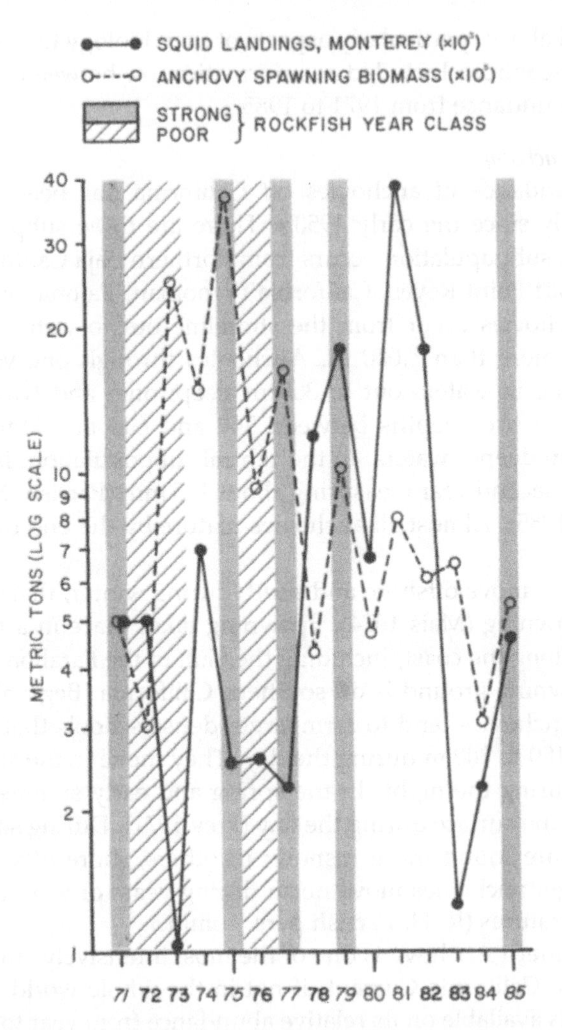

Figure 2.9. Relative strength of rockfish year classes, anchovy spawning biomass, and Market Squid landings in central California, 1971–85 (data sources identified in the text); years when juvenile rockfish abundance was not remarkably strong or weak, or for which data were unavailable (1980–82), are not shaded.

It is probable that fluctuations in spawning biomass were due to changes in spawning success and resultant year-class strengths, as well as to predation on older fish (Mais 1981). Strong year classes were produced from 1970 through 1973, so anchovies under two years old (smaller fish of the sizes consumed by birds) were abundant from 1971 through 1975. Poor year classes were produced in 1974,

1975, and 1977, and a class of only mediocre strength was produced in 1976. These poor year classes were, therefore, partly responsible for the lower levels of spawning biomass from 1978 on (Figure 2.9). The production of young fish, however, was fairly strong from 1978 through 1980, and again in 1982, but was weak in 1981 and 1983 (Mais 1981, Hewitt 1985). Information on the relative strengths of year classes beyond 1983 is not yet available.

During 1978 and 1982–84, commercial landings of anchovies were well below allowable takes. This was due to a number of reasons, including fish availability (Calif. Dept. Fish and Game 1985). Market conditions, which discouraged U.S. fishermen from taking anchovies, were certainly part of the story from 1982 to 1984 (A. D. MacCall pers. comm.)

Rockfish

In the offshore waters of central California, the species of rockfish most important to seabirds include the Blue, *Sebastes mystinus*, the Bocaccio, *S. paucispinis*, the Chilipepper, *S. goodei*, the Shortbelly, *S. jordani*, the Yellowtail, *S. flavidus*, and the Widow, *S. entomelas*. Among these, the most important is the Shortbelly Rockfish (PRBO, unpubl. data). Juveniles of all of these species occur in offshore waters of the continental shelf, although they are usually but not always most abundant within 15 to 30 km of land (MacGregor 1986). The relative composition of juveniles in the trawl samples taken by the National Marine Fisheries Service during the period 1983–86 was about 60% *S. jordani*, 30% *S. entomelas*, and 10% for the remainder of the above species (Tiburon Laboratory 1987, T. W. Wyllie Echeverria pers. comm.).

Juvenile rockfish are born alive and at parturition are about 3–7 mm long. Within a few months they grow to a length of about 75 mm. During their first few months they occur as scattered individuals or in loose schools at middle depths as well as in surface waters; they later settle near the bottom. An exception to this pattern is exhibited by the Shortbelly, the Widow, and sometimes the Yellowtail, which occur at middle depths as adults (Lenarz 1980, Lenarz and Moreland 1985, Lenarz and Gunderson 1987). Rockfish are born during winter and early spring, with the principal month of parturition being February for all species except the Blue and Chilipepper, for which the principal month is January (MacGregor 1986, Wyllie Echeverria 1987). Parturition in these months, along with a few months of subsequent growth, means that fish of a size suitable to be prey for smaller seabirds become available in March and April, and for larger seabirds from May to July. As rockfish settle to deeper waters beginning in late June and July, they become less available in surface waters (Tiburon Laboratory 1987), until only the very deepest-diving seabirds can reach them. During the spring and early summer, juvenile rockfish are exceedingly important prey to many marine predators, including, besides birds, salmon and marine mammals (Lenarz 1980, Jones 1981, Tiburon Laboratory 1987).

Commercial demand for rockfish has grown in the 1970's and 1980's, and juveniles are not targeted by fisheries. Consequently, long-term information on the annual variation in the abundance of juvenile rockfish does not exist. The abundance of juveniles has now been recognized as a possible indicator of future year-class strength, however, with the result that such data have been gathered since 1983 (Lenarz and Moreland 1985, E. Hobson et al. 1986, Tiburon Laboratory 1987, W. H. Lenarz pers. comm.) For prior years, indications of the annual abundance of juveniles can be approximated by the numbers of rockfish of specific ages caught commercially (see also MacCall 1986). Unfortunately (or perhaps fortunately for the marine food web), Shortbelly Rockfish are not fished, and thus only the more recent direct data are available for this species.

During our seabird study, year classes of most of the commercially important rockfish species were exceptionally strong in 1971, 1975, 1976, 1977 (not Chilipepper), 1979 (not Widow), and 1985 (1973 and 1978 were strong years only for Chilipeppers and Widows, respectively); year classes were exceptionally poor in 1972, 1973 (not Chilipepper), 1978 (not Widow), and 1983 (Hightower and Lenarz 1986, Henry 1986, Hobson et al. 1986, Thomas 1986). For the Shortbelly Rockfish, exceptionally strong years were 1971, 1975, 1977, 1979, and 1985 (MacCall 1986, Tiburon Laboratory 1987). The record for poor years is much more difficult to reconstruct, but 1972 and 1983 clearly were among such years, and 1984 and 1986 may prove to have been poor as well (MacGregor 1986, Tiburon Laboratory 1987, T. W. Wyllie Echeverria and D. Woodbury pers. comm.). These trends are summarized in Figure 2.9, with information after 1979 being less complete because rockfish born since then have not yet entered the fishery.

Pacific Herring

Herring occur in coastal waters throughout central California. They become available to seabirds especially during the winter, when they enter shallow coastal bays and lagoons to spawn. Important fisheries for herring exist in Tomales and, particularly, San Francisco bays, November to March. The spawning biomass of herring was particularly low in the winters of 1973–74, 1977–78, and 1983–84 (Spratt 1981a, Calif. Dept. Fish and Game 1983, 1984, 1985).

Market Squid

This species of cephalopod moves inshore to spawn in shallow waters (less than 100 m); otherwise it occurs widely in schools in California coastal and offshore waters. Schools remain in deep water during the day and approach the surface at night. A major West Coast spawning area is Monterey Bay, but some spawning occurs in the Gulf of the Farallones. The main spawning period is April to June, but some spawning takes place as late as November (Frey 1971, Recksiek and Frey 1978). Market Squid reach a maximum size of about 300 mm (including mantle and arms) in two years (Spratt 1981b); therefore, most eaten by seabirds are probably a year old or younger.

No estimates of spawning biomass exist for the Market Squid; squid landings by the fishery must suffice to indicate annual differences in availability, and to some degree they do, at least with regard to periods when the squid are unavailable. The squid fishery in Monterey Bay grew slowly until 1978, when demand began to increase rapidly. Thus, estimates of abundance must be separated by that year (Calif. Dept. Fish and Game 1982–86). Squid were notably few in 1973 and 1976 during the early part of the record and during 1980 and 1983–84 in the later part (Figure 2.9).

Euphausiids

Most of the research on zooplankton in the California Current has dealt with species in offshore waters; thus relatively little is known about *Thysanoessa spinifera* (S. E. Smith pers. comm.). This species occurs over the continental shelf and slope and is associated with centers of coastal upwelling. Usually occurring near the surface of deeper water is *Euphausia pacifica*, a species for which much more information is available. *E. pacifica* is one of the most abundant and ubiquitous euphausiids off California except over the continental shelf; it occurs year round but is most abundant during spring and summer. *T. spinifera*, on the other hand, seems to be abundant in surface waters mainly during the upwelling period (Brinton 1962, 1981). Both species, but especially *T. spinifera*, have been observed in swarms at the surface during the day (Smith and Adams 1988, J. T. Harvey ms). Surface swarms of *T. spinifera* appear to occur soon after strong northwesterly winds and upwelling events (pers. obs.).

Little is known about between-year variability in the abundance or availability of these two species. Brinton (1981) observed that *E. pacifica* becomes sparse during the early portion of sea-surface warming events. Most recently, this occurred from late 1977 to mid-1978 (Brinton 1981) and in 1983 (S. E. Smith pers. comm.). Even less is known about *T. spinifera*, and data on it are available for the 1980's only. In the Gulf of the Farallones it was sparse during 1983 and the first half of 1986 but was abundant in 1984 and especially 1985 (Smith pers. comm.).

To summarize these data, anchovies under two years old, of the sizes consumed by seabirds, appeared to be available throughout the period 1970–85, but abundance and fish size declined in the latter part of our study (1980's). Commercial fishermen were unable to find fish in 1978 and perhaps from late 1982 to early 1984, and this could well be true for seabirds, too. Juvenile rockfish (particularly the Shortbelly) were especially abundant in 1971, 1975, 1977, 1979, and 1985, and were less available in 1972–73, 1976, possibly 1978, and 1983. Market Squid were scarce in 1973, 1975–77, 1980, and 1983–84. Herring occurred in low numbers during their winter spawning periods in 1973–74, 1977–78, and 1983–84. Little is known about euphausiids except that warm water affects their abundance and distribution; such changes were noted in 1978, 1983, and 1986. Among these years of note, most frequently mentioned are the warm-water years, 1972–73, 1976, 1978, 1982–83, and 1986 (identified by **bold** type in the tables and figures of various chapters), when the seabirds' various prey were less available, and 1971, 1975, 1977, 1979, and 1985 (identified by *italics* in tables and figures), when the opposite condition of prey availability prevailed.

3

The Feeding Ecology of Farallon Seabirds

David G. Ainley, Craig S. Strong, Teresa M. Penniman, and Robert J. Boekelheide

The way in which two resources, space and food, are allocated plays a pivotal role in the structuring of animal communities (J. Diamond 1986, Roughgarden 1986; Chapter 1). For Farallon seabirds, we consider the allocation of nesting space in the chapters on breeding biology of individual species (Chapters 4–11), and here we will consider the way in which Farallon seabirds do or do not differ in their allocation of food resources. The allocation of food, of course, also involves spatial elements.

Coexisting species can divide available trophic resources by a number of means, and each can be considered a separate niche dimension or axis. These include allocation by prey species, by prey size, by foraging behavior, and by foraging habitat (or space). Individual species then may be plotted along these axes, in some cases perhaps overlapping with other species and in others perhaps not. All of these niche dimensions have been investigated many times in studies of seabirds, but, as detailed in the review below, few studies have considered all four dimensions for a particular community or species group.

We have organized this chapter by niche dimension rather than by a separate description of the foraging autecology of each species. First we consider foraging area and habitat; then foraging behavior, which is divided into prey-capture technique, temporal patterns, tendency to feed socially, and foraging effort; then diet composition, and finally prey size. Our data are largely from the breeding season, and, where possible, we include between-year variation in these niche dimensions.

SEASONAL PATTERNS OF TROPHIC DEMANDS

Because of the large sizes of the populations involved, it seems possible that seabirds compete for food in and near the Gulf of the Farallones (see Briggs et al. 1987). Many Farallon species are nonmigratory, residing within or near the Gulf year round. As discussed in Chapter 1 (see respective species chapters for more detail), these include the Ashy Storm-Petrel, Double-crested Cormorant, and, perhaps, Pelagic Cormorant, Common Murre, Cassin's Auklet, and Rhinoceros Auklet. Other species' populations disperse from the area entirely once breeding activities cease. These species include Leach's Storm-Petrel, Pigeon Guillemot, and Tufted Puffin. Finally, some portion of the Brandt's Cormorant and Western Gull populations remains in the Gulf year round while the remainder disperses along the West Coast. When the breeding populations of all these species are present, from mid-March to August, at least 300,000 individuals are exploiting food resources in the Gulf and are using the Farallones as their base. About 50,000 more breeding individuals from coastal sites can be added to the total (Sowls et al. 1980). In addition, during the breeding season, a large proportion of these species' nonbreeding populations also frequent breeding sites and thus feed in nearby waters. These nonbreeders are more free than breeding adults to exploit food farther afield, but little is known about whether or not they do. In tropical waters, Ashmole (1963) theorized that nonbreeders do feed elsewhere, but in the Ross Sea, Antarctica, Ainley, O'Connor, and Boekelheide (1984) found that this is not so. Only in the case of the Western Gull is there evidence to indicate that those *subadult* nonbreeders that visit the island feed on a resource different from that on which adult breeders and nonbreeders feed (L. B. Spear, unpubl. data; see gull section below). With the assumption that a significant portion of adult nonbreeders also frequents the Gulf during the breeding season, a minimum 400,000 individuals currently reside in or near the Gulf of the Farallones from mid-March through mid-August. Before the 1870's, this number was probably a million or more (Ainley and Lewis 1974).

In addition to these breeding species, at least one other seabird is important in the assemblage of species exploiting trophic resources in the Gulf of the Farallones. This is the Sooty Shearwater, a visitor from the Southern Hemisphere whose numbers in the Gulf probably exceed a million birds during portions of each summer, mainly May through July (Briggs and Chu 1986, Briggs et al. 1987). We will consider Sooty Shearwaters in this chapter because if seabird feeding does affect food availability, thus leading to competition, then certainly Sooty Shearwaters play a major role.

SEABIRD FEEDING STUDIES ELSEWHERE

Community-based studies of seabird diet and feeding ecology have been conducted in the Barents Sea (Belopol'skii 1961; diet, three to five breeding seasons, depending on species), in the Chukchi Sea (Swartz 1966; diet, one season), at Christmas Island (Ashmole and Ashmole 1967; diet and prey size, two seasons), on the Welsh coast (Pearson 1968; diet, prey size, feeding range and habitat, one season), at South Georgia (Croxall and Prince 1980; diet and foraging range, one season), at the Pribilof Islands (Hunt, Burgeson, and Sanger 1981, Schneider and Hunt 1984; diet and habitat, four seasons, but averaged into one season), at Kodiak Island (Sanger 1982; diet and prey size, one season), in the Gulf of Alaska (Baird et al. 1983; diet and behavior, two seasons), at Cousin and Aldabra atolls in the Indian Ocean (A. W. Diamond 1983; diet, prey size, and foraging range, one season), in the northwestern Hawaiian Islands (Harrison, Hida, and Seki 1983; diet, one season), and at Varanger, Norway (Furness and Barrett 1985; diet, prey size, and foraging range, one season). Trophic studies concerning major portions of breeding communities have been conducted on alcids at St. Lawrence Island (Bédard 1969a; diet and prey size, three seasons), on the Welsh coast (M. P. Harris 1970; diet and prey size, three seasons), and at the Olympic Peninsula (Cody 1973; diet, prey size, and foraging range, one season), on cormorants along the North American West Coast (Ainley, Anderson, and Kelly 1981; diet and foraging habitat, data for several years averaged), and on more diverse portions of communities in Oregon (J. M. Scott 1973, Wiens and Scott 1975; diet, prey size, and habitat, one season) and South Africa (Crawford and Shelton 1978; diet, several seasons). Feeding ecology studies of entire seabird communities at sea have been conducted in the high latitudes of the Antarctic by Bierman and Voous (1950; diet, one season) and by Ainley, O'Connor, and Boekelheide (1984; diet, prey size, behavior, and habitat, one season) and of partial communities (auks) by Bradstreet (1979, 1980, 1982; diet, prey size, and habitat, several seasons averaged) in the eastern Canadian Arctic.

Among these 20 studies, three characteristics are clear. First, most (14) were conducted in high-latitude, continental-shelf systems of relatively low, constant productivity rather than in highly but variably productive upwelling-dominated eastern boundary systems. Fundamental differences in the forces driving food-web structure have been identified in these two types of systems; in the former, predation has an important role in affecting species interactions, but in the latter, species interactions are submerged by the effects of environmental factors (Sherman and Alexander 1988). Of the remaining trophic studies of seabirds, three involved less productive subtropical and tropical oceanic systems, and only three involved more productive upwelling-based systems (West Coast of North America, Oregon, South Africa). Second, most (14) were conducted for only one year, or the data were averaged as if they covered only one year; and the majority (13) did not investigate within-year variability. Finally, only three studies, all at high latitudes, gathered data relating year-to-year variability in foraging ecology to changes in reproductive success for several species (Belopol'skii 1961; Hunt, Burgeson, and Sanger 1981, Hunt, Eppley, and Drury 1981, Schneider and Hunt 1984; Baird et al. 1983). A number of such studies exist, however, for single species.

A major common finding among the multispecies studies is that breeding seabirds at a given locality depend on only a few major species of prey, and diet overlap is high (A. W. Diamond 1983, Furness and Barrett 1985). The present consensus of opinion is that when segregation occurs it is on the basis of different sizes of common prey or differences in foraging behavior (e.g., range, habitat, or foraging depth). The underlying assumptions in most of these studies are that breeding seabirds compete for a limited amount of food and that food availability during the breeding season limits reproductive success and population size (Ashmole 1963, A. W. Diamond 1978, Hunt, Burgeson, and Sanger 1981, Hunt, Eppley, and Drury 1981, Hunt, Eppley, and Schneider 1986, Furness and Birkhead 1984). The opinion contrary to these assumptions has been expressed by Lack (1946, 1966), an opinion that seems lately to be falling out of favor. This opinion holds that because of a superabundance of a few prey species during the breeding season, seabirds do not generally compete for food then; rather, if competition occurs it should happen during periods when prey are scarce, namely, the nonbreeding season. Lack went further to hypothesize that scarcity of food during the nonbreeding season is the factor that limits seabird population size.

The relationship between feeding ecology, breeding success, and population size (or fluctuations in size) remains obscure, except in extreme situations where the total collapse or alteration of food webs has had dramatic effects on population size or breeding success (Murphy 1925, 1936, 1981, Crawford and Shelton 1978, Vermeer 1978, Vermeer, Cullen, and Porter 1979, Baird et al. 1983, Schreiber and Schreiber 1984). However, even in cases where total reproductive failure has been observed in a given year, the relationship of such events to population size has not been investigated. That breeding success is a function of food availability has been further supported by observations such as those by Bédard (1969a), who noted that auklets on St. Lawrence Island stopped breeding at the same time in late summer when they diversified their diet. This has also been observed for other species by Belopol'skii (1961) in the Barents Sea (see also Pearson 1968, Crawford and Shelton 1978). Such observations support the ideas that seabirds undertake breeding at the same time as (dependent upon?) the seasonal appearance of superabundant prey, and that breeding ceases when the prey become unavailable.

In further support of the idea that breeding success is dependent on food availability, a number of studies have noted a correspondence between interannual variability in diet composition and reproductive success (Anderson, Gress, and Mais 1982, Baird et al. 1983, Belopol'skii 1961, Bergman 1978, M. P. Harris 1984, Hunt and Butler 1980, Hunt, Burgeson, and Sanger 1981, Hunt, Eppley, and Drury 1981, Kuletz 1983, MacCall 1986, Schaffner 1982, and Vermeer 1980). Baird et al. (1983) and M. P. Harris (1984) felt that *interspecific* competition for prey reduces food availability for certain predators, leading in turn to lowered reproductive success; Birkhead and Furness (1985) and Hunt, Eppley, and Schneider (1986) felt that *intraspecific* competition has the same effect (see also Birt et al. 1987). In the other studies, authors ascribed diet and breeding variability to oceanic conditions independent of competitive interaction. Variation in competition and oceanographic factors are not mutually exclusive, but the paucity of studies relating reproductive success of seabirds to food-web dynamics, let alone to food availability as affected by predation or competition, and in turn relating this interaction to variation in population size, makes further discussion difficult, in other than the extreme cases (caused by environmental factors or human interference).

If we consider that most seabird community studies have been conducted (1) during the breeding season and (2) on (high-latitude) systems less productive and less variable than those in eastern boundary currents, it seems valid to ask whether or not our perceptions of the ultimate factors affecting seabird ecology and populations are not accordingly biased. Or, at the least, to ask how far we can go in generalizing these perceptions. Furness and Monaghan (1987) concluded that generalization is still premature. In the last chapter of this book, we will propose that in productive, upwelling-based systems, seabird populations are not directly limited by food during the breeding season, an opinion already voiced by Murphy (1981) and Duffy (1983a) for the Peru Current, a system analogous to the California Current. An additional proposal will be made that for seabirds of the California Current upwelling domain, ecological segregation of food resources is most evident during nonbreeding periods. The implication is that generalizations of patterns of resource segregation, and their relationship to population and breeding biology, among all breeding seabird communities, are unlikely to be valid, particularly in the case of upwelling systems.

METHODS

Foraging areas and habitat

During April and June in both 1985 and 1986, we crisscrossed the Gulf of the Farallones and neighboring waters by ship in a study (with K. T. Briggs) to relate seabird foraging to ephemeral oceanographic features. In the process of censusing seabirds we were able to determine feeding areas of Farallon breeding species. Cruise dates were 11 to 16 April and 31 May to 19 June 1985 and 14 to 22 April and 4 to 19 June 1986. The first of these years was one of the most productive for Farallon seabirds since our study began, with chicks fledged per breeding pair exceeding the 1971–82 average for all species (Appendix 3.1). In contrast, during the second, which was an ENSO year, breeding was late in all species except storm-petrels and reproductive success was well below the 1971–82 average; in fact, Pelagic Cormorants and Pigeon Guillemots failed in their nesting attempts. Thus, we consider 1985 representative more of conditions during the average or better years of seabird reproduction described in the various species' chapters, and we consider 1986 representative of the warm-water years, such as 1973, 1976, 1978, and 1982–83.

Census methods were the same as those described by Ainley, O'Connor, and Boekelheide (1984). This work supplemented the information provided by Briggs et al. (1987), who censused seabirds by airplane off the northern California coast from 1980 to 1982. Their census lines were perpendicular to the coast and more widely spaced than our cruise tracks.

To construct the 1985 and 1986 maps of foraging areas, we divided the Gulf of the Farallones and vicinity into blocks measuring three minutes of latitude by four minutes of longitude for the April (prebreeding season) and June (breeding season) cruises. Each block was thus about 5 km on a side (ca. 27.9 km^2). If a census segment (each one-half hour or 8.8 km long) intersected a block, then each species' density for that segment was assigned to that block. In cases where more than one segment intersected a block, values were averaged. When a segment intersected more than one block, its density values were assigned to each. Finally, when an uncensused block was bordered by at least two censused blocks, the uncensused block was given a value equal to the average densities of the bordering blocks. We considered only birds on the water or actively feeding.

We assessed change in the habitat use of species between 1985 and 1986 by comparing blocks occupied in the two years. This was done by means of Cole's (1949) coefficient of association and census blocks common to both years, where

$$C = \frac{(ad - bc)}{(a + b)(b + d)},$$

and the variance

$$s^2 = \frac{(a + c)(c + d)}{n(a + b)(b + d)}$$

for most cases. In some instances where $ad < bc$, a variation of this formula is used (see Cole 1949). In the formula, a is the number of blocks in which both species are present, b is the number of blocks in which species A (the less abundant of the two species being compared) is present in the absence of species B, c is the number of blocks in which B is present in the absence of A, and d is the number in which neither occurs; n equals the sum of the four variables. In the case of abundant, widespread species (Sooty Shearwater, Common Murre, and Cassin's Auklet), we compared only blocks where density was greater than or equal to one bird per square kilometer; for less abundant species (Brandt's Cormorant, Western Gull, and Rhinoceros Auklet), densities had to exceed 0.1 bird per square kilometer for inclusion; and for the remaining species, we included all blocks where they were observed. In this way we eliminated from analysis the incidental occurrence of isolated individuals. Using the same technique we also assessed habitat overlap among the species and whether overlap changed between the warm and the cool year.

We were able to derive a further assessment of between-year variation in foraging areas by censusing seabirds along the supply route between the Golden Gate and Southeast Farallon Island. These censuses were conducted from mid-1972 until early 1981 aboard the small craft that ferried us to and from the island, compliments of the Oceanic Society Farallon Patrol (see Acknowledgments). This census effort proved useful for determing variability in foraging areas only because, as indicated by observations from the island, a significant proportion of Brandt's Cormorants, Western Gulls, and Common Murres on their foraging trips during the breeding season disappeared toward or appeared from the east or southeast, the direction along which the supply craft traveled. Subsequently, the aerial censuses of Briggs et al. (1987) and the 1985–86 shipboard surveys showed that a major foraging area in the Gulf of the Farallones (especially in warm years) is in the coastal waters from southern Marin County to southern San Francisco County, i.e., east or southeast of the island. This area is influenced by the plume of waters that flow out of San Francisco Bay. Within it, the interface between estuarine waters and shelf waters, which is usually a marked, visually observable boundary, is often an area of high bird numbers (see also Storer 1952). In essence then, by good fortune and the quirks of geography, the Golden Gate–to–Farallon route provided the means to index the more distant foraging efforts of a large proportion of Farallon cormorants, gulls, and murres.

On all our trips, only birds that occurred within an estimated 300 m of one side of the boat were counted; observers were usually about 5 m above the sea surface on supply boats and 15 m above on our 1985–86 censuses. Supply boats traveled at 5 to 8 knots, and the vessel used in the 1985–86 work traveled at 10 knots. The supply route was partitioned into four units (see Figure 3.2): segment I, Golden Gate Bridge to Point Bonita, the outermost point of the Marin Headlands (a 4.4-km stretch); segment II, Point Bonita to the outermost shipping-channel buoy (6.2 km); segment III, outermost buoy to San Francisco Pilot Buoy (8.7 km); and segment IV, Pilot Buoy to about 2 km off Southeast Farallon (24.6 km). When winds were above

25 knots we were unable to census effectively; usually, if winds exceeded that level, the supply boats were canceled and the larger vessel anchored in the lee of the Point Reyes headland. During the eight years of supply-run censuses, we were able to make 220 censuses with an exceedingly constant effort of one to three censuses per month from April to September.

Foraging behavior

Several previous studies have suggested that much insight into the foraging ecology of seabirds can be gained by measuring dive sequences and characteristics (Dow 1964, Stonehouse 1967, K. A. Hobson and Sealy 1985, Cooper 1986). We investigated the diving capabilities of Farallon seabirds by using stopwatches to determine how long birds remained under water and the amount of time between successive dives. As in the previous studies, we did not include dives that resulted in known capture of prey or the handling of prey at the surface, and we timed only those pauses terminated by another dive. Thus, measured dives and pauses were not prematurely ended or unduly prolonged. From the duration of dives and pauses between dives, we calculated the dive–pause (*D*/*P*) ratio. All birds were timed within the same 100-m^2 area off East Landing at Southeast Farallon; water depth there is about 20 m. All the species observed can certainly dive deeper than 20 m, but we felt it important to compare them in the same habitat.

We also investigated the tendency of species to feed in flocks, including mixed-species flocks, as well as seasonal and year-to-year differences in such tendencies. We recorded the presence of feeding flocks that occurred within 3 km of the South Farallones. The effort from 1972 to 1978 was semicasual, so only large flocks were recorded then, but from 1979 to 1983 the effort was more concerted and probably most flocks were noted. We logged time of day, direction from the island, species composition, size of flock, and persistence at a given location. To analyze flocking tendencies, we used Cole's (1949) coefficient of association with flocks being the unit of comparison rather than the census blocks used above.

We assessed another aspect of foraging behavior—foraging effort—by conducting all-day watches of selected species' nests during the breeding season. We assumed that the time birds spent away from nests, i.e., the interval between successive nest reliefs, was a measure of the ease with which birds located food (A. W. Diamond 1983). An all-day watch was conducted every three to five days from blinds.

From 1972 to 1978, both cormorants and murres were observed from the blind atop Shubrick Point, which overlooked the nests of several species (Figure 1.3). Samples of about two to five Pelagic Cormorant nests, 15 Brandt's Cormorant nests, and 75 murre sites within close (10 m), easy view were identified on colony photos. From 1979 to 1983, Brandt's Cormorant watches were moved to the blind overlooking the large colony at Sea Lion Cove, but murre watches were continued at Shubrick Point. Pelagic Cormorant watches were conducted only during 1973–77. The sites observed for all three species were among those we followed to assess reproductive effort and success (detailed in later chapters). During watches, we noted the times of arrival and departure of the birds occupying these nests. Biologists spelled one another every hour to prevent fatigue; three to five biologists participated in the rotation each day.

We assessed time of day when foraging occurred by combining observations of feeding flocks (above), the all-day watch data just mentioned, and information on the daily patterns of attendance of each species (results mainly in individual species' chapters).

Diet composition

We investigated the diet of Farallon seabirds during the nestling period by a variety of means, the method depending on the biology of respective species. As described by Ainley, Anderson, and Kelly (1981), we collected pellets from Double-crested, Brandt's, and Pelagic cormorants just after chicks fledged, 1973 to 1977. During 1977, we were also able to collect Brandt's Cormorant pellets during the nest-building period, and this provided the opportunity to compare early- and late-season diet during that year. The qualities of pellets as a means for assessing cormorant diet are discussed by Ainley, Anderson, and Kelly (1981); basically they are quite satisfactory and have been used by a number of workers (e.g., Jordán 1959, Schlatter and Moreno 1976). Within pellets, fish can be identified by their otoliths, and invertebrates can be identified by such hard parts as beaks for cephalopods and carapaces for decapods. A number of studies indicate that counts of hard parts may overrepresent cephalopods and decapods at the expense of fish (Bradstreet 1980, Ainley, O'Connor, and Boekelheide 1984, Duffy and Jackson 1986), but in our study squid beaks were so tiny and delicate that for cephalopods this may not have been so.

Four different sources of information were used to quantify the diet of Western Gulls:

1. Pellets. Most pellets were collected from the concrete water-catchment pad on the island's southwest side. Once per week, weather permitting, biologists walked over the area, broke open all pellets encountered, and recorded the predominant food type in each. Otoliths were collected for identification. This technique was used from March to August 1973–78, and in March and April 1979; a total of 11,490 pellets was inspected.

2. Feces. In the same catchment basin discrete feces were classified according to major components. Fecal identification was done from May to August 1976–78; a total of 1,024 fecal samples was inspected.

3. Chick regurgitations. In the course of our annual chick-banding effort, during a two-week period in June 1974–83, we collected food regurgitated by chicks. A total of 325 regurgitations was collected.

4. Adult regurgitations. Observations of courtship feeding or adults feeding chicks were made opportunistically from one of our study blinds in the course of other work. Most identifications required use of a binocular. Information from 1974 was extrapolated from Pierotti (1981), and L. Spear (unpubl. data) contributed his many observations from 1978 to 1983. A total of 1,468 meals was observed.

In the case of the Common Murre, we determined only the diet (including prey size) fed to chicks during the nestling period (Chapter 8). As indicated by J. M. Scott (1973), D. H. Varoujean (unpubl. data), and Bradstreet and Brown (1985), the diet fed to murre chicks differs somewhat from that retained by parents during the period of chick dependence. In general, adults feed their offspring items that can be easily carried lengthwise in the bill, as well as perhaps those of high caloric value (M. P. Harris 1984). These are usually fish or squid from 4 to 15 cm long. In addition to fish, adults eat small invertebrates such as euphausiids, which would be too energetically costly and nutritionally insufficient to carry one by one to the offspring. In 1985 and 1986, when we analyzed diet of adults at sea as well as diet fed to chicks, we found that adults were eating the same species of fish fed to chicks and were not then feeding on invertebrates; whether or not adults ate different-sized fish awaits future analysis. During the all-day watches mentioned above (also Chapter 8), we noted the prey brought to specific pairs, as well as the time of day. Fish were identified to the lowest taxon possible, which for most was to species or species group (in the case of rockfish).

We quantified the diet fed to Pigeon Guillemot chicks in a fashion similar to that used for the murre, although we did not conduct all-day watches. Each day, at about 1400 when we made daily rounds of the guillemot study plot (see Chapter 9), we noted prey size and species brought to chicks. As long as we were present, adults refused to enter their burrow with the meal, but instead waited on their roost near the burrow entrance. Thus, we usually had a good opportunity to observe each prey item. In Black Guillemots, *Cepphus grylle,* studied in the high Arctic, the diet of adults differed somewhat from that fed to chicks in much the same way as in murres (Gaston et al. 1985, Bradstreet and Brown 1985).

Observations of prey brought to the chicks of Tufted Puffins were gathered on an opportunistic basis. Several puffin pairs nested in the vicinity of the Shubrick Point blind. When they arrived with prey in their bills and paused long enough at the burrow entrance, we were able to identify many prey to species. On many occasions, however, they were so quick to enter their burrow that we could see only that the prey was fish rather than squid or vice versa. During the years when we observed puffin diet, we were unable to collect information on the diet of the Rhinoceros Auklet.

Cassin's Auklets—like cormorants and gulls, but unlike murres, guillemots, or puffins—feed their chicks by regurgitation. Unlike other seabirds that feed their chicks in this way, auklets have gular pouches in which they store the food until they make their once-daily trip to the nest. The food consumed by adults, on the other hand, is swallowed directly into the alimentary tract. Cassin's Auklet adults feed their chicks on the same prey they catch for themselves (PRBO, unpubl. data 1985 and 1986). During the chick-rearing periods of 1977 and 1979–81, we captured weekly samples of adults just as they arrived to feed their chicks, and we allowed them to regurgitate into plastic vials. These samples were then analyzed to determine diet.

For most species, we calculated diet diversity by using the Shannon-Weiner formula (Hurtubia 1973):

$$H = -\Sigma p \ln p,$$

where p is the proportion of the total diet contributed by each prey species.

We also measured the extent of overlap in the diet composition between species and within species between years. To do this we used Morisita's (1959) index:

$$C = \frac{2\Sigma x^i y^i}{(\Sigma x^i)^2 + (\Sigma y^i)^2},$$

where x and y are the numerical proportions of various prey in the two diets being compared.

Prey size

To investigate overlap in prey size, we calculated the lengths of fish prey that were common to several species' diets. In the case of cormorants, standard lengths of fish were determined by regression of otolith diameters. Many of the regressions have been determined by us (PRBO, unpubl. data), but a few, especially for commercially important fish such as anchovies and herring, are available in the literature. In the case of murres, guillemots, and puffins, fish length was estimated crudely by a comparison of the fish carried in the bill to bill length. For the more abundant prey species, we were able to convert estimated length to weight by using regressions that we had determined or, in a few cases, that were available in the literature.

FORAGING AREAS AND HABITAT

Overall patterns

As mentioned earlier, we know that waters along the mainland coast out to the oceanic boundary of the San Francisco Bay plume from southern Marin County to southern San Francisco County (Point San Pedro) are exceedingly important. This is apparent in the data of Briggs et al. (1987), who censused seabirds by air monthly from 1980 to 1982 within 70 km of the northern and central California coast. Equally important are waters of the continental slope, where the islands actually lie. These patterns were confirmed during 1985 and 1986, when we censused seabirds by ship in both April and June. The former month represents the prelaying period and the latter the nestling period for most Farallon seabirds. These data, in addition, provided comparison of foraging habitat use during years of exceedingly high (1985) and exceedingly low (1986) reproductive success (see above). According to preliminary analysis of trawl data taken by the National Marine Fisheries Service on these cruises (P. Adams, S. Smith, and T. Wyllie Echeverria, pers. comm.), juvenile rockfish and euphausiids (especially *Thysanoessa spinifera*) were much less abundant in and near the Gulf of the Farallones during 1986 than during 1985. The data gathered on the Golden Gate–to–Farallon supply run, 1972–80, were also instructive in providing further perspective on between-year variability.

A general feeling for the location of Farallon species' feeding areas

TABLE 3.1
Sightings of Farallon seabird species along the Golden Gate–to–Farallon census track, by segment, 1972–81

Species	Segment I	II	III	IV
Leach's Storm-Petrel	0	0	0	1
Ashy Storm-Petrel	0	0	1	3
Double-crested Cormorant	19	6	5	9
Brandt's Cormorant	169	183	196	217
Pelagic Cormorant	27	14	7	26
Western Gull	127	134	147	147
Common Murre	145	184	216	222
Pigeon Guillemot	6	12	14	43
Cassin's Auklet	1	0	2	74
Tufted Puffin	0	0	1	36
Rhinoceros Auklet	0	0	2	47

NOTE: Data show number of transects in which each species was seen. Segment I is nearest the coast, segment IV nearest the islands; see Figure 3.2.

can be derived by comparing the relative frequency with which they were observed on the four segments of the Golden Gate–to–Farallon track (Table 3.1). Storm-petrels were seen rarely, and then only in deeper waters near the islands. Double-crested Cormorants were seen near San Francisco Bay, flying rather than feeding. Brandt's Cormorants were quite evenly distributed among the four census segments, but we know that during some years they prefer inshore and during other years they prefer offshore waters (see below). The Pelagic Cormorants encountered in segment IV nest at the Farallones, and those encountered in segments I and II nest near Point Bonita (see Figure 3.2 below). The distributions of the Western Gull and murre resemble that of Brandt's Cormorant. The remaining four species, all alcids, were encountered primarily in the deeper waters near the islands.

These patterns are supported closely by information provided by Briggs et al. (1987), and the 1985–86 censuses. Leach's Storm-Petrels reside mainly in the warmer, blue waters of lower productivity west of the continental slope and of the Gulf of the Farallones (Figure 3.1). These are waters influenced little by coastal upwelling. Ashy Storm-Petrels, on the other hand, occur over the same waters but also nearer the continental shelf break, particularly in the vicinity of Cordell Bank 50 km to the north (Figure 3.1). These are waters influenced strongly by coastal upwelling. At a minimum, then, Leach's Storm-Petrels feed

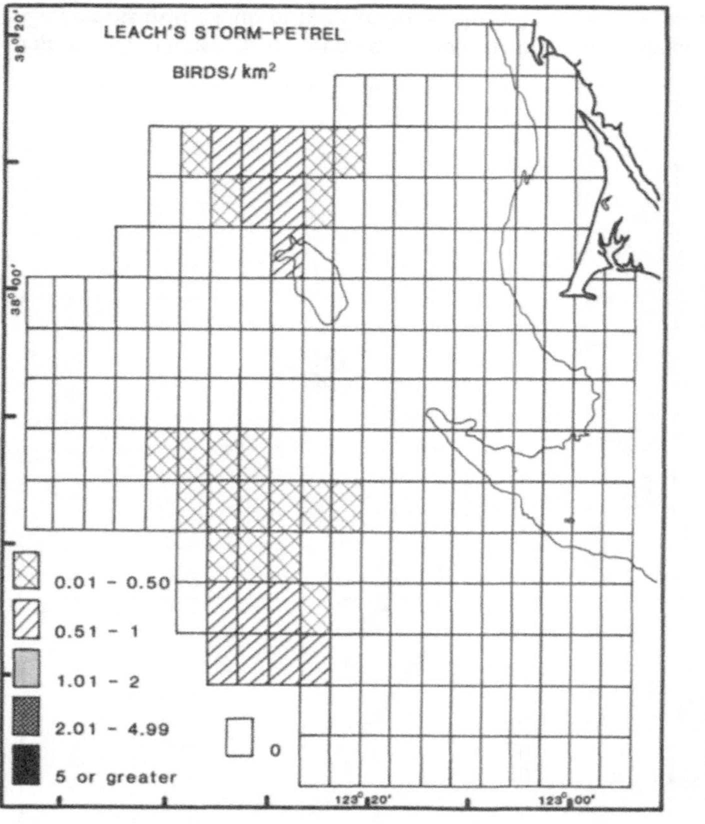

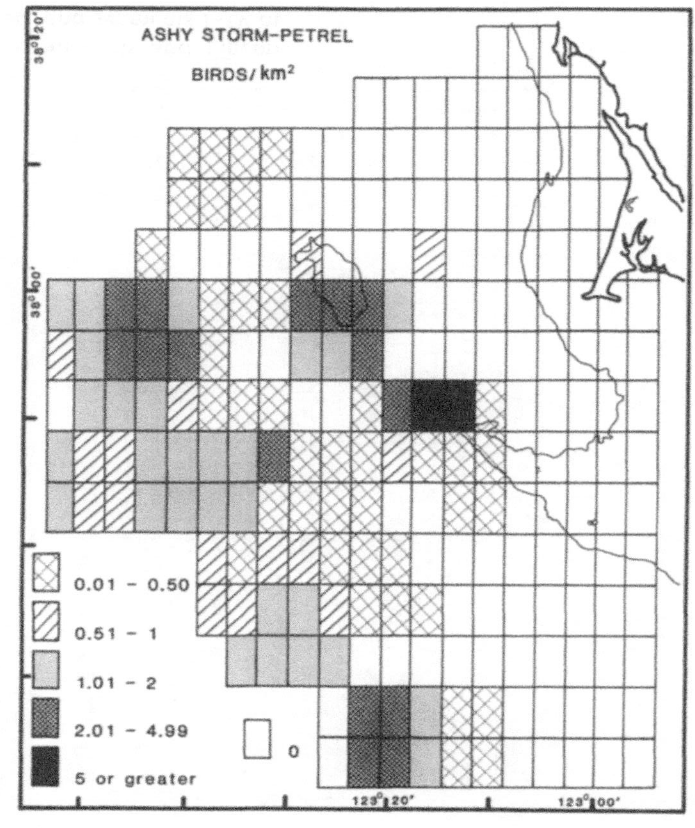

Figure 3.1. Densities of all Leach's and Ashy storm-petrels along cruise tracks, April and June 1985–86, combined.

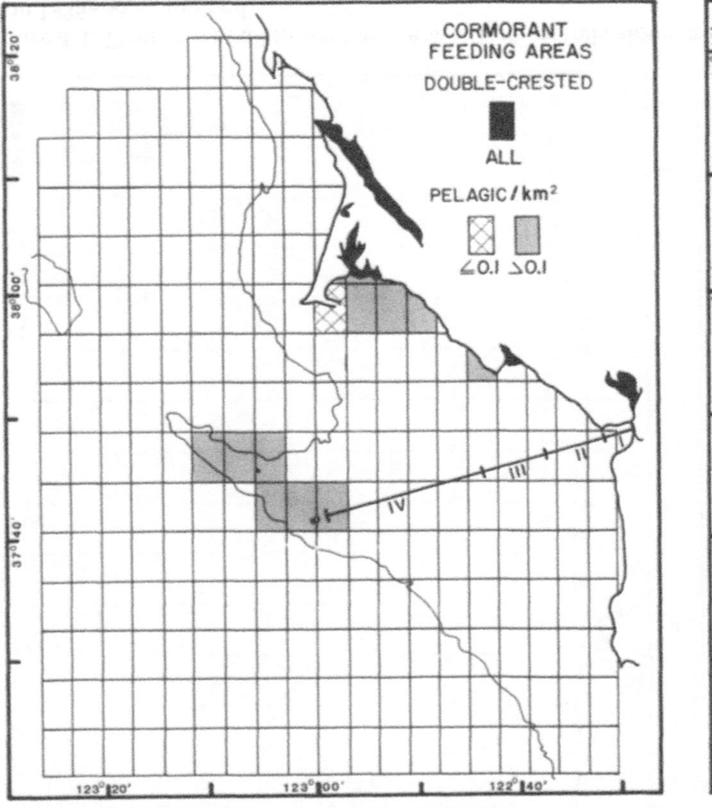

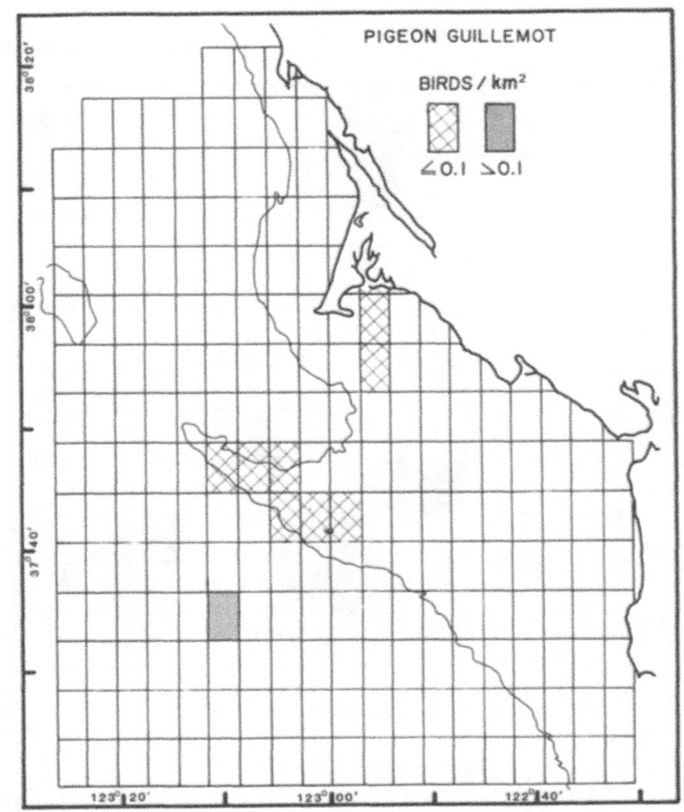

Figure 3.2. Densities of feeding Double-crested and Pelagic cormorants and Pigeon Guillemots in the Gulf of the Farallones, June 1985–86, combined, and segments I–IV of the Farallon–to–Golden Gate census track (see text).

at least 50 km from the Farallones, and Ashy Storm-Petrels feed at least 25 km away. At maximum, individuals of both species probably feed much farther afield, the Leach's more to the west and the Ashy perhaps more to the north and south of the Farallones. Briggs et al. (1987) reported an association between storm-petrels and convergence lines visible at the sea surface. These boundaries along eddies and water types are generally quite ephemeral in this region.

Largely on the basis of observations of color-banded birds, we know that Double-crested Cormorants feed principally in Bolinas Lagoon, Drake's Bay, and Limantour Estero (Figure 3.2). In earlier years, when many thousands of Double-cresteds nested at the Farallones (Ainley and Lewis 1974), large numbers fed in San Francisco Bay (Bartholomew 1943). The extent to which they feed there now is not known, but at times perhaps they do (Table 3.1). Similarly, the extent to which Farallon individuals use Tomales Bay is not known, but we suspect it is an important area. The waters where Double-crested Cormorants feed are no more than 10 m deep, and they overlie flat sand or mud. During the nesting season Double-crested Cormorants feed no closer than 30 km and perhaps more than 80 km from Southeast Farallon.

Early in the spring, Brandt's Cormorants tend to feed within San Francisco Bay, sometimes as far as the bay's east shore 80 km from the Farallones, as well as in nearshore waters of the Gulf of the Farallones (Figure 3.3). Later in the season, depending on year, they feed either near the islands or in coastal waters (Figure 3.4). Waters in the latter area are turbid and are influenced markedly by the outflow from San Francisco Bay. Thus, Brandt's Cormorants feed 0–80 km from Southeast Farallon during the breeding season. Most of the habitat along the mainland coast is composed of flat sand or mud bottoms 10–60 m deep; the habitat offshore includes rocky bottoms as well and is up to 120 m deep.

Virtually the entire breeding population of the Pelagic Cormorant probably feeds within a few kilometers of the South Farallones (Figure 3.2). In this area are many submerged, rocky reefs, the feeding habitat preferred by this species. Pelagic Cormorants exhibit remarkably little plasticity in this choice of feeding area. We have observed them in waters up to 120 m deep.

The Western Gull probably exploits more of the Gulf of the Farallones and vicinity than any other species (Figures 3.5 and 3.6). This is perhaps a function of its being confined to feeding at the surface (see below). During some years gulls feed farther from the island

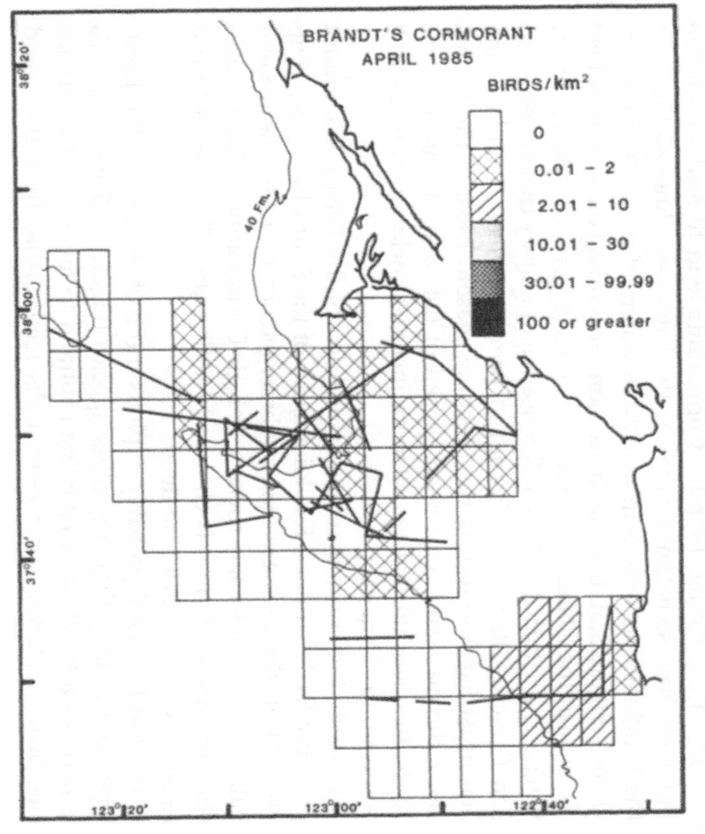

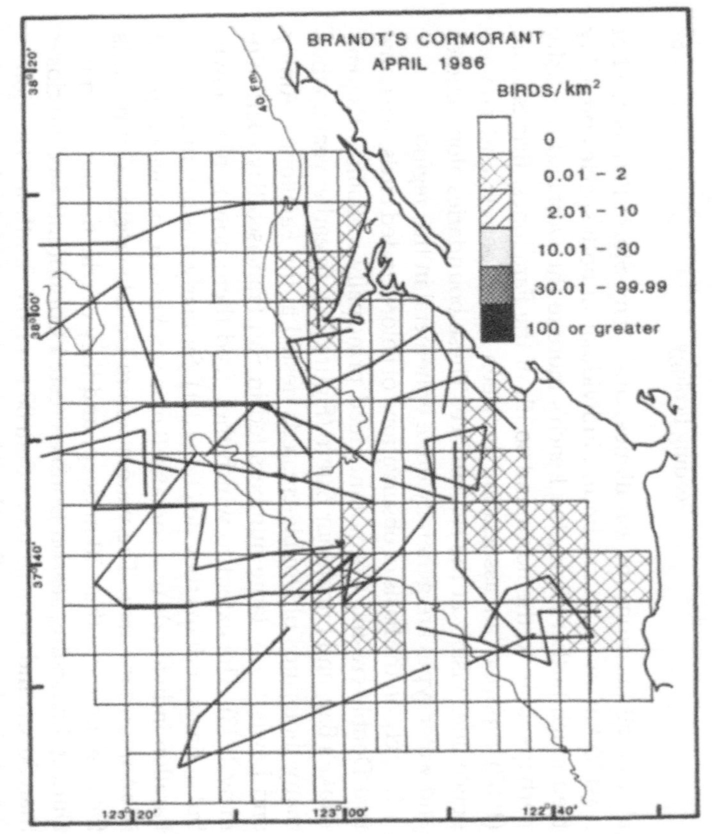

Figure 3.3. Densities of feeding Brandt's Cormorants along cruise tracks in April 1985 and 1986; tracks indicated by bold lines.

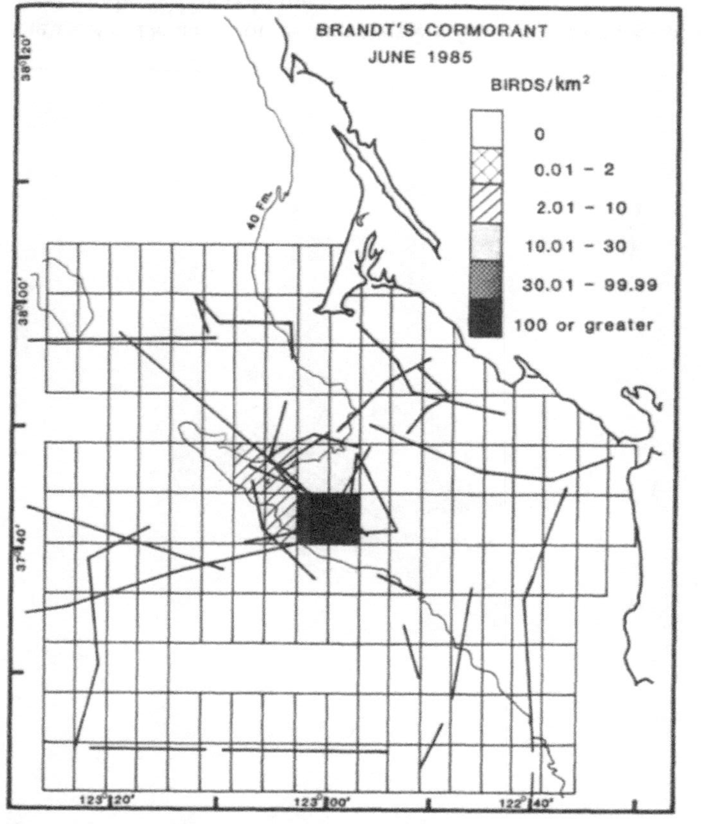

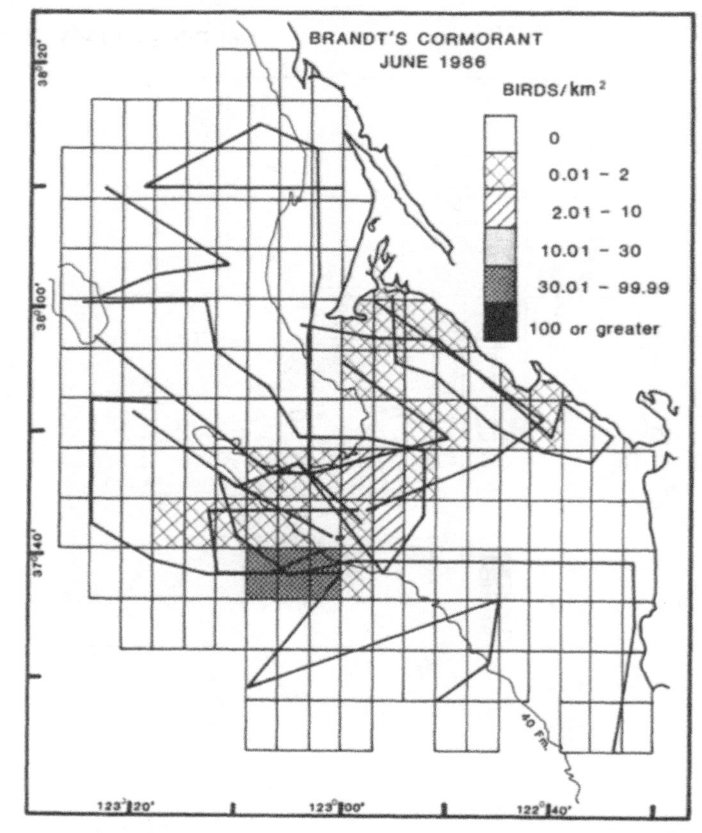

Figure 3.4. Densities of feeding Brandt's Cormorants along cruise tracks in June 1985 and 1986; tracks indicated by bold lines.

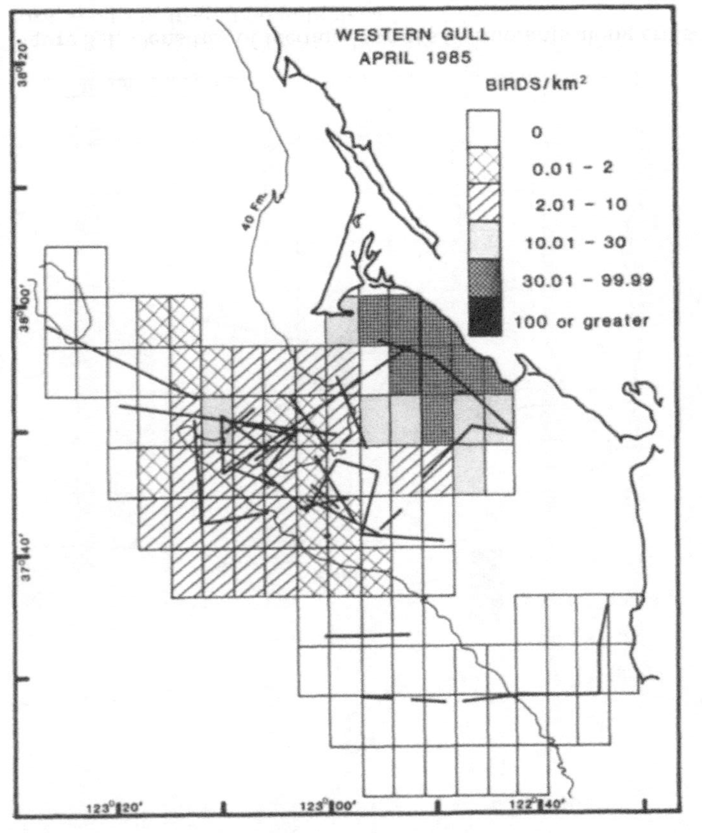

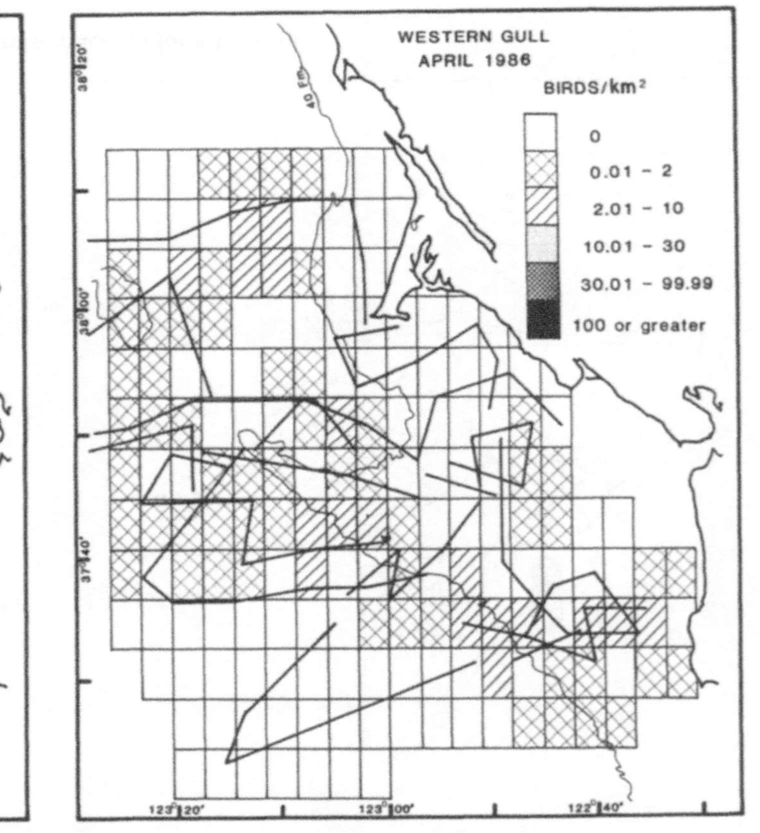

Figure 3.5. Densities of feeding Western Gulls along cruise tracks in April 1985 and 1986; tracks indicated by bold lines.

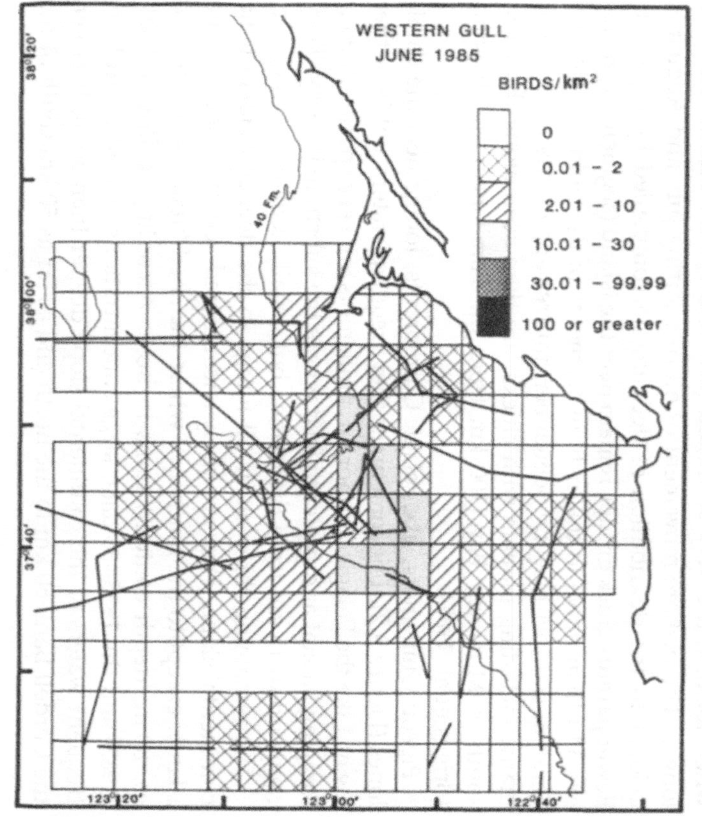

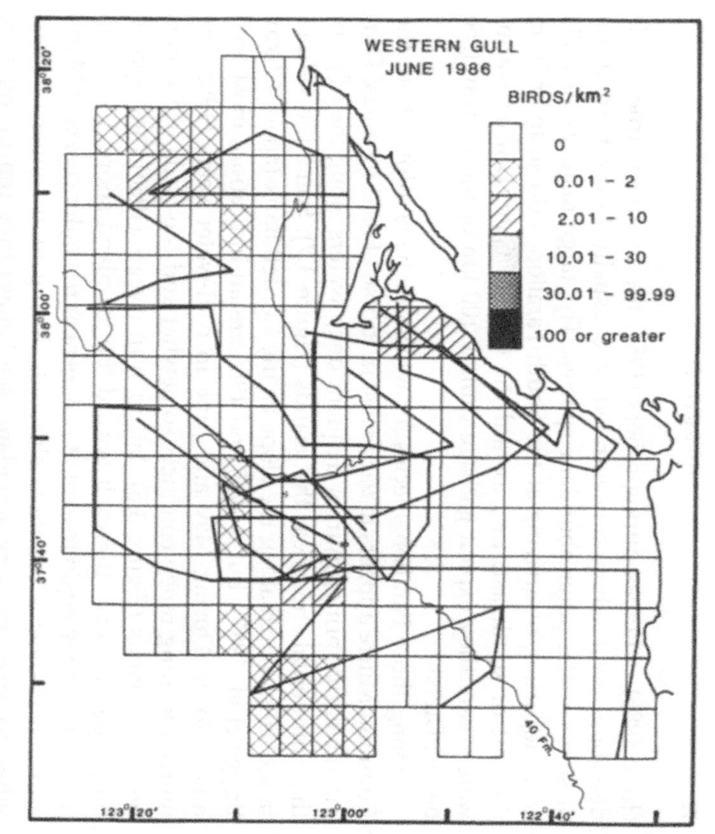

Figure 3.6. Densities of feeding Western Gulls along cruise tracks in June 1985 and 1986; tracks indicated by bold lines.

(June 1986), whereas during others (1985) the reverse is true. Their densities, even when concentrated, are usually lower than those of some of the other abundant species. The large majority of gulls seen departing or arriving at the South Farallones do so to or from the east or southeast, i.e., toward the Golden Gate. During warm-water years they exploit garbage dumps 100 km or more away (see below, Diets).

Among those Farallon species for which we have information, the Common Murre appears to vary most in its use of feeding areas. During the early spring, murres frequent deep waters along the edge of the continental shelf near the islands (Figure 3.7). Some individuals range as far north along the slope as the Cordell Bank (about 60 km; Briggs et al. 1987), but if 1985 and 1986 are an indication, most feed much closer. During May and June in cool-water years (1985) the murre's feeding range contracts somewhat and the birds use waters near the islands (Figure 3.8), but in warm years (1986) they spread out, especially over the shelf toward the mainland shore. By July in many years they begin to exploit waters near the outer coasts of Marin and San Francisco counties, although during midsummer of years when rockfish are very abundant (see below) they remain offshore longer. The inshore movement of murres during July and August is evident in the distribution of fledglings accompanied by a parent. Fewer parent–chick groups remain near the island (segment IV in Figure 3.9) than move to the inshore segments of the Golden Gate–to–Farallon transect line. By September, murres apparently spread out; many, including parents with chicks, are then seen along the coast south to Monterey Bay (100 km from the Farallones) and beyond (Briggs et al. 1987).

Pigeon Guillemots, like Pelagic Cormorants, forage near the Farallones (Figure 3.2), but sometimes they frequent waters a little farther away. Like the Pelagic Cormorant, guillemots are very specific in their foraging habitat and prefer to feed where the substrate is rocky. They range up to 15 km from the islands but usually remain much closer.

The remaining three alcids all forage in the deeper waters of the continental slope. They are rarely encountered very far inshore of the islands except to the northeast, where the shelf break (= 40-fathom, ca. 80-m, depth contour) turns inshore. According to Briggs et al. (1987) and the 1985–86 data, large concentrations of Cassin's Auklets frequent waters of the continental slope deeper than 80 m, between the Cordell Bank 60 km to the north and as far as 50 km south of the

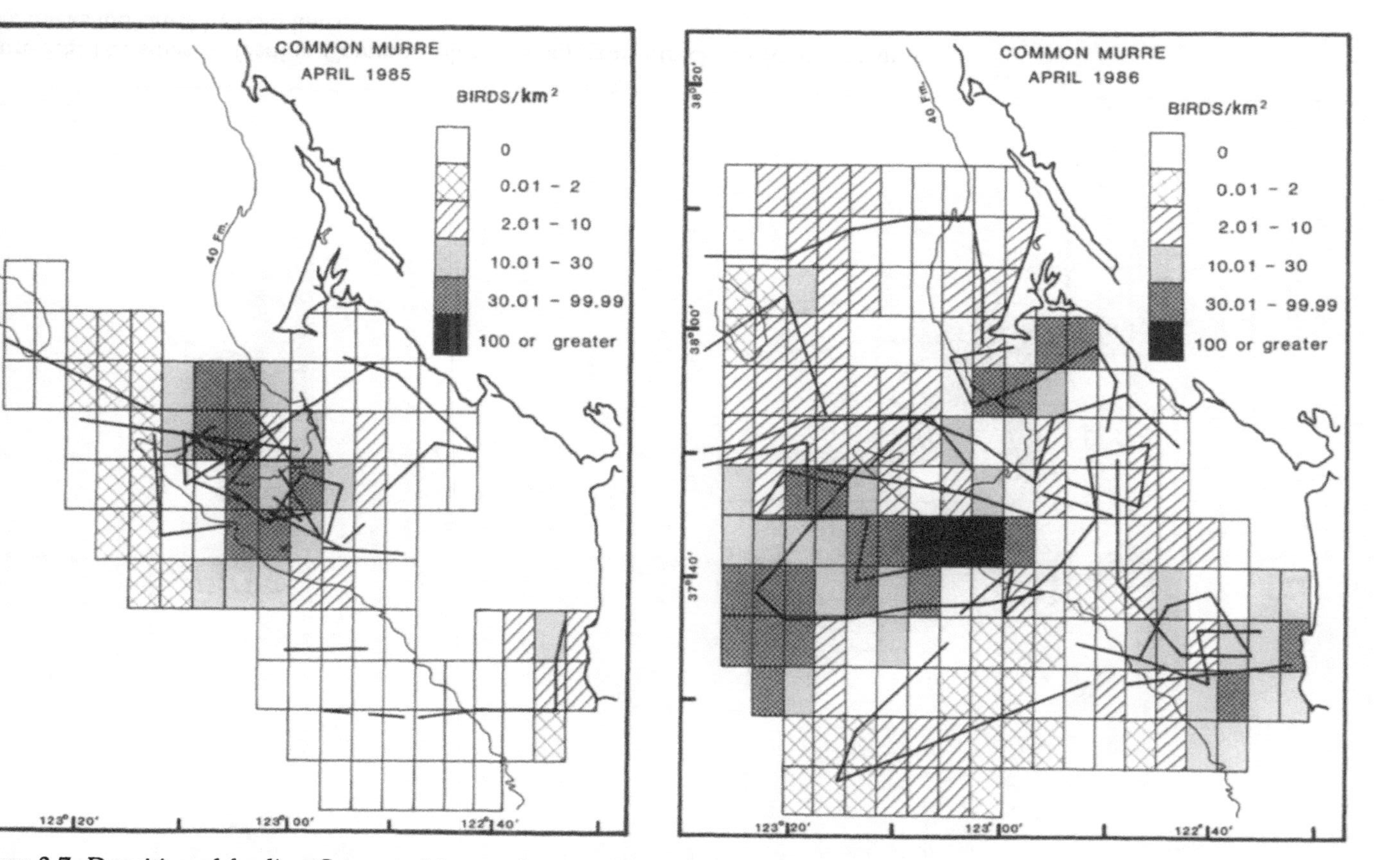

Figure 3.7. Densities of feeding Common Murres along cruise tracks in April 1985 and 1986; tracks indicated by bold lines.

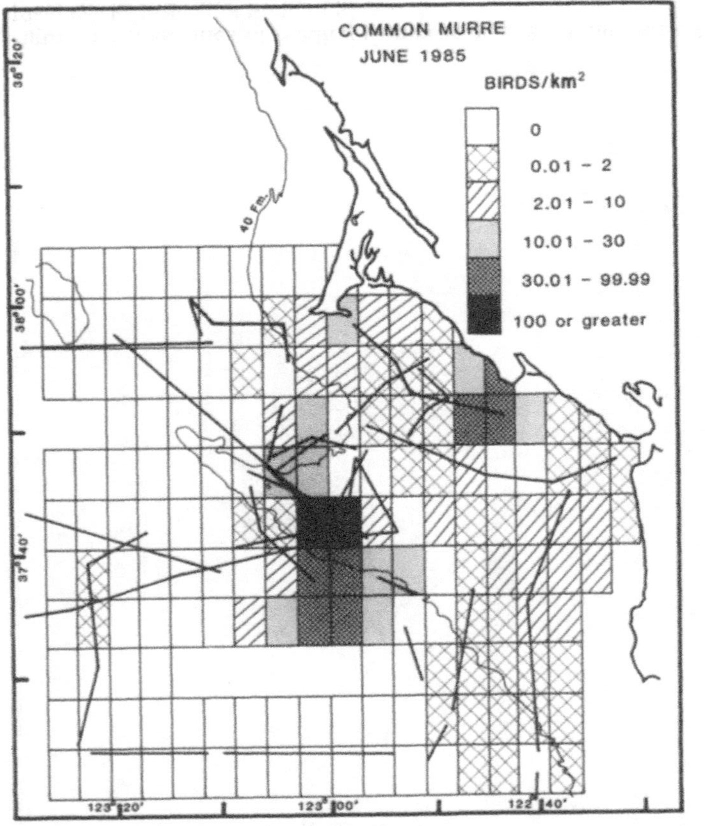

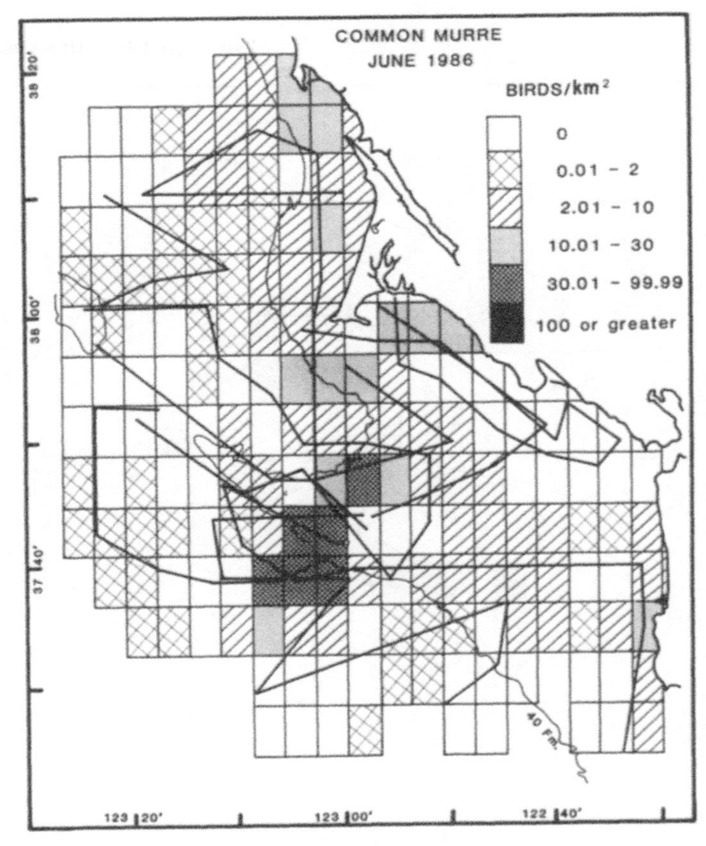

Figure 3.8. Densities of feeding Common Murres along cruise tracks in June 1985 and 1986; tracks indicated by bold lines.

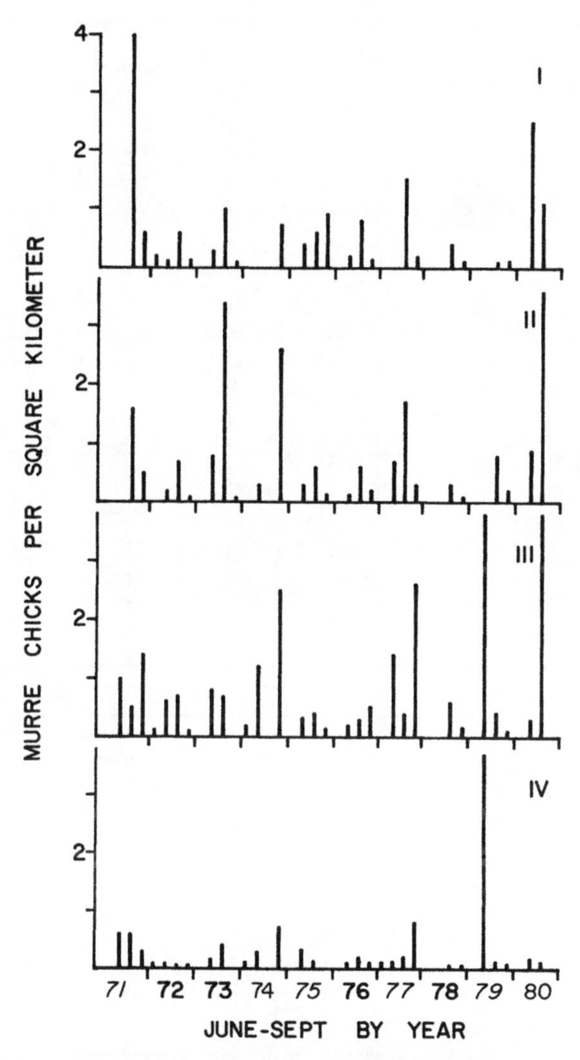

Figure 3.9. Average numbers of murre chicks (with parents) seen within the four segments of the Farallon–to–Golden Gate cruise track, by year, 1971–80.

island (Figures 3.10 and 3.11). Rhinoceros Auklets (Figures 3.12, 3.13) and Tufted Puffins (Figures 3.14, 3.15) also occur most frequently in waters over these same slopes.

Sooty Shearwaters can be found throughout the shelf and slope as well as the deeper waters of the Gulf and vicinity (Briggs and Chu 1986). Depending on year, after arriving in late April, they may con-

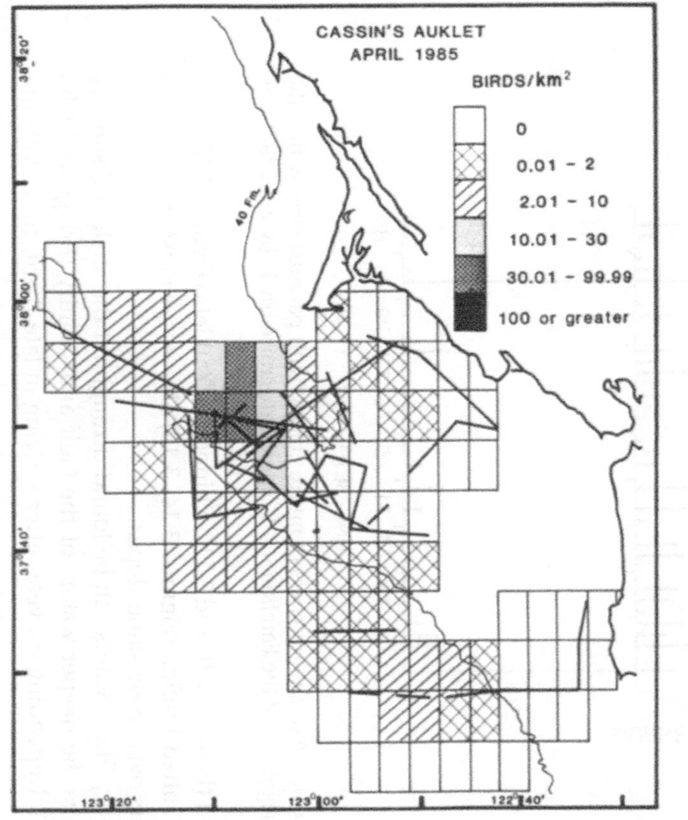

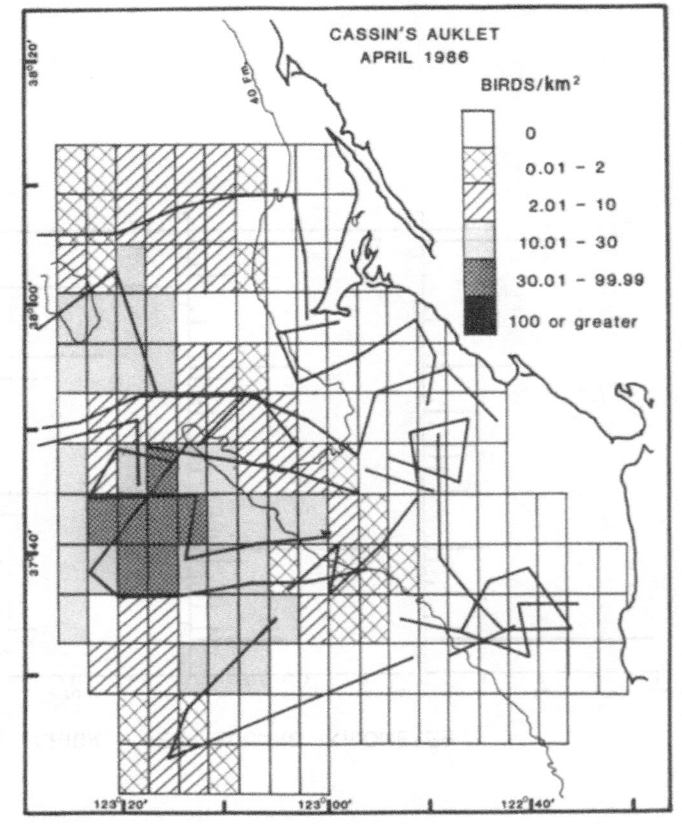

Figure 3.10. Densities of feeding Cassin's Auklets along cruise tracks in April 1985 and 1986; tracks indicated by bold lines.

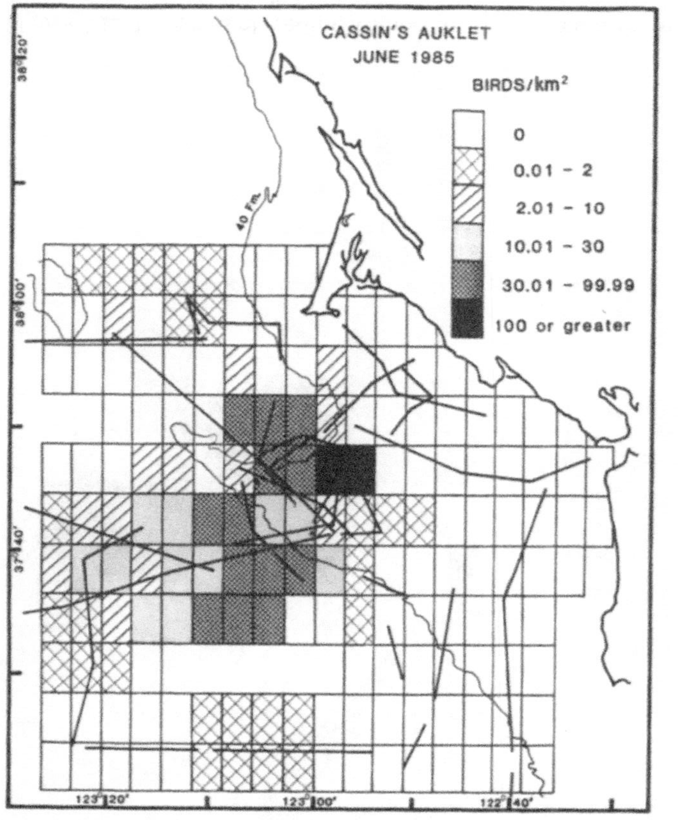

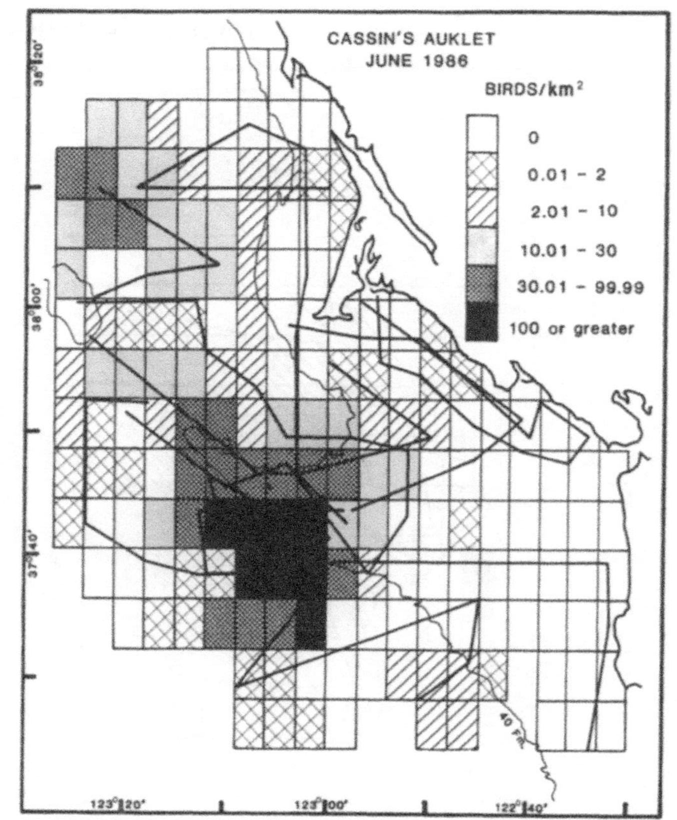

Figure 3.11. Densities of feeding Cassin's Auklets along cruise tracks in June 1985 and 1986; tracks indicated by bold lines.

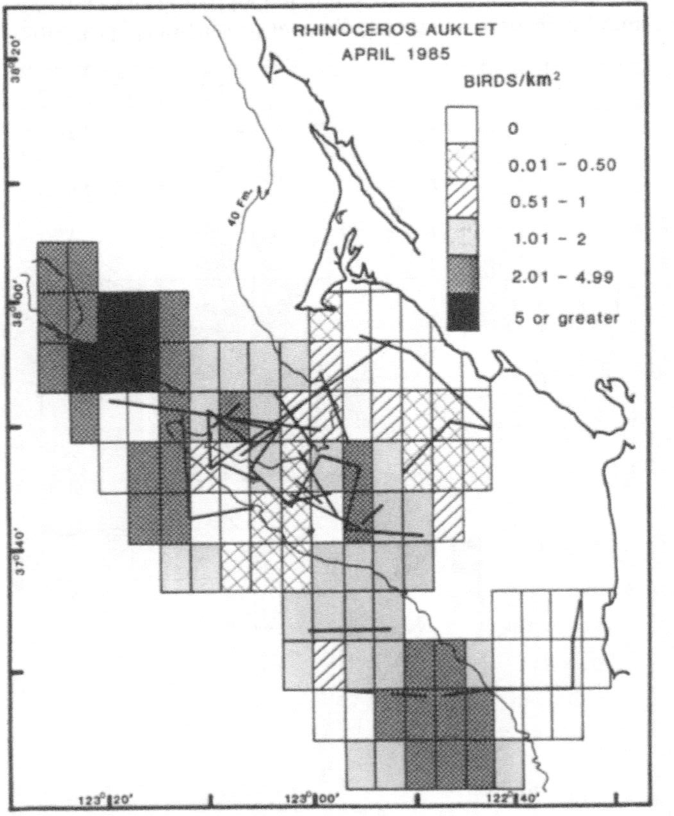

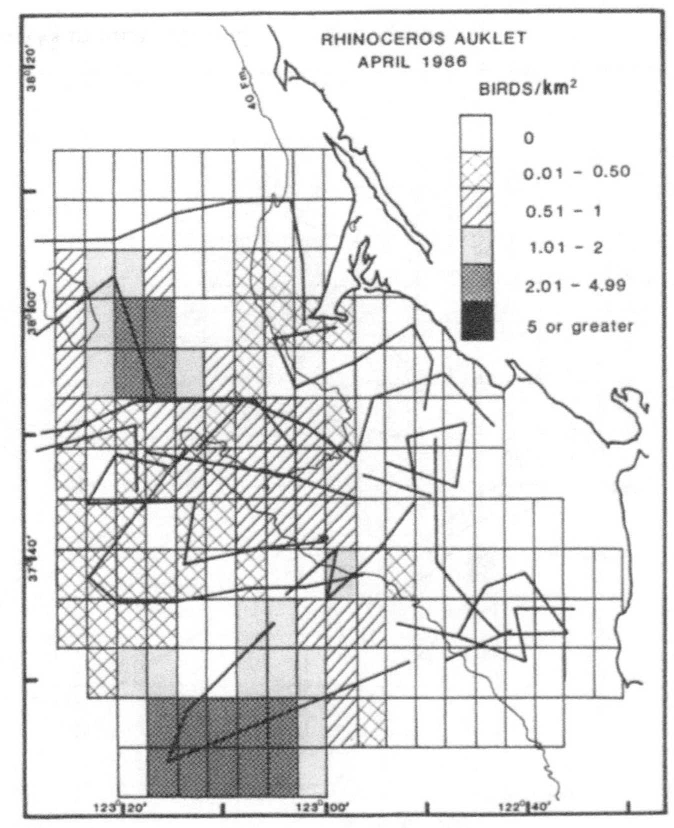

Figure 3.12. Densities of feeding Rhinoceros Auklets along cruise tracks in April 1985 and 1986; tracks indicated by bold lines.

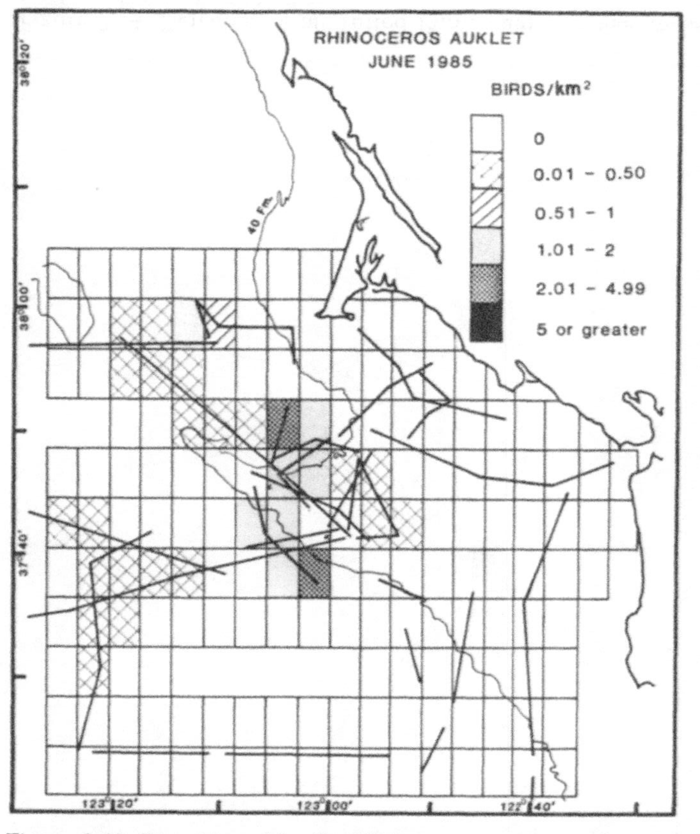

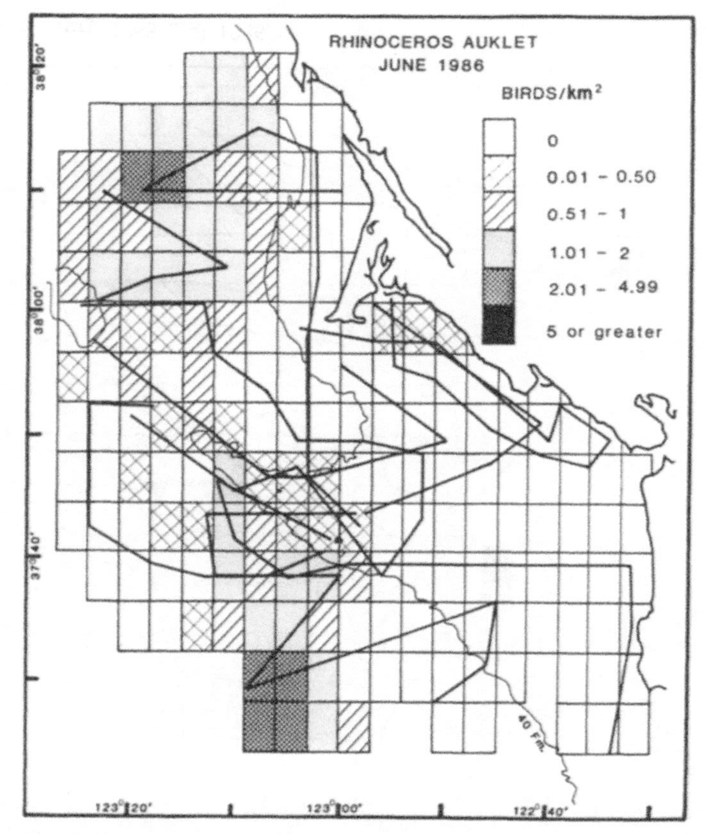

Figure 3.13. Densities of feeding Rhinoceros Auklets along cruise tracks in June 1985 and 1986; tracks indicated by bold lines.

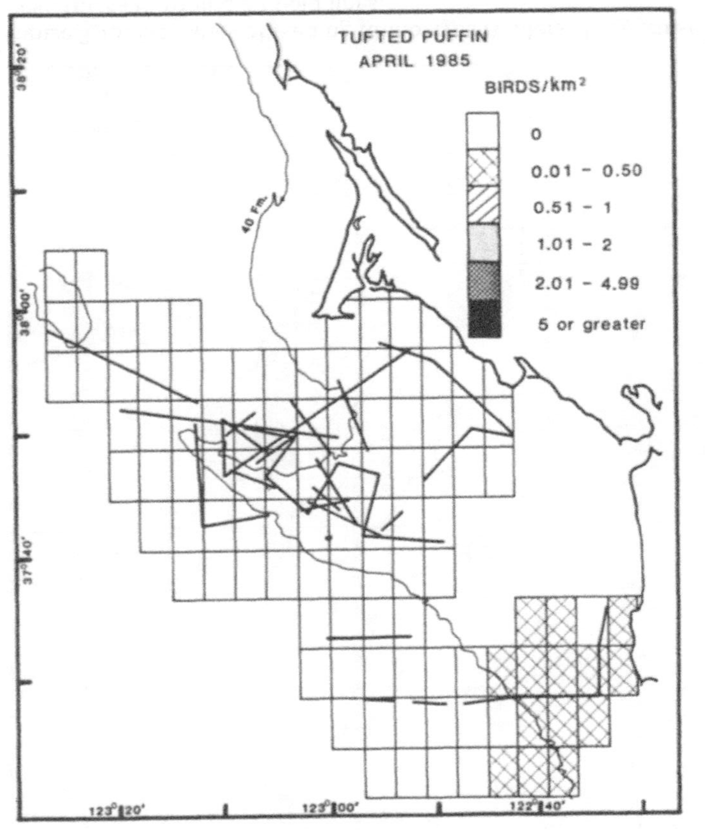

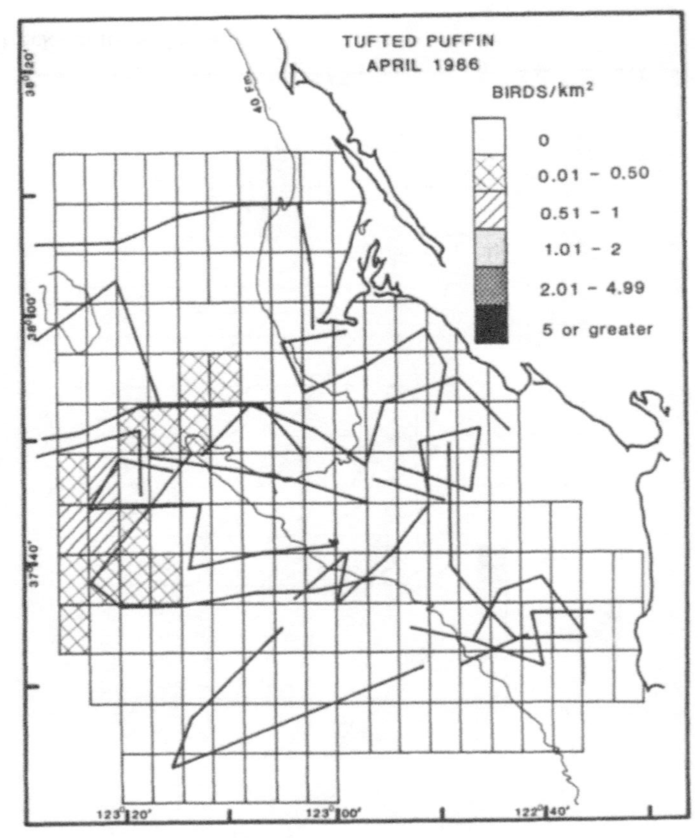

Figure 3.14. Densities of all Tufted Puffins along cruise tracks in April 1985 and 1986; tracks indicated by bold lines.

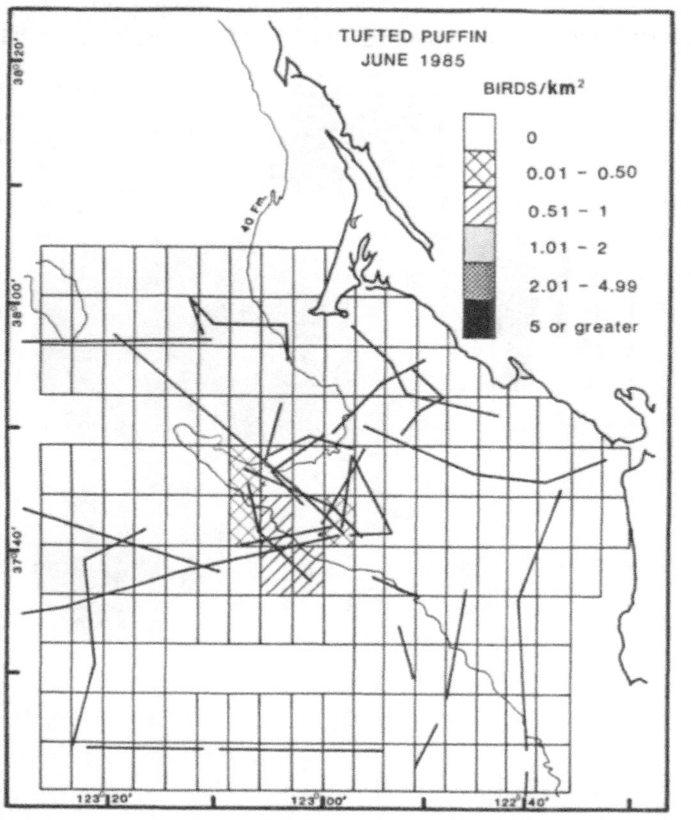

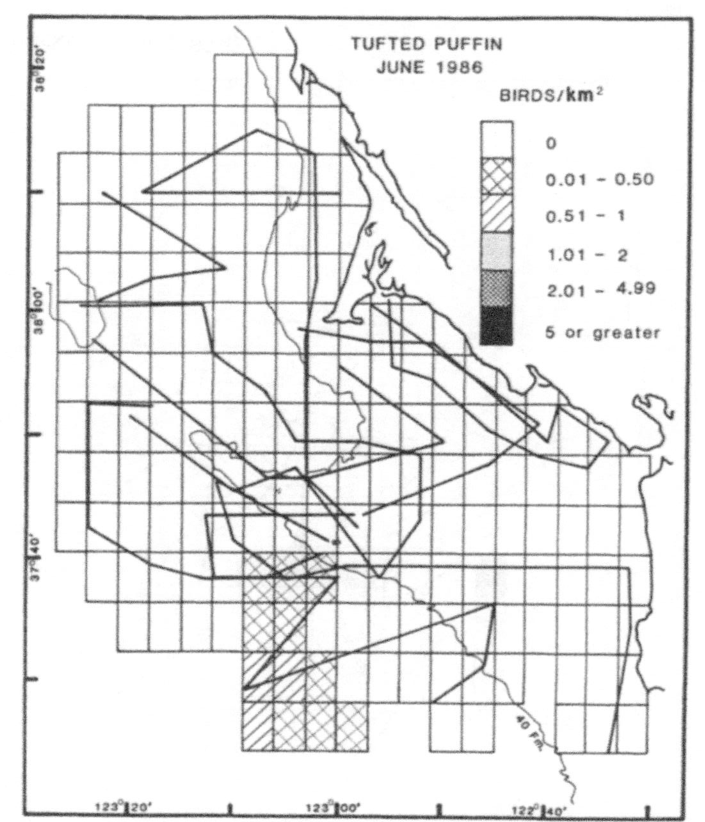

Figure 3.15. Densities of all Tufted Puffins along cruise tracks in June 1985 and 1986; tracks indicated by bold lines.

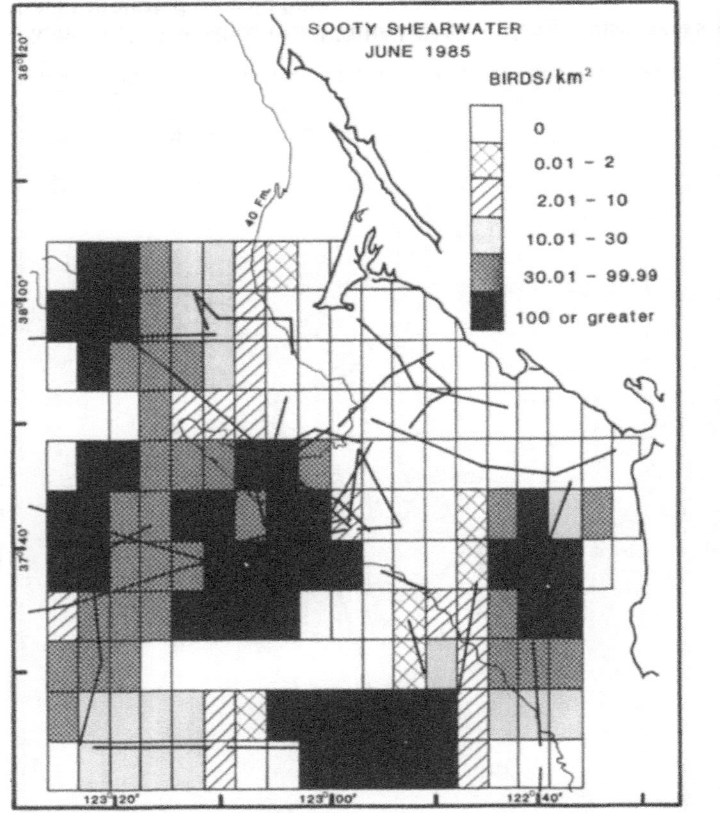

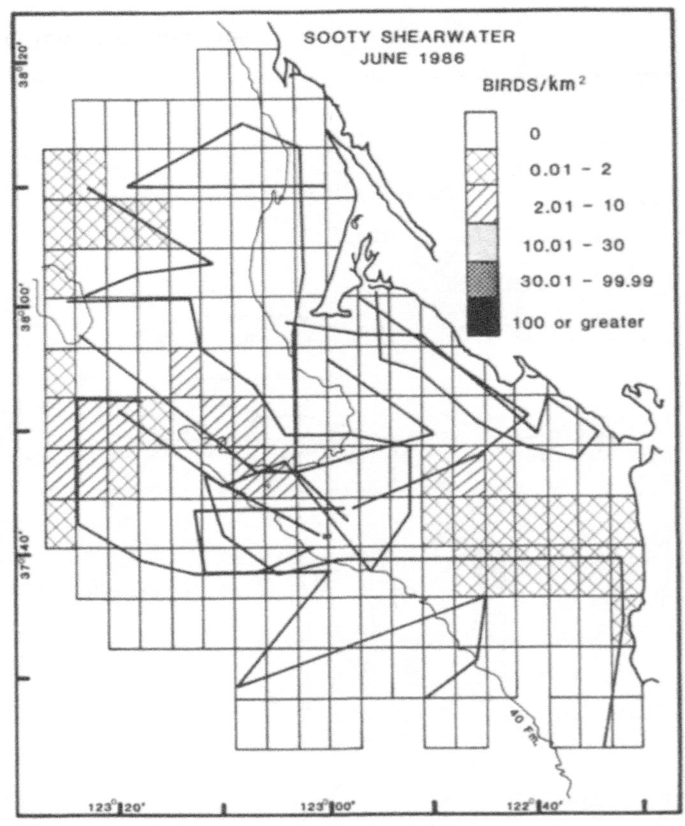

Figure 3.16. Densities of feeding Sooty Shearwaters along cruise tracks in June 1985 and 1986; tracks indicated by bold lines.

centrate more inshore than offshore or vice versa (Figure 3.16). In June 1985, immense numbers occurred around the Farallones, particularly over the deeper waters to the west, but also over shallow waters to the southeast. The large concentrations of this species near the Farallones in 1985, presumably a response to food availability, indicate that the feeding ranges of Farallon species, also concentrating there, were a function of where the prey were located rather than a constraint of nesting duties. Conversely, few Sooty Shearwaters were observed in June of the warmer year, 1986, particularly over the continental shelf.

Between-year variation

We have few data to indicate what annual variability might exist in the foraging ranges and habitats of the storm-petrels. Their densities were too low to allow adequate comparison of between-year variation. In the case of Double-crested and Pelagic cormorants and Pigeon Guillemots, the data available (Briggs et al. 1987; our 1985–86 data) indicate little variability at the height of the breeding season (in June) from one year to the next. Values of the coefficient of association between habitat blocks occupied June 1985 compared to those occupied June 1986 equaled 1.00 ± 0.01. In contrast, the habitat occupied by Cassin's and Rhinoceros auklets changed appreciably, with a slight shift to slope waters (C = 0.52 ± 0.01 for both). The habitats of other species, however, shifted even more from 1985 to 1986. Between-year C-values for Brandt's Cormorant and Western Gull were 0.21 ± 0.01 and 0.16 ± 0.01, respectively, and were associated with a shift toward the mainland. These values indicate little correspondence in habitat use between the two years, but even so the values are artificially high because they do not incorporate the many individuals feeding near the Golden Gate and in San Francisco Bay where our ship could not go (because of heavy vessel traffic). Tufted Puffins and Sooty Shearwaters remained in slope waters during the two years but spread farther from the islands in 1986 (C = 0.23 and 0.36, respectively; ±0.01). A large shift in habitat was exhibited also by the murre with its switch from slope waters in 1985 to inshore waters in 1986 (C = 0.38 ± 0.01). These habitat changes are discussed more qualitatively in the following paragraphs.

During April in the 1985 nesting season, Common Murres, Cassin's Auklets, and Sooty Shearwaters fed in large concentrations within 10 km of the Farallones and Fanny Shoal, 10 km north. Brandt's Cormorants fed inshore near the coast. Habitat occupancy in June exhibited little change, except that cormorants had moved to feed at the islands as well (Figures 3.3 to 3.11). Many cormorants and murres, in fact, fed within several hundred meters of island shores, and our density estimates are accordingly biased downward because the survey vessel could not pass closer to the island than about 1 km. The only cormorants we observed, other than those from mainland nesting colonies feeding near the mainland, were flocks of Double-crested Cormorants in transit to and from Drake's Estero. Similarly, concentrations of murres also occurred around mainland nesting areas. Gulls were "concentrated" within about 15 km of the islands (Figures 3.5 and 3.6). Rhinoceros Auklets and puffins were relatively numerous in slope waters adjacent to the Farallones (Figures 3.12 to 3.15).

During the 1986 nesting season, when warm-water conditions prevailed, the patterns were markedly different. Although in April murres and Cassin's Auklets concentrated near the North Farallones, by early June murres began to feed in mainland coastal waters and auklets spread more widely along the slope (Figures 3.8 to 3.11). Some Brandt's Cormorants, and a few murres, fed 5 km northwest of Southeast Farallon in a flock that persisted for several days in early June. Otherwise, most Brandt's Cormorants undertook long flights to waters near the mainland coast. Gulls and Rhinoceros Auklets were dispersed in much lower densities all over the Gulf and over the deep waters to seaward, respectively. Tufted Puffins also occurred in slope waters, but farther from the islands than in 1985.

These patterns indicate broader overlap of feeding habitat among species in 1985 than in 1986, with some notable exceptions. In a comparison of coefficients of association between species by census block, 33 of a possible 45 species combinations showed significantly greater overlap in habitat use during 1985 than during 1986; 8 combinations (six negative ones involving the Ashy Storm-Petrel) did not change (Appendix 3.2). Only four species combinations, three of which involved the Common Murre, exhibited greater overlap in 1986. The fourth involved the Rhinoceros Auklet and Western Gull. What factor affected the increased association of Common Murres with the habitat used by the Sooty Shearwater, Pigeon Guillemot, and Cassin's Auklet during 1986 is not known, but given the large difference in food availability between the two years, as noted above, it is likely that the occurrence patterns of prey were somehow involved.

Our Farallon–to–Golden Gate transects 1972–80, as well as observations of feeding flocks near the island, reveal the frequency at which the above two sets of conditions may have alternated during

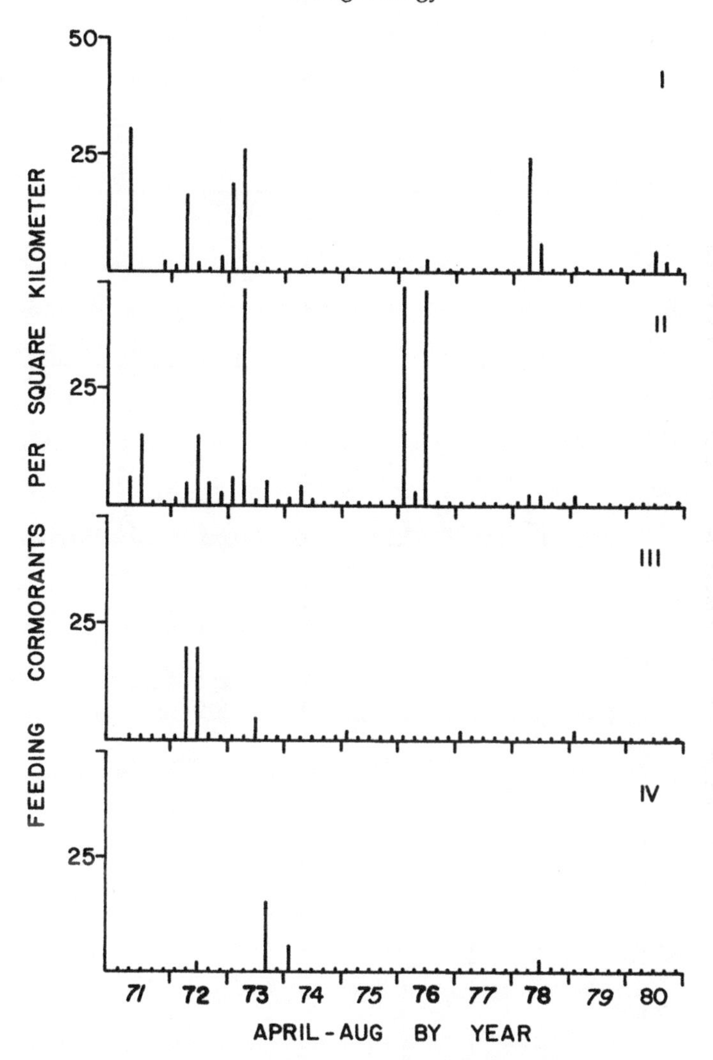

Figure 3.17. Mean number of feeding Brandt's Cormorants observed along the Golden Gate–to–Farallon track, by track segment and year, April to August, 1971–80.

the study period. For purposes of the following review, recall from Chapter 2 that 1973, 1976, 1978, 1982–83, and to a lesser extent 1980 were years of warm water and low marine production. In four of five of these warm-water years, i.e., early 1971, 1973, 1976, 1978, and to a lesser degree 1980, large numbers of Brandt's Cormorants fed near the coast (transect segments I and II, Figure 3.17). Conversely, cormorant

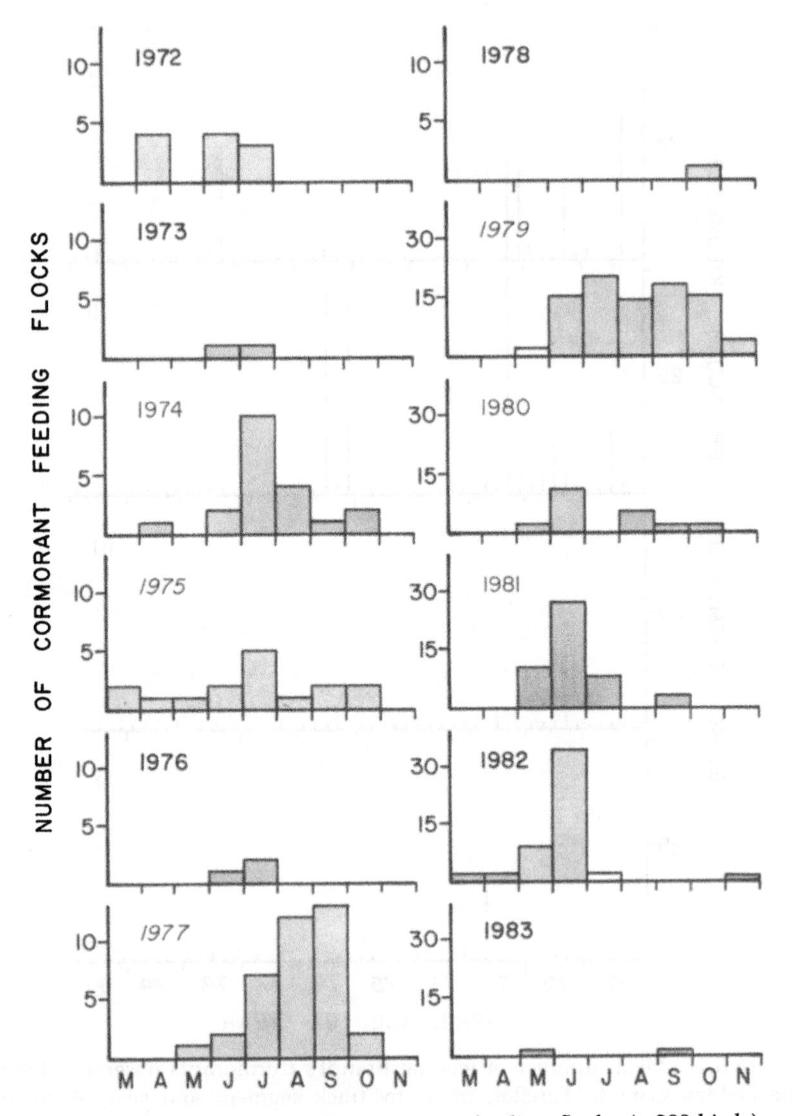

Figure 3.18. Number of Brandt's Cormorant feeding flocks (>200 birds) per month within 3 km of Southeast Farallon, by year, 1972–83.

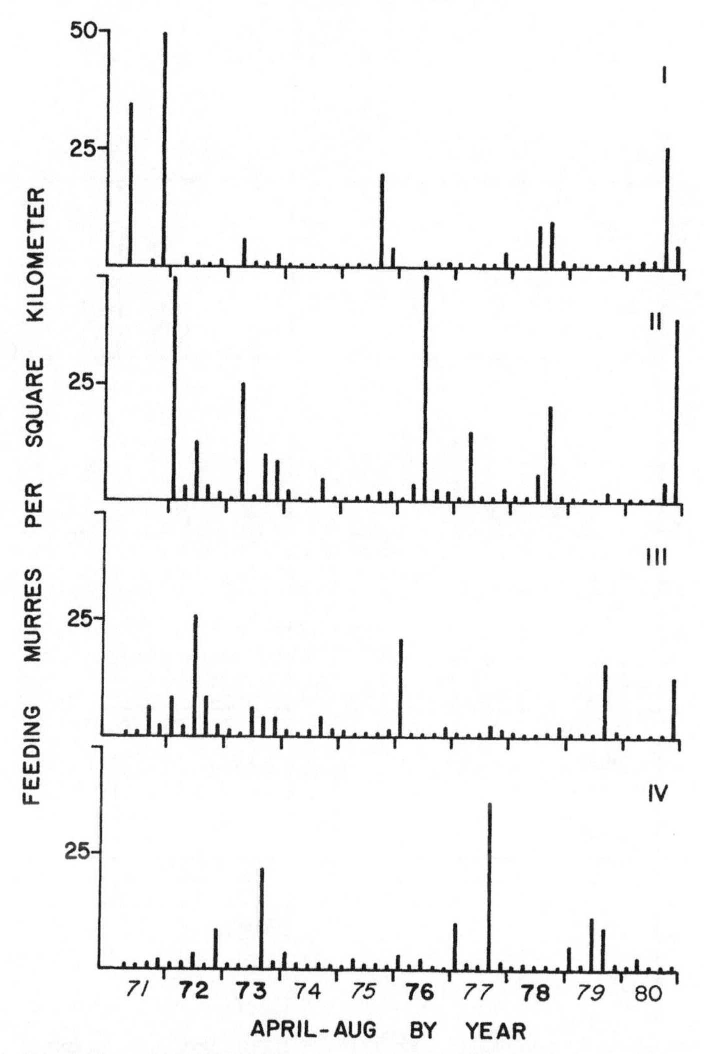

Figure 3.19. Mean number of feeding murres observed along the Golden Gate–to–Farallon track, by track segment and year, April to August, 1971–80.

flocks were especially prevalent near the islands in 1974, 1975, 1977, 1979, and to a lesser extent 1981 and 1982. Very few Brandt's Cormorants were observed feeding near the islands in 1983 (Figure 3.18).

Common Murres fed inshore near the coast mainly during the warm-water years, late 1971 to early 1972, 1973, late 1975 to 1976, 1978, and 1980 (Figure 3.19). Murre feeding flocks were prevalent

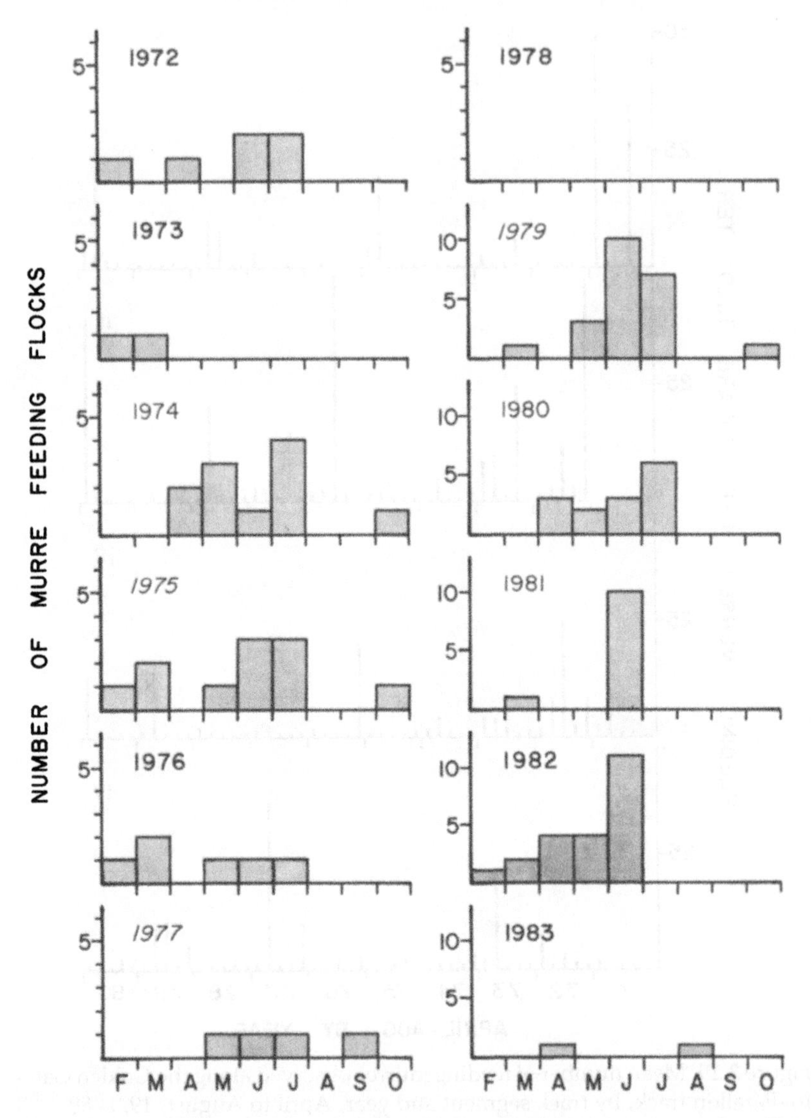

Figure 3.20. Number of murre feeding flocks (>200 birds) per month within 3 km of Southeast Farallon, by year, 1972–83.

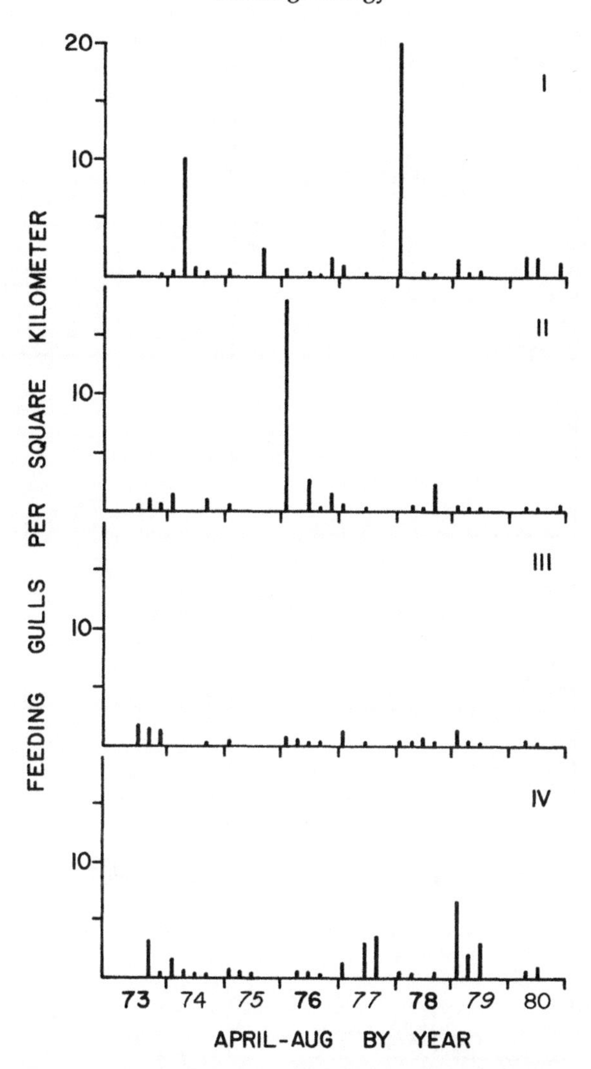

Figure 3.21. Mean number of feeding Western Gulls observed along the Golden Gate–to–Farallon track, by track segment and year, April to August, 1973–80.

near the islands during 1974, 1975, 1979, and 1982. Few were seen near Southeast Farallon in 1972, 1973, 1976, 1977, and 1983, and none in 1978 (Figure 3.20).

Gulls were especially prevalent in inshore coastal waters in the two warm years 1976 and 1978 (Figure 3.21). They fed close to the islands in 1974, early 1975, 1977, and from 1979 through early 1981 (Figure 3.22).

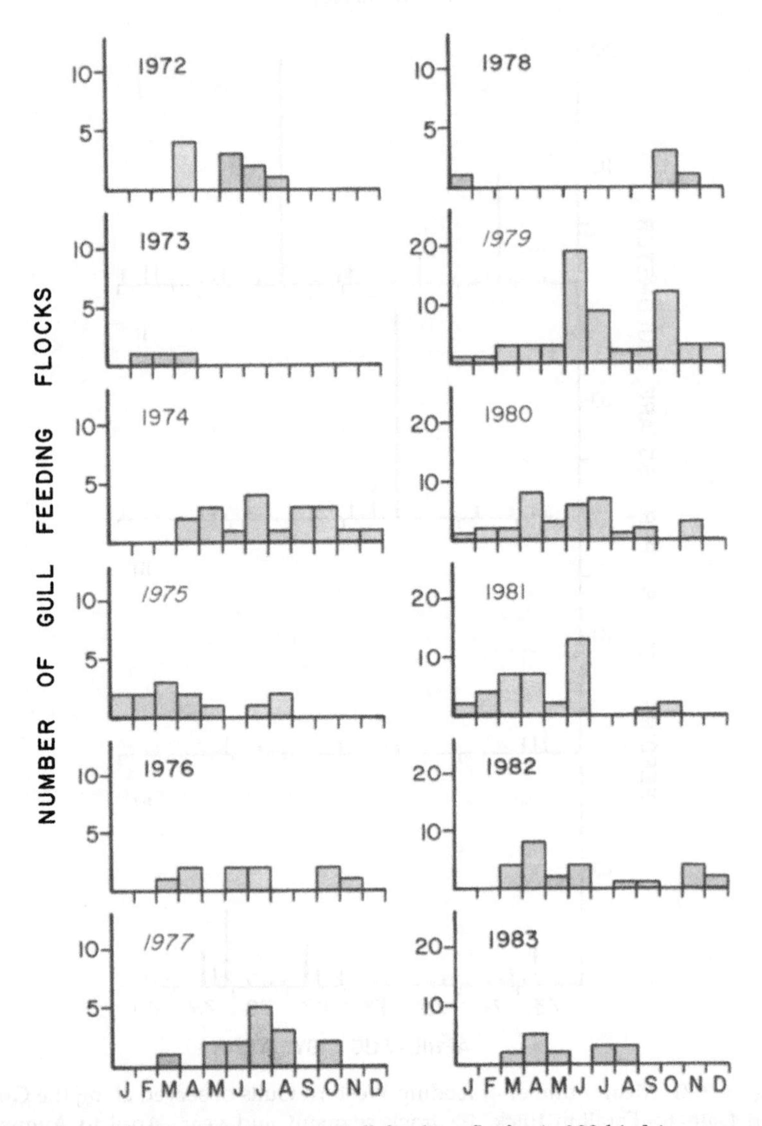

Figure 3.22. Number of Western Gull feeding flocks (>200 birds) per month within 3 km of Southeast Farallon, by year, 1972–83.

FORAGING BEHAVIOR

Depth of foraging

Storm-petrels and gulls feed at the surface by dipping, seizing, or making shallow plunges in which they hardly submerge themselves. They are thus restricted to acquiring prey that occur at the surface. Sooty Shearwaters feed by pursuit plunging to depths up to 10 m (R. G. B. Brown et al. 1981, A. Baldridge pers. comm.). All of these species are thus capable of exploiting mobile prey and capable of searching a large area. The remaining eight species of Farallon seabirds feed by diving; cormorants propel themselves underwater by using their large webbed feet, whereas alcids propel themselves with their wings. Diving species are more heavy-bodied than the shallow-feeding species. Thus, only at great energetic cost (relative to the others) can diving species search large areas for food. Rather, they require food that is predictable in occurrence or feeding habitats that are habitually productive (Ainley 1977, Ainley, O'Connor, and Boekelheide 1984, Crawford and Shelton 1978).

It is intriguing that eight species of diving seabirds can coexist in the same region. As part of our effort to investigate resource partitioning by Farallon seabirds, we compared diving capabilities by timing dives and pauses between dives and by calculating dive-to-pause ratios (Dow 1964, Stonehouse 1967, K. A. Hobson and Sealy 1985). We were able to measure these ratios on birds feeding in the same plot, where depths were shallow (20 m). All the species observed are probably capable of reaching depths of at least 50 m, and most can probably swim deeper (as deep as 180 m in murres; Piatt and Nettleship 1985). Were it possible, it would have been instructive to observe them all feeding in deep waters (180 m, for example). Nevertheless, although our analysis does not prove the maximum capabilities of any of the species, we believe the comparison is instructive.

The Double-crested Cormorant remained submerged the shortest time at 25 seconds per dive, and the Pigeon Guillemot remained submerged the longest time at 75 seconds per dive (Figure 3.23). Both these extremes were significantly different from the values of other species, which in order of time submerged per dive ranged downward as follows: Tufted Puffin, Common Murre, Brandt's Cormorant, Pelagic Cormorant, Cassin's Auklet, Horned Puffin (*Fratercula corniculata*, a nonbreeding species), and Rhinoceros Auklet. Pigeon Guillemots took the most time between successive dives and Tufted Puffins the least (Figure 3.23).

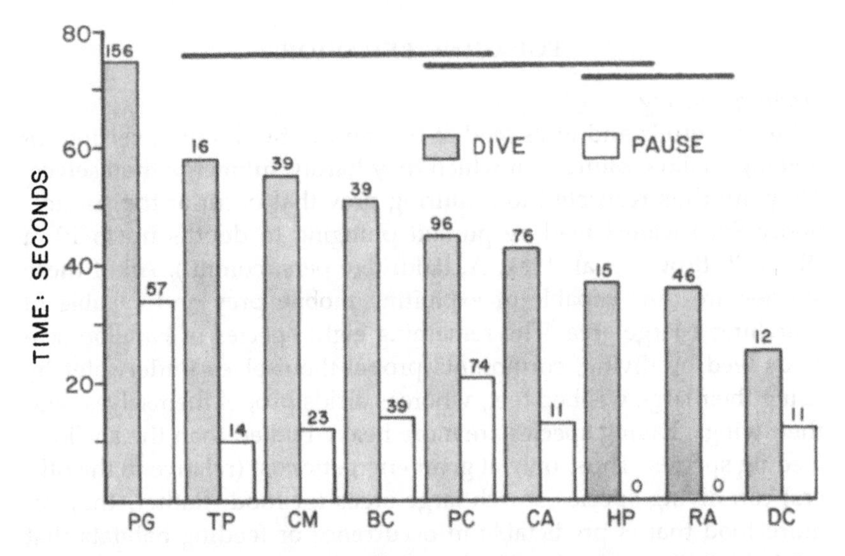

Figure 3.23. Average number of seconds that species remained submerged or paused between dives during feeding: numbers are sample sizes; horizontal lines at top connect similar diving times (SNK test, $p > .05$). PG, Pigeon Guillemot; TP, Tufted Puffin; CM, Common Murre; BC, Brandt's Cormorant; PC, Pelagic Cormorant; CA, Cassin's Auklet; HP, Horned Puffin; RA, Rhinoceros Auklet; DC, Double-crested Cormorant. No pause data for HP or RA.

Our measurements of diving times for some species agreed but for others disagreed with those reported in the literature. Among cormorants our values for the Pelagic (mean 45 sec, maximum 70 sec, $n = 96$) exceeded most values summarized by Cooper (1986) and were almost double those derived by J. M. Scott (1973); for Brandt's (mean 51 sec, maximum 95 sec, $n = 39$) our values also exceeded Scott's by a wide margin; and for the Double-crested (mean 25 sec, maximum 35, $n = 12$) our values were similar to those reviewed by Cooper (1986). If diving times are indeed an effective indication of diving depths, then Brandt's and Pelagic cormorants are among the deepest divers of the 29 cormorant species addressed by Cooper (1986). Little direct information is available on the actual diving depths of cormorants.

Among auks, information for Atlantic species was reviewed by Bradstreet and Brown (1985); additional information for Pacific alcids is contained in J. M. Scott (1973) and Cody (1973). The mean and maximum values determined by us for the murre exceeded Cody's values but were exceeded in Scott's study (mean 55 vs. their 41 and 71 sec; maximum 70 vs. their 71 and 140 sec). The mean and maximum values determined for the Pigeon Guillemot exceeded both Cody's and Scott's (mean 75 vs. their 41 and 36 sec; maximum 110 vs. their 68 and 69 sec), as well as the values determined by Bradstreet (1982) for the Black Guillemot (mean 60, maximum 146 sec). Our values for Cassin's Auklet exceeded those calculated by Cody (mean 43 vs. his 10 sec; maximum 70 vs. his 24 sec), as did our values for the Tufted Puffin (mean 58 vs. his 37 sec; maximum 80 vs. his 60 sec). No comparison is available for the Horned Puffin.

Bradstreet and Brown (1985) concluded that on the basis of dive times, murres and guillemots can dive much deeper than other Atlantic alcids. Their conclusions for the murre were supported by known diving depths (Piatt and Nettleship 1985). The latter authors recorded depths of 50 m for guillemots, but on the basis of circumstantial evidence Bradstreet and Brown concluded that guillemots can dive in excess of 100 m (see also Follett and Ainley 1976). In the Pacific, it appears that Tufted Puffin can be added to the deep-diver category, along with Brandt's and Pelagic cormorants. Cooper (1986) and others have stated that diving depth is a function of body size, which may be true within bird families but certainly is not evident when alcids and cormorants are compared (if dive times are an indication; cf. Figure 3.23). The apparently shallow dives of the large-bodied Double-crested Cormorant, as well as the deep dives of the small-bodied Pelagic Cormorant, appear to be rather enigmatic, or the relationship between body size and diving depth is a weak one.

Another measure that has been used to compare indirectly the diving capabilities of marine birds is the dive-time:pause-time ratio (*D*/*P*). Dewar (1924) and Stonehouse (1967) concluded that those species with higher values are capable of diving deeper, or at least more efficiently, than those with smaller values. A comparison of the *D*/*P* ratios we calculated indicates that Tufted Puffins clearly outperformed the other species (Figure 3.24). This means that per unit measure of time submerged, they required the least amount of time for recovery. Among the various species considered here, the *D*/*P* values we measured in most cases are similar to those measured by other authors for similar groups (Dewar 1924, Cooper 1986), although those for the Pelagic Cormorant were a third again higher than measurements in other studies (e.g., K. A. Hobson and Sealy 1985).

For cormorants, Cooper (1986) noted a direct relationship between *D*/*P* ratios and body mass, although for the many fewer species we compared, an indirect relationship is apparent. J. M. Scott (1973) observed that the *D*/*P* ratio is also to some degree a function of water

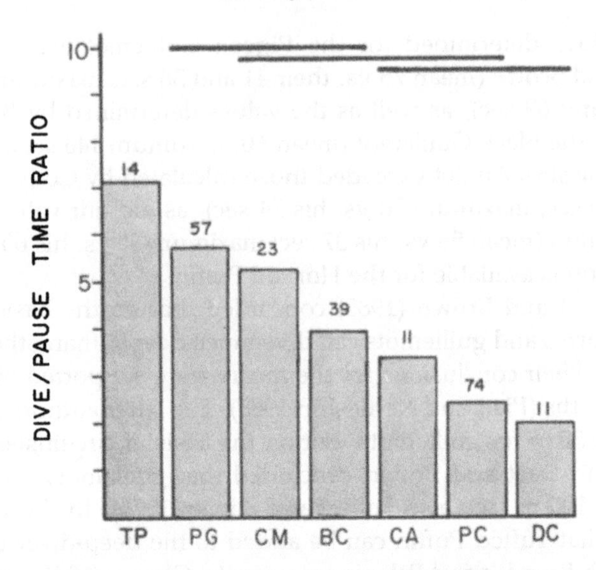

Figure 3.24. Average dive-to-pause ratio in the feeding of various species: numbers are sample sizes; horizontal lines at top connect similar averages (SNK test, $p > .05$). For abbreviations, see Figure 3.23.

depth, which in our case was the same for all species. Might it also be possible that the diving capabilities of a seabird are related to the type of prey it is designed to catch? The three Farallon species with highest *D/P* ratios are those that feed almost exclusively on midwater-schooling prey (as we shall see later), and the remainder feed a great deal, and in some cases exclusively, on benthic prey. This is logical. A seabird that feeds on mobile prey should spend little time resting on the surface once it locates a school. Otherwise the prey may move away while the bird is recuperating, and this would increase energetic costs by requiring it to relocate the school. On the other hand, seabirds that feed on prey that hide near, on, or in the substrate can afford to rest at the surface, because their quarry is not likely to move away. In the case of the Pelagic Cormorant and Pigeon Guillemot, both benthic foragers with long dive times and long rest times, it would seem beneficial to maximize time spent under water to investigate the nooks and crannies where their prey hide.

Dive times and *D/P* ratios could also be a function of age, as pointed out by Bradstreet and Brown (1985). Confirming this, diving times and *D/P* ratios averaged 75 ± 16 sec (n = 156) and 2.5 ± 0.8 (n = 57) for adult Pigeon Guillemots, compared to 62 ± 14 sec (n = 120) and 2.0 ± 0.6 (n = 36) for recently fledged juveniles. The

Double-crested Cormorants in our sample were also recently fledged, and, therefore, our values for that species, although consistent with other studies (ages of subjects not specified), could be an underestimate as well.

To summarize our information and that of others reviewed above on diving capabilities, we seem to have three groups of species in the Gulf of the Farallones community. The deep divers, capable of exceeding 100 m depth, comprise Brandt's and Pelagic cormorants, Common Murre, Pigeon Guillemot, and Tufted Puffin; intermediate-depth divers, which probably dive 20 to 80 m, are the Double-crested Cormorant, Cassin's Auklet, and Rhinoceros Auklet (as well as the Horned Puffin); and the only shallow diver is the Sooty Shearwater (to 10 m). The remaining Farallon species feed at the surface.

Tendency to feed socially

Seabirds differed in their tendency to feed socially. The most solitary feeders among Farallon species were the Pelagic Cormorant and the Pigeon Guillemot. During the period April through August 1979–83 these two species were seen in only four and 41 presumed feeding flocks, respectively (< 5% of 738 flocks); in fact, some of the guillemot flocks could have been social gatherings (Storer 1952). Both species associated more frequently with multispecies flocks during years of superabundant prey. Otherwise, they fed alone at submerged reefs and hunted for the solitary prey that hide in the rocks (Ainley, Anderson, and Kelly 1981, Follett and Ainley 1976). Even when in flocks, rather few individuals were usually involved (Figure 3.25). The Rhinoceros Auklet and Tufted Puffin also tended to feed alone or occasionally in small flocks (PRBO, unpubl. obs. 1985 and 1986, Briggs et al. 1987, Sealy 1973a, Hoffman, Heinemann, and Wiens 1981, Grover and Olla 1983). Double-crested Cormorants are highly social in their feeding (Bartholomew 1942, Ainley pers. obs.), but as indicated above, they do not feed near the island. Cassin's Auklet feeds in dense flocks, but not usually in asociation with other species except the Sooty Shearwater (PRBO, unpubl. obs. 1985 and 1986, Briggs et al. 1987) and rarely in sight of the island.

The Farallon species that we most often observed feeding in flocks were Brandt's Cormorant, Western Gull, and Common Murre, along with the abundant nonbreeding visitor, the Sooty Shearwater (Table 3.2). Among the latter four species, the cormorant and gull were observed in 63 and 51%, respectively, of the 738 flocks observed April to August 1979–83. Numbers of cormorants in flocks tended to be

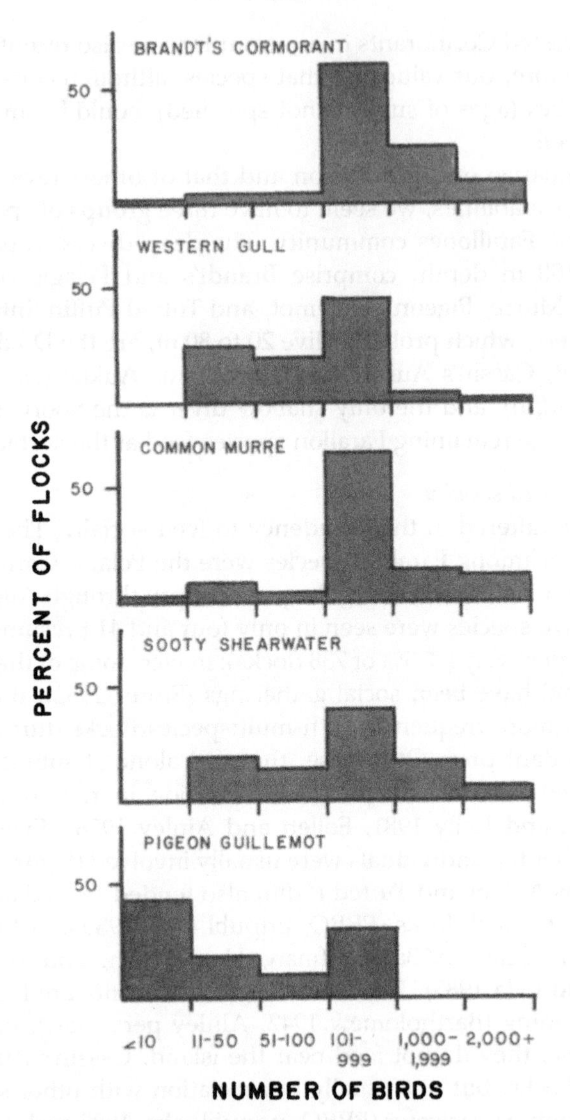

Figure 3.25. Frequency distributions showing by species the number of birds participating in feeding flocks seen within 3 km of the island, April to August, 1979–83 (n = 738 total flocks).

much greater than numbers of gulls in flocks (Figure 3.25). Both were most often observed feeding in unispecies flocks (i.e., three or more birds together), and their tendency to feed in flocks containing only their own species increased during years of warm oceanic conditions when food was presumably less available. This is consistent with results from diet analyses (presented later) and is also consistent with a lessened tendency to feed in association with other species (Table 3.2).

According to several studies on the benefits of feeding in multispecies flocks, or in any flocks for that matter, feeding socially should increase an individual bird's chances of finding food (e.g., Sealy 1973b, Hoffman, Heinemann, and Wiens 1981, Porter and Sealy 1982, C. R. Brown 1986). On the other hand, our observations seem to indicate that abundant food brings individuals of different species together, while conditions of sparse feeding opportunities drive species apart. This is perhaps consistent with the ideas of Hoffman, Heinemann, and Wiens (1981), who observed that when feeding socially, seabird species play different roles: catalysts (gulls, some alcids), divers (shearwaters, cormorants, alcids), kleptoparasites (gulls), and suppressors (shearwaters, cormorants). When food is abundant, species in their various roles accrue benefits from multispecies flocks, but when it is not, the benefits may be fewer. For instance, the patterns we observed, in warm-water years vs. cool-water years (Table 3.2), could be explained largely by some species' avoidance of others that dominate feeding situations by their feeding behavior or success (suppressors). Also, although during warm-water years we observed plenty of gull (catalyst) flocks, other species seemed to "know" that their existence did not necessarily indicate the presence of appropri-

TABLE 3.2

Tendency of near-island Farallon seabird species to forage in mixed-species flocks, 1971–83

Species and year	Flocks	Proportion flocking alone	Cole's coefficient		
			BC	WG	CM
Warm-water years					
Brandt's Cormorant (BC)	139	0.50[a]			
Western Gull (WG)	126	0.54[a]	−0.92[a]		
Common Murre (CM)	28	0.07	0.59[a]	0.28	
Sooty Shearwater	12	0.00	−0.14	0.75[a]	−0.46[a]
Other years					
Brandt's Cormorant	391	0.37[b]			
Western Gull	227	0.21[b]	−0.56[a,b]		
Common Murre	99	0.08	0.10[b]	0.39[a]	
Sooty Shearwater	15	0.20	−0.03	0.17[b]	−0.27[b]

NOTE: "Near-island species" in this table are those foraging within 3 km of the island. Warm-water years: **1973, 1976, 1978,** 1980, **1983.**

[a] These are statistically significant values ($p < .05$).

[b] Comparison of species' percentages for the two groups of years showing statistical significance ($p < .05$, paired t-test).

ate food. An exception to this was the Sooty Shearwater, which increased its tendency to feed with gulls in flocks near the island. This might suggest that when food is scarce, the apparent feeding behavior of a catalyst species becomes even more attractive to species sensitive to the catalyst (suppressors). Brandt's Cormorants and Common Murres also tended to feed more together near the island in the warm years, although on a larger scale they overlapped much less in habitat use (as noted above). The significance of this is not clear.

Murres and shearwaters were observed in 15 and 5%, respectively, of the flocks we observed. Like cormorants, numbers of murres in flocks tended to be large; shearwaters were quite variable in the numbers participating in flocks (Figure 3.25).

Most feeding flocks we observed (principally gulls and cormorants) near the Farallones were downcurrent, i.e., to the southwest, south, and southeast. The frequency with which feeding flocks occurred was a function of year. Few flocks were observed near the island during warm-water years: 1973, 1976, 1978, 1983, and, to a lesser extent, 1980. This pattern was apparent when we looked at the frequency of feeding flocks by month as well as by year. For instance, Brandt's Cormorants tended to feed in flocks near the island only early in the 1981 and 1982 breeding seasons (Figure 3.18); in 1977, more flocks occurred late in the season than early; and in 1980, flocks were equally abundant early and late. These changing patterns of feeding, as we shall see, have important effects on feeding effort and on differing rates of breeding success.

We saw murres in far fewer flocks than we saw of cormorants or gulls. Except in warm-water years, murre flocks occurred near the island most consistently during June and July, the chick-rearing period (Figure 3.20). We rarely saw murre flocks nearby during April and May, when they tended to feed north of the island (Figure 3.7). Compared to cormorants, Western Gulls appeared to be less erratic in the frequency with which they fed in flocks near the island (Figure 3.22). Notwithstanding the warm-water years, only during 1981 was the observed difference in the number of gull flocks between early and late in the nesting season particularly marked.

Foraging effort

We measured foraging effort for Pelagic Cormorants, Brandt's Cormorants, and murres by determining the amount of time they spent away from their nest or the number of trips they made per day between sea and nest site. Our measure of effort included the time required to capture prey as well as that required to fly to and from the feeding area. As an index to "foraging difficulty," our measure is imperfect for a number of reasons. First, a bird could spend as much time gathering prey from an abundant source far away as from a situation where prey are everywhere scarce. Second, for cormorants, for which we measured the time between nest reliefs (we recorded actual arrival and departure times in only a few years, and interpretation of these is difficult because of trips for nesting material, bathing, etc.), the interval also included the brooding tendencies of the incubating bird—though a mate has returned from feeding at sea, the incubating bird can "elect" to remain on the nest. Thus, in years when food was abundant and parents were not hungry, we might still observe as long an interval between nest reliefs as during years when foraging individuals were absent for a long time. Third, we had to assume that any trip away from the nest site (except for those obviously for bringing back nest material) was for feeding. This was usually true, but cormorants also take a swim each day in order to bathe, and murre probably do so as well. Finally, this method does not take into consideration actual time spent looking for prey, which is the factor we are really interested in measuring, as opposed to other components of foraging effort such as prey handling or capture time.

Not surprisingly, the interval between nest reliefs for Brandt's Cormorant during incubation (289 $\pm$ 125 min, or about 4.8 hours; $n = 1{,}200$) was longer than during the nestling period (204 $\pm$ 101 min, or about 3.4 hours; $n = 1{,}871$, $t = 20.8$, $p < .001$). Certainly the difference was due at least in part to the fact that parents were not pressed to feed chicks during the incubation period, but it may also have been due to seasonal shifts in the location of available prey. As indicated previously, during April and May large numbers of Brandt's Cormorants feed by the Golden Gate; in fact, many at that time feed within San Francisco Bay itself. Nest-relief intervals were usually longest during the early season regardless of whether birds were tending eggs or chicks (Figure 3.26). Brandt's Cormorants fed that far away, i.e., on the order of 50 to 80 km, early in the breeding season during at least five of the ten years when we censused birds along the Golden Gate–to–Farallon transect—1971, 1972, 1973, 1976, and 1978 (Figure 3.17). If during other years they were foraging away from our transect line, then we would not have known. This probably was the case in most of the other years, because cormorant feeding flocks were not observed close to the island early in the nesting season, except during 1981 and 1982 (Figure 3.18). We can begin to see, then, that nest relief interval does, in fact, indicate "foraging effort" to some degree.

A comparison of mean nest-relief intervals during incubation (Table

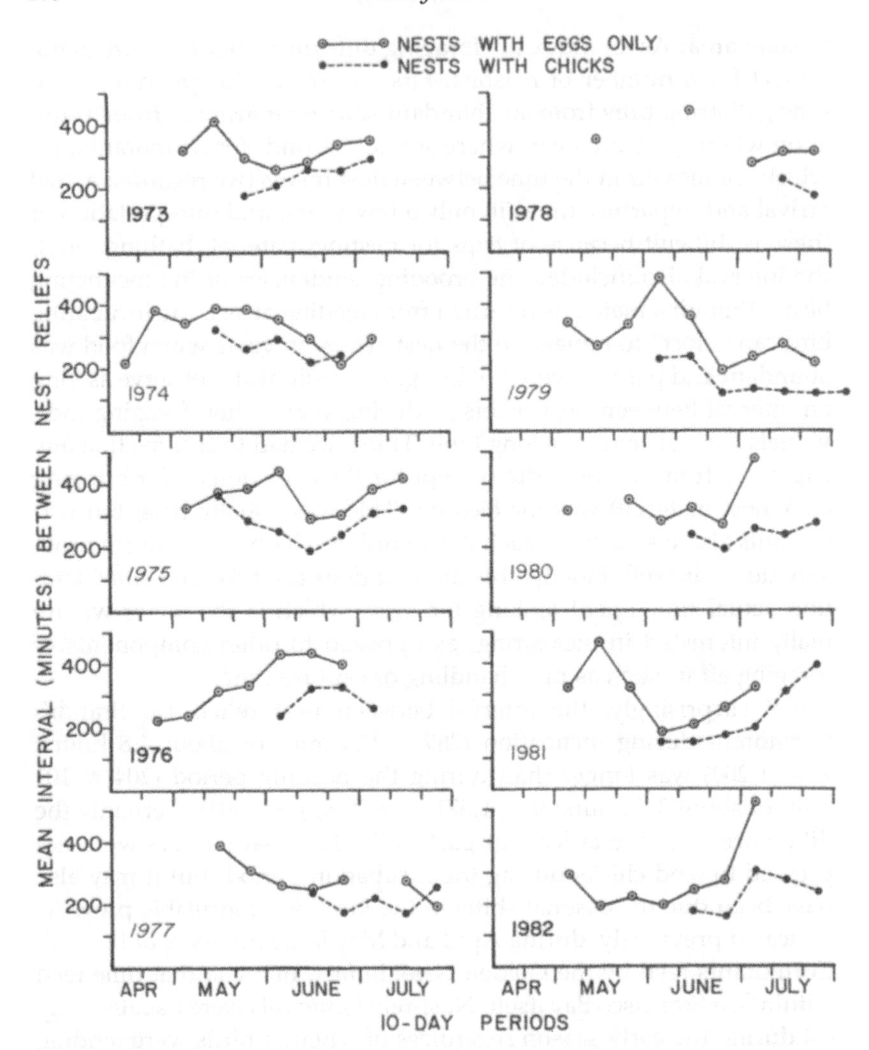

Figure 3.26. Within-year variation in the time interval between nest reliefs by Brandt's Cormorants, 1973–82. Lines connect consecutive data only.

3.3) revealed that intervals were shortest during 1981 and 1982. Nest-relief intervals during incubation were longest in 1974, 1975, and all the warm-water years (no nesting in 1983): 1976, 1978, 1980, and 1973. What was happening in 1974 and 1975 is not known; Brandt's Cormorants fed neither near the Golden Gate nor to the south or east of the island during the early part of those years (Figures 3.17 and 3.18).

During the nestling period, Brandt's Cormorants took longer to return from feeding during all warm-water years; the cool years of 1974 and 1975 were again enigmatic in this regard (Table 3.3). The position of 1978 relative to other years could be an artifact of the small sample size and the possibility that the few adults feeding chicks during this especially poor year were atypical, i.e., older, more experienced individuals. The nest-relief interval of birds feeding chicks lengthened dramatically late in 1981 (Figure 3.26), which is consistent with the disappearance of cormorant feeding flocks near the island at that time (Figure 3.18). Nest-relief intervals during incubation were correlated with those during the chick phase. That is, years with long intervals early tended to have long intervals late ($r_S = .6606$, $t = 2.49$, DF = 8, $p < .05$). Exceptional were 1978 and 1982 (Table 3.3).

During the nestling period, Brandt's Cormorants showed a close correlation between the number of feeding trips per day and the nest-relief interval ($r_S = .9030$, $t = 5.94$, DF = 8, $p < .05$; Table 3.3). Two years, 1977 and 1979, were clearly outstanding in the high frequency of feeding trips during the chick period. These years were also outstanding in the large number of cormorant feeding flocks observed close to the island late in the nesting season (Figure 3.18).

Patterns of Pelagic Cormorants were similar to those of Brandt's Cormorants. During the egg stage, foraging time was shortest in 1975;

TABLE 3.3

Mean interval between nest reliefs, and number of feeding trips per day, Brandt's Cormorants, 1973–82

Minutes between reliefs, nests with eggs

1975	1974	**1976**	**1978**	1980	**1973**	*1977*	*1979*	1981	**1982**
355	340	334	329	322	297	277	260	248	233

Minutes between reliefs, nests with chicks

1976	*1975*	**1973**	1974	1980	1981	*1977*	**1982**	**1978**	*1979*
303	258	244	241	231	201	193	188	182	132

Trips per day per nest[a]

1976	**1973**	*1975*	1974	**1982**	1980	**1978**	1981	*1977*	*1979*
2.6	3.5	3.8	4.0	4.3	4.6	5.2	5.9	6.7	8.6

NOTE: Lines connect similar means; SNK test, $p > .05$.
[a] Includes only nests with chicks between 11 and 35 days old.

during the nestling stage, foraging time was shortest in 1975 and 1977 (Table 3.4). The longer trips early in 1977 coincided with the late start to nesting that year (Chapter 6). Trip times were long during the two ENSO/warm-water years, 1973 and 1976, as well as in 1974. Pelagic Cormorants made the most frequent feeding trips during 1974, 1975, and 1977, and the fewest during 1973 and 1976 (Table 3.4).

The patterns for murres were similar to those for cormorants, although we did not have data for as many years. The amount of time between arrivals of parents with fish was greatest during the two warm-water years, 1976 and 1983 (Table 3.5). Corresponding to the longer feeding absences, fewer fish were brought each day to chicks during those years. Except for 1977, the year of the highest return rate, the ranking of years by the length of time between feeds is consistent with the frequency of feeding flocks containing murres within 3 km of the island. A comparison of Figure 3.20 and Table 3.5, for example, indicates few murre feeding flocks nearby during June 1974, 1976, and 1983, when absences were long, but many flocks nearby in June 1975 and 1982, when absences were short. In 1977, though few murres were feeding within 3 km, many were doing so within segment IV of our transect corridor, or within 15 km of the colony (Figure 3.19). The latter distance is still relatively close.

We have few data with which to compare feeding effort among years for other species. Results of the Golden Gate–to–Farallon censuses indicate that Western Gulls fed close to the Marin coast during the early parts of the warm-water years, 1976 and 1978, and to a lesser extent during early 1974 (Figure 3.21). Late in the nesting season dur-

TABLE 3.4

Mean intervals away from nests, and number of feeding trips per day, Pelagic Cormorants, 1973–77

Nests with eggs					Nests with chicks[a]			
Minutes away								
1976	*1977*	1974	**1973**	*1975*	1974	**1973**	*1975*	*1977*
248	201	191	179	123	191	134	96	91
Trips per day per nest								
1976	**1973**	*1977*	1974	*1975*	1974	**1973**	*1977*	*1975*
3.3	3.6	3.6	4.6	5.4	5.0	6.4	7.9	8.6

NOTE: Lines connect similar means; SNK test, $p > .05$.

[a] No nests had chicks in **1976**.

TABLE 3.5

Mean intervals between feeds, and number of feeds per day, Common Murre chicks, 1973–83

Minutes between feeds						
1982	*1977*	1974	*1975*	**1976**	**1983**	
141.2	142.6	147.4	151.6	194.4	217.9	
Feeds per day						
1977	**1982**	*1975*	1974	**1976**	**1983**	**1973**
5.0	4.8	4.6	3.8	3.0	2.8	2.2

NOTE: Lines connect similar means; SNK test, $p > .05$.

ing the cooler-water years of 1977 and 1979, they apparently fed near the island (also evident in Figure 3.22). These patterns are consistent with those of the cormorants and murre. Thus, we would expect feeding trips of longer duration during the warm-water years.

Temporal patterns of foraging

A mass of circumstantial and direct evidence indicates that most Farallon seabirds are diurnal foragers, although feeding activity is greatest during or near twilight and is least between about 1000 and 1500 hours. This evidence includes (1) all-day watches of cormorants, gulls, murres, and puffins (data presented above; Pierotti 1976; Chapters 5, 6, 7, 8, 11), (2) observations of gulls and cormorants at night through a night-vision scope (Pierotti 1976), (3) direct observations of Rhinoceros Auklet arrivals (Sander 1986), (4) electronic monitoring of Ashy Storm-Petrel and Cassin's Auklet arrivals (Chapters 4, 10), (5) collections of murres and auklets (see Diets below; PRBO, unpubl. data), and (6) observations to determine temporal patterns of territory occupancy (Chapters 4–11).

Only the storm-petrels, which depart their nests and the island in the early morning hours (0100–0300; Chapter 4), likely forage extensively at night, probably in the predawn hours. In the data derived from collections of several storm-petrel species at sea, but not off California, night foraging is clearly indicated (PRBO, unpubl. data). Cormorants and gulls definitely roost on the island during the night (Chapters 5–7), but members of pairs not incubating eggs or tending small chicks depart at the first twilight of dawn and/or arrive back just at dark. The cormorants' diet contains several species that are active in the water column at night and are looking for hiding places in the substrate at dawn (Ainley, Anderson, and Kelly 1981, J. Fitch pers. comm.). Murres and guillemots also roost on the island at night and feed their chicks at the peak rate from dawn to about 1000 (Chapters 8, 9), as well as in the late afternoon and evening. Cassin's Auklets feed at the same time in the morning as these other two alcids, judging from collections of birds at sea (PRBO, unpubl. data); they spend the night ashore and arrive and depart the island just after dark and just before dawn, respectively (Chapter 10). Tufted Puffins remain in their burrows at night (Chapter 11) and usually bring fish to their chicks in the early morning (ca. 0800–0900) and just before dark. Finally, Rhinoceros Auklets, which also spend the night on the island, arrive with fresh fish within the twilight hour just after sunset (Sander 1986).

DIETS OF FARALLON SEABIRDS DURING THE BREEDING SEASON

We know little about the diet of the storm-petrels that nest on the Farallones. Presumably, they feed on the zooplankton and micronekton that occur near the surface in waters seaward of the continental shelf. Similarly, we know little about the diet of the Rhinoceros Auklet at the Farallones; elsewhere it feeds heavily on zooplankton and feeds fish to its young (Ainley and Sanger 1979, Vermeer 1978, 1980). Collections of items brought to chicks in 1986 indicated a diet that year dominated by three fish species, juvenile rockfish, Pacific Saury (*Cololabis saira*), and juvenile Black Cod (*Anoplopoma fimbria*) (Sander 1986). Except for the Tufted Puffin (see below), no other Farallon seabird, to our knowledge, feeds substantially on the latter two species, both of which occur in waters seaward of the continental shelf. The Rhinoceros Auklet's diet is thus consistent with the observed pattern in its foraging area and habitat.

For the remaining Farallon species, we have amassed a considerable amount of information on the within-season and between-year variability of the diet fed to chicks. In the case of murres and puffins, diets of the adults, which we did not sample, may differ from those of the chicks (Bradstreet and Brown 1985), but in the case of other species the chick and adult diets are probably similar. Previous work on the trophic relationships of seabirds in the Gulf of the Farallones is restricted to five studies: Ainley, Anderson, and Kelly (1981) for cormorants, Pierotti (1981) for Western Gulls, Follett and Ainley (1976) for Pigeon Guillemots, and Thoresen (1964) and Manuwal (1974b) for

Cassin's Auklets. The information from all these publications, except that for auklets, is incorporated into the data discussed below. The earlier data for auklets were not collected or reported in a way that allows comparison with ours. In the following discussions, we are concerned largely with the period from about late May through August, the nestling period, for each species. We will first consider diets in what A. W. Diamond (1983) termed a "level-three" analysis, that is, by determining diet overlap among years for each species, as well as between species in different years, on the basis of numerical composition only. In level-three analysis, Diamond used prey categories based on taxonomic families, whereas we use the lowest taxa possible. Many food items in our study could be identified to species. We also look at dietary overlaps after combining data on numerical composition with data on prey weight (a function of size). Such a procedure was considered to be a "level-one" (size at lowest possible taxon) or "level-two" (size at family level) analysis. Our analysis of number and size is thus a combination of these two levels. To be consistent with Diamond's analysis, we considered an overlap index greater than or equal to 0.8 (80%) to indicate diet similarity.

Cormorants

The diet of the Double-crested Cormorant was investigated during four summers, 1974–77. A level-three analysis indicated virtually total overlap in diet composition among the four years (Appendix 3.3). This is not surprising given the greater stability of the estuarine habitat where these birds feed over the upwelling-influenced waters offshore, where greater dietary variability might be expected. The major prey was surfperch, particularly the Shiner Surfperch, *Cymatogaster aggregata* (Table 3.6). Most of the other prey species consumed are, like the surfperch, typical of quiet, shallow, inshore waters. Most are schooling species that live above flat sandy bottoms. Prey diversity was greater during 1976, a warm-water year, than during the other three years. Because of the lack of a significant difference in the size of the surfperch eaten (Appendix 3.4), an analysis of diet overlap that included prey size did not alter the results.

The diet of the Pelagic Cormorant was investigated during three summers, 1975–77. Diets were virtually identical in 1975 and 1977 but were largely different in the warm year, 1976 (Appendix 3.3). Prey were characteristic of submerged, rocky reefs and were dominated by organisms that hide in the substrate. Predominating were several species of sculpin (cottids), juvenile rockfish, mostly *Sebastes flavidus* and *S. jordani*, and a mysid shrimp, *Spirontocaris* sp. (Table

TABLE 3.6

Composition of diet among three cormorant species, 1973–77

(Percent)

Prey[a]	Double-crested, 1974–77		Pelagic 1975, 1977		Pelagic 1976		Brandt's 1973–75, 1977		Brandt's 1976	
	n	*w*	*n*	*w*	*n*	*w*	*n*	*w*	*n*	*w*
Porichthys notatus							2%	1%	6%	1%
Citharichthys sordidus	1%						2	13	14	25
Citharichthys stigmaeus									1	2
Clupea pallasi									1	1
Cottids	3	1%	27%	24%	78%	61%	3	1	12	4
Cymatogaster aggregata	91	98					1	1	10	19
Phanerodon furcatus	1									
Engraulis mordax							1	4	14	10
Microgadus proximus	1						21	22	8	2
Acanthogobius flavimanus									1	1
Coryphopterus nicholsii			1	1	1	1				
Chilara taylori			1	1			3	6	6	3
Glyptocephalus zachirus								1	1	2
Parophrys vetulus							1	8	10	20
Sebastes spp.	1		65	72	20	37	66	42	13	10
Spirontocaris sp.			5	2						
Loligo opalescens									2	3
Total items	2,815		5,673		1,166		10,125		1,065	
Total weight (gm)	104,394		11,832		2,517		54,644		21,950	
Mean diversity (*n*)	0.359 ± 0.153		0.830 ± 0.077		0.462		1.013 ± 0.597		2.379	
Mean diversity (*w*)	0.066 ± 0.121		0.749 ± 0.063		0.715		1.647 ± 0.439		2.157	
Prey/sample	16.5 ± 2		40.5 ± 1		20		44 ± 20		14	

NOTE: Years are grouped according to overlaps identified in Appendix 3.3; means are shown ± SD. *n*, number; *w*, weight. Total figures are not percentages.
[a] Only prey species contributing at least 1% of the total diet are shown.

3.6). Certainly, the species of scorpaenids and cottids listed by Follett and Ainley (1976) as guillemot prey included the major species eaten by this cormorant. Few of the prey could be considered schooling species. Because most of the prey items were 5–10 cm long, their otoliths were too tiny to distinguish species within the rockfish or sculpin families. As with the Double-crested Cormorant, diet diversity based on numerical composition (but not weight) differed, but was lower, in the warm-water year of 1976. That year, there were also fewer items per sample. The lack of any between-year differences in size of prey (Appendix 3.4) resulted in no change of dietary overlaps among years.

The diet of Brandt's Cormorant was investigated during five summers, 1973–77. In a comparison of diet overlap among all five summers (excluding the early spring collection from the 1977 sample; see below), diets proved to be largely similar except for the warm year, 1976 (Appendix 3.3). The principal prey were rockfish, mainly *Sebastes flavidus* and *S. jordani* (Table 3.6). Other important prey were flatfishes (bothids and pleuronectids), Pacific Tomcod, *Microgadus proximus*, Plainfin Midshipman, *Porichthys notatus*, and Spotted Cusk-Eel, *Chilara taylori*. On a weight basis, the importance of flatfishes increased dramatically. This assemblage of fishes, except for the juvenile rockfish, is typical of waters near or on the bottom over flat relief (mud, sand). The rockfish and tomcod are schooling species, but the others are not. Diet diversity was exceedingly low during most years as a result of the preponderance of juvenile rockfish; on the other hand, diversity was exceedingly high during 1976 (Table 3.6). The number of prey per sample was low during 1976 as well. If the larger rockfish and sculpins eaten by Brandt's Cormorants during 1976 are taken into account (probably year-1 instead of year-0 fish; Appendix 3.4), diet overlap was even lower between that and the other years than Appendix 3.3 indicates.

During 1977, we were able to collect some samples from Brandt's Cormorants just before egg laying for comparison with diet during the chick period (Table 3.7). The early diet was far more diverse than that eaten later, and there was little overlap in species composition. The menu of species was similar to that of the previous summer, 1976, but the contributions of anchovies, flatfishes, surfperch, and cusk-eels were quite different.

For each of the three cormorant species, we combined data for those years in which the birds ate similar diets: all years were the same for the Double-crested, 1976 differed from 1975-plus-1977 for

TABLE 3.7

Composition of Brandt's Cormorant diet early and late in the nesting season, 1977

Prey	Percent of diet	
	Mar.–Apr.	Jul.–Aug.
Ammodytes hexapterus	< 1%	
Atherinopsis californiensis	1	
Porichthys notatus	3	
Citharichthys stigmaeus	2	
Hemilepidotus spp.	10	
Leptocottus armatus	2	
Phanerodon furcatus	6	
Engraulis mordax	29	
Microgadus proximus	< 1	
Merluccius productus	< 1	
Chilara taylori	25	
Isopsetta isolepis	2	
Parophrys vetulus	8	
Sebastes spp.	12	99%
Loligo opalescens		1
Total items	105	2,811
Diversity	1.993	0.056

NOTE: Numerical data only.

the Pelagic, and 1976 differed from all other years for the Brandt's (Appendix 3.3, Table 3.6). Judged on the basis of diversity indexes, the diet of Brandt's Cormorant was much more complex than that of the other two species. There was little overlap of the Double-crested's diet with that of the other two species (Appendix 3.5) except in 1976, when overlap with the Brandt's diet was slightly greater. This was likely due to Brandt's Cormorants feeding more inshore (even within estuaries?) that year, as discussed in a previous section. In contrast, the diets of Pelagic and Brandt's cormorants overlapped to a remarkable degree during all years except the warm one, 1976. This overlap was due largely to the preponderance of juvenile rockfish in both cormorants' diet during those years. On the basis of weight, the between-year similarity of the Pelagic Cormorant's diet increased slightly. Otherwise, overlap indexes based on weight were generally lower than numerical overlaps, a tendency also observed by A. W. Diamond (1983).

Western Gull

Because we used several different techniques to assess the Western Gull's diet, it is first necessary to evaluate each method with respect to the others. The various methods have been employed in numerous studies of the diet of gulls, but a comparison of the methods' results has not been performed previously.

We employed two techniques, pellet examination and fecal examination, to sample the diet of gulls that roosted on the catchment basin (Table 3.8). The percentages of fish in samples of each type were similar, even in 1977 (*t* test after arcsin transformation, $p < .5$). If only the total of marine invertebrates is considered, the species breakdown reveals different prey compositions: euphausiids did not occur in pellets but did in feces, and barnacles clearly predominated in the pellets (Table 3.9). Terrestrial invertebrates, primarily beetles, were equally represented in both samples. Bird remains occurred in up to 7% of pellets but not in feces. Garbage was also present in pellets; less so in feces. These differences were thus likely due to digestive processes. Accordingly, a large percentage of unidentifiable fecal remains was likely composed mostly of garbage and bird components. The overall composition of pellets did not change even with a reduced sample and sampling period. This suggests only slight within-season variability in the diet of the gulls that frequent these roosts, a large portion of which are nonbreeders (Spear 1988a).

We also employed two techniques, collection and observation of chick regurgitations, to sample the prey fed to chicks. The percentage of fish was consistently higher in collected than in observed samples,

TABLE 3.8
Composition of gull diets, 1976–78
(*Percent*)

	1976			1977		1978		
Prey	F	P	P-all	F	P	F	P	P-all
Fish	63%	65%	61%	55%	39%	27%	31%	30%
Euphausiids[a]	(10)		(< 1)	(14)	(< 1)	(4)	(< 1)	(< 1)
Invertebrates								
Marine[a]	10	7	6	15	13	8	17	18
Other[b]	1	1	1	< 1	< 1		< 1	< 1
Birds		4	4		7	< 1	5	5
Garbage	5	23	28	1	41	10	46	47
Unidentified	21	—	—	28		55	< 1	< 1
Total items	354	2,350	3,189	354	2,025	359	2,589	2,877

NOTE: Data derived from feces (*F*) and pellets (*P*). Total figures are not percentages. *P* columns include data only from period also covered by fecal data; *P-all* columns include all pellets.
[a] Euphausiids are also included in marine invertebrate totals; therefore, euphausiid values are in parentheses.
[b] Includes tenebrionid beetles.

TABLE 3.9
Marine invertebrates in Western Gull diet, 1975–78
(Percent)

	1975	1976		1977		1978	
Prey[a]	P	P	F	P	F	P	F
Arthropods						1%	
Lepas sp.	72%	82%	3%	89%	6%	91	50%
Mysids			6				
Euphausiids		1	91	1	94		50
Decapods	3	3		5			
Mollusks							
Squid	6	3					
Acmaeids[b]	5	3				1	
Mytilus sp.	8	2				2	
Unidentified	2	1				1	
Total items	151	209	35	256	53	509	28

NOTE: Data derived from feces (*F*) and pellets (*P*). Total figures are not percentages.
[a] Only items contributing at least 1% of the total diet are included.
[b] Species of limpets identified in pellets: *Collisella digitalis, C. pelta, C. scabra,* and *Lottia gigantea. Collisella pelta* accounted for 79% of those identified ($n = 28$).

except during 1983, when the percentages of fish were equal (Table 3.10). These differences may be due to (1) the collected sample being smaller, (2) difficulty in identifying fish at a distance, as perhaps reflected in the category of unidentified fish, or (3) within-season variability. In regard to the first and last points, regurgitations were collected only during a short period when most chicks were ten to 20 days old, the age of maximum food demand (Coulter 1973) and a time when improved diet quality (i.e., more fish) may be required. Most likely, however, the differences were due to the differences in methods. First, we could not compare invertebrates (marine or terrestrial) because of different procedures for recording observations from 1978 to 1982. Second, in 1983, few regurgitations could be collected ($n = 5$), in spite of increased efforts. Food was not easily available that year, and comparisons with 1983 data are unfortunately difficult. Third, birds did not appear as prey in collections. This may be due to (1) the relative unimportance of birds in chick diets (1–3%), (2) lower-quality items not being fed during the chick-banding period, as mentioned above, or (3) the high visibility of regurgitations containing birds, which increases their likelihood of being sighted.

During years with adequate samples (1978, 1979, 1981, and 1982), the percentages of garbage were similar. In the two years with the lowest reproductive success—1978 and 1983 (warm-water years; Chap-

TABLE 3.10

Composition of Western Gull chick diet, 1974–83

(Percent)

Prey[a]	1974		1978		1979		1980		1981		1982		1983	
	CR	OB	CR	OB	CR	OB	CR	OB	CR	OB	CR	OB	CR	OB
Fish	86%	74%	58%	51%	93%	78%	62%	58%	87%	64%	78%	73%	57%	57%
Euphausiids[b]	(8)	(12)	(1)	(9)	(< 1)	(8)	(24)	(5)	(2)	(8)	(3)			(4)
Invertebrates														
Marine[b]	12	13	2	9	2	10	38	7	3	10	9	4		9
Other[c]			1		< 1									
Birds		2		3		1		1		2		2		2
Garbage	1		39	28	4	6		24	10	16	13	15	43	24
Unidentified		11		9		5		10		8		6		7
Total items	7	805	112	43	59	184	17	68	62	83	19	53	5	232

NOTE: Data derived from collected (*CR*) and observed (*OB*) regurgitations. Total figures are not percentages.

[a] All observations (*OB*) in 1974 are extrapolated from Pierotti 1981; observations (*OB*) 1978 to 1983 from Spear, unpubl. data.

[b] Euphausiids are also included in marine invertebrate totals; therefore, euphausiid values are in parentheses.

[c] Includes tenebrionid beetles.

Table 3.11

Composition of Western Gull diet, 1975–79

(Percent)

Prey	1975			1976					1978						1979		
	P-all	P	CR	F-all	F	P-all	P	CR	F-all	F	P-all	P	CR	OB	P	CR	OB
Fish	51%	40%	77%	63%	64%	61%	61%	61%	27%	43%	30%	26%	58%	51%	24%	93%	78%
Euphausiids[a]			(2)	(10)	(4)	(< 1)		(4)	(4)		(< 1)		(1)	(9)		(< 1)	(8)
Invertebrates																	
Marine[a]	6	12	19	10	6	6	10	11	8		18	30	2	9	12	2	10
Other[b]	1	1		1	3	< 1	< 1				< 1	< 1	1				
Birds	7	14				4	4		< 1		5	5		3	1	< 1	1
Garbage	34	33	4	5	4	28	24	28	10	4	47	43	39	28	62	4	6
Unidentified				21	23				55	53	< 1			9			5
Total items	2,245	434	26	354	112	3,189	615	19	359	47	2,877	277	112	43	525	59	184

NOTE: Data derived from feces (*F*), pellets (*P*), and collected (*CR*) and observed (*OB*) regurgitations. Total figures are not percentages. *P* and *F* columns include samples only from periods covered by regurgitations; *P-all* and *F-all* include all pellets and all feces, respectively. No data for 1977.

[a] Euphausiids are also included in marine invertebrate totals; therefore, euphausiid values are in parentheses.

[b] Includes tenebrionid beetles.

ter 7)—garbage was more prevalent in collected than in observed regurgitations. This difference arose from differences in sampling period; for 1983, during the chick-banding period only, the difference in prevalence of garbage was negligible (28% in observations vs. 24% in collections). Thus, sample size, fraction of unidentified items, and time of season may all have affected differences in diet estimation by these two techniques.

The gull populations whose diets were sampled by pellets, rather than by regurgitations, may have been exploiting different resources. The percentage of fish was much higher in regurgitations than in pellets, except in 1976, when percentages were equal (Table 3.11). In 1975 and 1978, the difference between pellets and regurgitations sampled at the same time increased. Pellets then showed an even lower percentage of fish.

Barnacles were not detected in regurgitations, either collected or observed, again suggesting that pellets sampled a different subpopulation of gulls. Euphausiids and squid were also lacking in the pellet samples, but as they were present in the feces, this difference was probably due to the methods employed. The difference in the garbage component between the two methods was greatest during 1975 and 1979. Reproductive success during those years was especially high (Chapter 7), and not surprisingly the diet contained a low percentage of garbage. It is therefore interesting that garbage remained a constant, important feature in the diet of individuals that roosted on the catchment basin, but this is consistent with Spear's (unpubl. data) findings that these birds tend to feed more at coastal sites. Similarity in percentage of garbage for all methods in the warm years, 1976 and 1978, indicates that breeding adults fed more on garbage when fish were scarce, and hence exploited the same resource as the roosting, nonbreeding gulls during those years.

Fecal and pellet collections probably sampled the diet of nonbreeders, so the results of these two techniques should be similar. Indeed they were, with the differences noted above. The presence of euphausiids was quite similar for the two methods, especially in samples from the same periods (Table 3.11).

To judge from the analysis of otoliths in pellets, the major fish eaten by gulls were juvenile rockfish 39%, cusk-eels 21%, Pacific Hake, *Merluccius productus*, 12%, and midshipmen 11% (Table 3.12). The importance of rockfish is also evident in chick regurgitations, of which rockfish constituted 47% of the fish eaten. No other fish approached this level of importance. Juvenile rockfish were much smaller than the

TABLE 3.12

Fish species in Western Gull diet, 1973–75

(Percent)

Prey[a]	1973	1974	1975	Total
Atherinopsis californiensis	4%	1%	4%	4%
Porichthys notatus	29	13	9	11
Citharichthys sordidus	4	3	3	3
Brosmophycis marginata		1		
Cymatogaster aggregata			1	1
Zalembius rosaceus	6	2	1	2
Engraulis mordax	6	2		1
Microgadus proximus		2	3	2
Merluccius productus	8	9	13	10
Chilara taylori	17	21	21	21
Glyptocephalus zachirus	1	2	1	1
Microstomus pacificus		1		
Genyonemus lineatus	4	6	4	5
Sebastes spp.	21	36	40	39
Total fish	72	533	1,513	2,118
Diversity	0.855	0.852	0.773	0.956

NOTE: Fish species identified by otoliths in pellets.

[a] *Clupea pallasi, Hemilepidotus* spp., *Leptocottus armatus, Hexagrammos decagrammus, Lyopsetta exilis,* and *Parophrys vetulus* represent < 1% each.

other fishes prominent in the Western Gull's diet, but even on a percent weight basis they remained the major fish prey; samples containing *Sebastes* had more individual fish than samples without. Hake can be considered "offal" and so represent garbage foraging, but at sea rather than at land dumps. This fish is frequently discarded by boats fishing for salmon, and up to 1978 there was a large commercial trawl fishery for hake in waters near the islands. Where the cusk-eels and midshipmen came from is not clear, because both these prey tend to avoid the surface.

Overall, fish accounted for 60–80% of the total diet of gull chicks. Marine invertebrates contributed 5–7% of the total mass, although they occurred in 10% of samples. Garbage was certainly supplemental in the diet, and it is clear that the Farallon gull breeding population primarily exploits the marine environment.

Although gulls are often considered to be scavengers, this is not the case for the Western Gull at the Farallones (see also Hunt and Hunt 1975). Breeding individuals apparently feed in surface waters on live marine organisms. This was dramatically evident in 1983, a severe warm-water year when chick production reached an all-time low. If the Farallon population had been able to switch effectively to feeding on garbage, chick production should have been little affected (see Chapter 7).

Because of the variety of methods employed to sample the gulls' diet, we did not consider it valid to calculate diet diversity or overlaps with other species. Relative to the other seabirds, however, some points are worth noting. First, many of the fish species eaten by gulls generally occur in deeper waters than gulls should be able to exploit (as noted above). The negative association of gulls with cormorants, as discussed in a previous section (Table 3.2), may account for the existence of these species in the gulls' diet. Either the gulls sometimes parasitize cormorants, or they are scavenging fish that cormorants regurgitate in order to take off (sometimes cormorants eat too much), or the cormorants (or other predators) are driving some of these fishes to the surface. Another source of these fish for gulls is trawler spoils, and certainly this is so for hake. Second, gull diets appeared to be more diverse during warm-water years, including the increase in use of garbage. During these years—1976, 1978, and 1983—the preponderance of fish was lower as well (Tables 3.10, 3.11). In all habitats, terrestrial, intertidal, or oceanic, gulls can be considered surface feeders.

Alcids

The diet fed to Common Murre chicks was studied during 11 years, 1973–83 (Appendix 3.6, Table 3.13). Principal prey were juvenile rockfish, anchovies, Nightsmelt, *Spirinchus starksi*, and Market Squid. These are all midwater-schooling organisms of the shelf and slope. During years when rockfish (*Sebastes flavidus* and *S. jordani*) composed relatively little of the diet, i.e., the warm-water years 1973, 1976, 1978, and 1983, they were replaced by anchovies and diet diversity was high. These years of diverse diet and the high anchovy component were also years when murres fed inshore the most (Figures 3.7, 3.8, 3.19, and 3.20). The greater negative rockfish anomaly in the diet during warm-water years, compared to the positive anchovy anomaly (Figure 3.27), suggests that anchovies and other species may have been "replacement" prey. A comparison of the diet among all 11 years indicates that (1) diet during 1976 overlapped little with diet during other years, but most with another warm-water year, 1978, (2) diets during the warm-water years 1973, 1978, and 1983 were similar, and (3) diets in the remaining years were similar largely because of domination by rockfish (Appendix 3.6, Table 3.13). During the atypical warm-water year of 1976 (Chapter 2), squid and smelt replaced

TABLE 3.13

Composition of the diet of two large alcids, 1972–83

(Percent)

	Common Murre						Pigeon Guillemot							
	1973, 1978, 1983		1976		1974, 1975, 1977, 1979–82		1972, 1976		1973		1978		1974–75, 1977, 1979–82	
Prey	*n*	*w*	*n*	*w*	*n*	*w*	*n*	*w*	*n*	*w*	*n*	*w*	*n*	*w*
Porichthys notatus		1%	1%	1%										
Citharichthys sordidus	4%	2	1	1	1%	1%	15%	16%	4%	8%	49%	60%	3%	6%
Clinids							2	1	3	3			1	1
Cottids							41	24	26	22	35	22	15	13
Engraulis mordax	62	71	22	20	10	26								
Spirinchus starksi	8	7	20	14	2	3								
Pholids							6	4	9	8	8	6	2	2
Salmonids	1	1												
Cololabis saira	1	1												
Sebastes spp.	19	4	36	10	86	63	13	8	54	46	8	5	77	78
Stichaeids							1	1						
Peprilus simillimus	3	9												
Loligo opalescens	3	6	21	54	1	8								
Octopus rufescens							22	46	4	12	2	7		
Total items	3,672		2,033		14,722		766		192		26		1,602	
Total weight (gm)	30,224		19,822		70,393		2,537		446		80		3,475	
Mean diversity (*n*)	1.105 ± 0.109		1.281		0.531 ± 0.249		1.646 ± 0.006		1.467		0.371		0.938 ± 0.302	
Mean diversity (*w*)	1.068 ± 0.209		1.298		0.995 ± 0.416		1.416 ± 0.005		1.454		0.926		0.994 ± 0.330	

NOTE: Years grouped according to overlaps identified in Appendixes 3.6 and 3.8. *n*, number; *w*, weight. Total figures are not percentages.

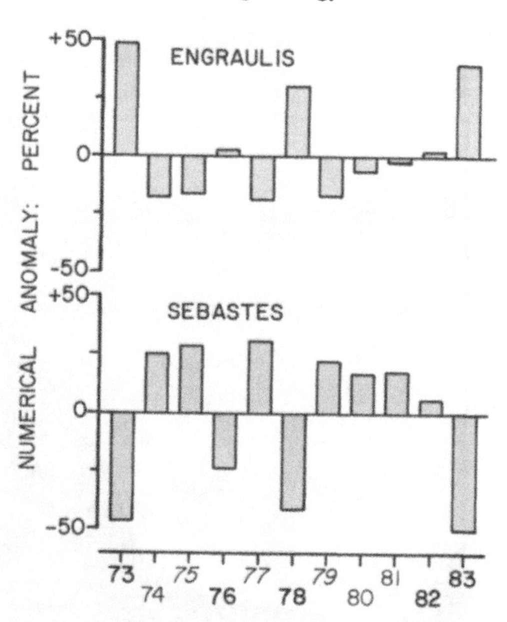

Figure 3.27. Switching between anchovies (*Engraulis*) and juvenile rockfish (*Sebastes*) by Common Murres, 1973–83; for each prey group, annual prevalence is compared against its 11-year mean contribution to the diet.

rockfish much more than in the other warm-water years, when anchovy was the principal prey.

Murres ate larger Pacific Sanddabs, *Citharichthys sordidus*, in 1977, and the length of anchovies eaten increased gradually during the study period (Appendix 3.7). The large size of anchovies relative to rockfish increased the importance of anchovies slightly when diet is considered on a weight basis (Table 3.13).

We were able to look at seasonal shifts in the murre's diet (Figure 3.28). All years combined, diet diversity increased from 0.581 during the first ten days of the nestling period to 1.068 in the seventh and last ten-day period ($r_S = .786$, $p < .05$). This trend, however, was largely a function of warm-water years, when murre nesting was late. In those years, parents were feeding chicks on the ledges from about 20 June into the first week of August (Chapter 8), and during that period the contribution of rockfish decreased from about 20% at the start to 13% near the end, while the contribution of anchovies changed from 40 to 70%. The prevalence of other species increased as well. In other years, when nestlings were being fed from 10 June to about 19 July, the contributions of rockfish and anchovies remained relatively stable at about 85 and 7%, respectively. The shift to ancho-

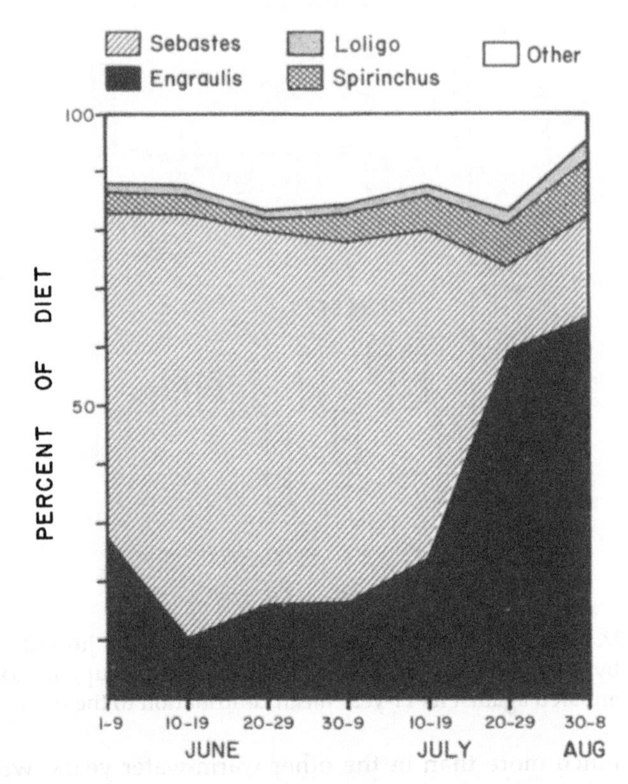

Figure 3.28. Temporal change in the numerical proportions of prey in the Common Murre's diet, all years, 1973–83.

vies and smelt late in the chick period of warm years is consistent with the inshore movement of murres during that time, as discussed previously. As many parents with chicks eventually move inshore even in years other than warm ones, Figure 3.28 illustrates fairly well the general trend in the seasonal shift of the murre's diet.

The diet fed to Pigeon Guillemot chicks was investigated for 11 years, 1972–82. We would have collected observations during 1983, but the species failed to breed that year (Chapter 9). Primary prey were the same as for the Pelagic Cormorant, i.e., juvenile rockfish and cottids, but guillemots ate more octopus, *Octopus rufescens*, and fewer mysid shrimp. When the contribution of rockfish was low during a given year, cottids and other species were more prevalent (Figure 3.29), and diet diversity was much higher (Table 3.13). Curiously, rockfish were also less prevalent in the guillemot's diet during years preceding ENSO's. This indicates that rockfish may have been generally less available during these years, and, consequently, with their much less variable foraging range, guillemots had a harder time finding them than, for example, murres did. A comparison of diet among the 11 years (Appendix 3.8) produced groupings of years approximately similar to those of the murre, except that 1978 was the most unusual. When groupings of years were compared to those of the murre and the cormorants, little overlap occurred during the warm-water years, but much overlap occurred in other years when rockfish were prevalent in the diet of all seabirds (except Double-crested Cormorant; Appendix 3.5). During "rockfish years," the diets of the guillemot and Pelagic Cormorant were virtually identical, which is interesting because the two species were both restricted to feeding in the same habitat close to the island. Even on the basis of prey size (weight), overlap was almost complete. During 1978, poor prey availability in the rocky habitat apparently forced the guillemot to feed more on sandy bottoms, though probably near rocks (note high incidence of sanddabs that year; Table 3.13). Interestingly, few guillemots attempted to breed during 1978 (Chapter 9). During the warm-water years, what little diet overlap was evident on the basis of species composition was reduced 50% more on the basis of prey size.

Like the murre, guillemots diversified their diet toward the end of

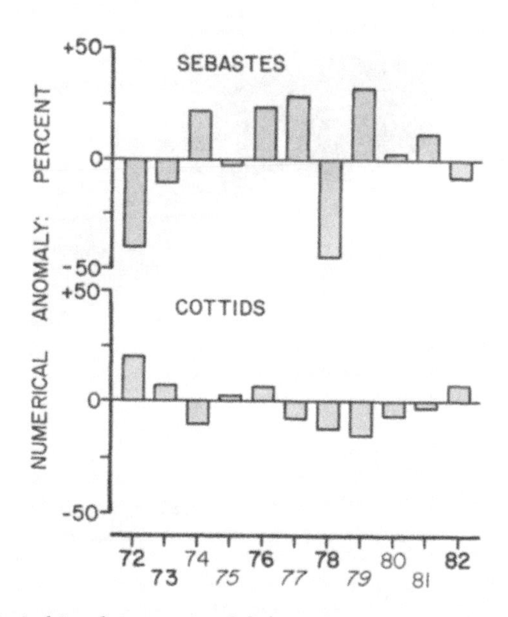

Figure 3.29. Switching between rockfish and cottids by Pigeon Guillemots, 1972–82; for each prey group, annual prevalence is compared against its 11-year mean contribution to the diet.

the chick period, when rockfish decreased and other species increased in importance (Figure 3.30). Diet diversity changed from 0.500 early in the nestling period to 1.619 at the end (eight 10-day periods; $r_S = .988$, $p < .05$). Unlike the situation with the murre, the decrease in importance of rockfish (and increase of other species) occurred in all years. In warm years, the contribution of rockfish to the diet changed from 50 to 9%, and in others it changed from 80 to 50% during the course of the nestling period.

We investigated the diet of Cassin's Auklet during four years, 1977 and 1979–81 (Table 3.14). A lack of data during the warm-water year of 1978 was partly the result of having few breeding auklets available to sample (Chapter 10). Auklet prey were all midwater-schooling organisms of the outer shelf and slope. Unlike all other Farallon species, auklets fed primarily on zooplankton, principally euphausiids of two species, *Thysanoessa spinifera* and *Euphausia pacifica*. The principal fish

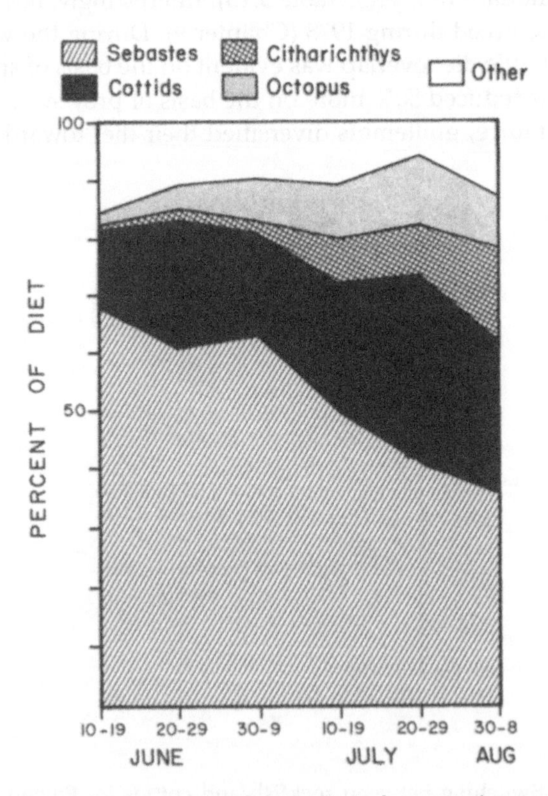

Figure 3.30. Temporal change in the numerical proportions of prey in the Pigeon Guillemot's diet, all years, 1972–82.

TABLE 3.14

Composition of Cassin's Auklet diet, 1971–81

(Percent)

Prey	1971[a]	1977		1979		1980		1981	
	n	*n*	*w*	*n*	*w*	*n*	*w*	*n*	*w*
Fish[b]	1%	1%	2%	6%	16%	3%	21%	8%	34%
Crustaceans									
Amphipods									
Gammarids	—					11	2		
Hyperiids	—			19	6	38	3		
Subtotal	74%			19%	6%	50%	5%		
Cyprids				4%	1%				
Mysids				1	1			9%	2%
Euphausiids[c]	22%	99%	98%	66	74	46%	72%	82	64
Squid	3			4	1	2	1		
Total items	8,083	966		10,907		7,031		6,005	
Total weight (gm)		271		2,963		1,337		1,270	
Diversity (*n*)		0.0980		0.1012		1.1510		0.5814	
Diversity (*w*)		0.0560		0.8230		0.7937		0.7307	

NOTE: *n*, number; *w*, weight. Total figures are not percentages.
[a] Data from Manuwal 1974b; most samples from latter third of nestling period.
[b] Almost all fish were *Sebastes* spp.; a few larval *Citharichthys* spp.
[c] Percent contribution of *Thysanoessa spinifera*, by year = 85%, 68%, 6%, and 55%, respectively, 1977–81 (no data 1971).

in the diet were *Sebastes* spp., and their contribution on a weight basis was much higher than by numerical composition. Our results were similar to those of Manuwal (1974b), who studied the diet during 1971, but because most of his data were from late in the nestling period they are not really comparable. The diet shown in Table 3.14 is similar to the diet during April (incubation period); the adults' diet is the same as that of chicks (PRBO, unpubl. data 1985–86). We also know that adult murres and puffins, and to some extent Western Gulls, have a diet similar to that of the auklet during March and April (1985–86 data). At that time the cliffs turn reddish from the murres' guano, another indication that they are eating euphausiids.

Among the few years sampled, we detected little year-to-year variability in the auklet's diet. Within our sample of years, 1980 was actually somewhat different from the others for auklets because *E. pacifica* instead of *T. spinifera* dominated the euphausiids, and amphipods became much more important. During another warm-water year, 1986, this same shift in diet occurred (PRBO, unpubl. data). As with the two other alcids discussed above, the auklet's diet diversified

Table 3.15

Composition of Tufted Puffin diet, 1973–82

(Percent)

Prey	1973		1974		1975		1976		1977		1978		1979		1980		1981		1982	
	n	*w*	*n*	*w*	*n*	*w*	*n*	*w*	*n*	*w*	*n*	*w*	*n*	*w*	*n*	*w*	*n*	*w*	*n*	*w*
Engraulis mordax	94%	87%	88%	94%	21%	71%	55%	71%	44%	83%	46%	57%			22%	48%	15%	56%	4%	3%
Cololabis saira					3	3	4	2			9	2								
Sebastes spp.			11	4	65	19	30	4	56	17			100%	100%	73	27	85	44	48	47
Peprilus simillimus											9	2								
Loligo opalescens	6	13	1	2	2	7	11	23			36	39			4	25			48	50
Decapods					9															
Total items	285		124		76		56		32		34		7		45		13		56	
Total weight (gm)	4,888		1,653		554		976		267		324		18		307		64		1,135	
Unidentified items[a]	5		14		525		52		156		236		6		3		4		49	

NOTE: *n*, number; *w*, weight. Total figures are not percentages.

[a] Not included in percent composition calculations.

toward the end of the chick period, a trend also evident in Manuwal's (1974b) data. The diversity index increased from 0.551 early in the nestling period to 1.259 at the end (nine periods; $r_S = .768$, $p < .05$). The pattern occurred in all years; the late-season diet is similar to that of the warmer years.

We were able to collect information on the diet fed to Tufted Puffin chicks during ten summers, 1973 to 1982. As with the guillemot, we would have gathered data during 1983 had the species bred (Chapter 11). Unfortunately, we could identify fewer than 50% of the prey, because many times puffins entered burrows too quickly or did so near dark (see Methods). Because so many prey were not identified we calculated neither diet diversity nor an index of overlap. Among the prey identified, anchovies, rockfish, and squid predominated (Table 3.15). Squid apparently replaced fish during the warm-water years 1976, 1978, and 1982. The contribution of rockfish was greatest during years when they were especially abundant in the diet of other species, i.e., 1975, 1977, 1979, 1980, and 1981. On a weight basis, however, rockfish were important only in 1979 and 1981.

The puffin's diet diversified as the nestling period progressed, the diversity index changing from 0.423 early to 1.153 late (five 10-day periods; $r_S = .800$, $p < .05$). This change was largely a function of anchovies decreasing, squid and unidentified fish increasing, and the prevalence of rockfish staying about the same (Figure 3.31). This trend was just the opposite of that of the other seabirds. We suspect many of the unidentified fish were Pacific Saury, a species of slope and pelagic waters.

The puffin's pattern with regard to anchovies is intriguing in that it is so different from that of the murre. Anchovies apparently move inshore late in the birds' breeding season (Chapter 2) and are followed there by murres; by remaining offshore, the puffin no longer has access to them. The fact that rockfish do not decrease in importance in the puffin's diet as the summer progresses can perhaps be accounted for by the deep-diving abilities of this seabird. As the juvenile rockfish grow and settle deeper in the water column (Lenarz 1980), the puffin is still able to feed on them. Unlike the pattern of anchovy consumption by murres, the proportion of the generally larger anchovies eaten by puffins did not increase as the study progressed (Appendix 3.7).

No data are available on the diet of Sooty Shearwaters in the Gulf of the Farallones, but anchovies, squid, euphausiids, and especially juvenile rockfish are important in their diet in the southern portion of

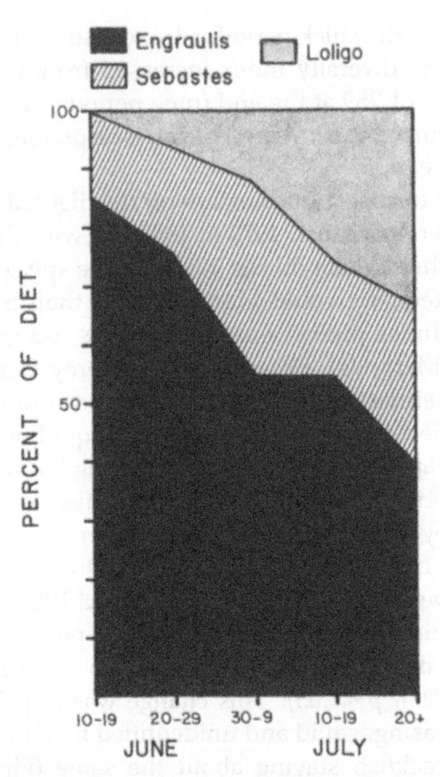

Figure 3.31. Temporal change in the numerical proportions of prey in the Tufted Puffin's diet, all years, 1973–82.

central California waters during early summer (Chu 1984). On fortuitous occasions in the Gulf we have observed them feeding in large numbers on these prey, with individuals regurgitating their catch prior to taking wing. According to Chu (1984), Sooty Shearwaters switch from rockfish to anchovies in August. Thus, their diet and the seasonal changes therein appear to be similar to those of several seabird species breeding in the Gulf of the Farallones. Chu felt that the switch to anchovies was a behavior to take advantage of a more energy-rich food source just before the shearwaters' long transequatorial migration. The information presented here and in Chapter 2, however, indicates that breeding seabirds switch because rockfish become unavailable when they settle closer to the bottom. Thus, it would appear that Chu's idea requires more data on the relative availability of various prey at the time of the switch to determine whether or not energetic considerations are a part of the story.

PREY SIZE

We used data on fish and squid size in these analyses to investigate diet composition by weight. We now make interspecific comparisons to investigate whether or not trophic segregation by prey size was an important factor in this community. We were able to compare the sizes of six fish species common to the diets of the three cormorants: juvenile rockfish and sculpins (cottids) were common to the diets of all three cormorants, while sanddabs, anchovies, Shiner Surfperch, and Pacific Tomcod occurred mainly in the diets of the Double-crested and Brandt's. Among the four years for which data were available, 1974–77, the smaller-billed Pelagic Cormorant (see Figure 1.1) consistently took smaller *Sebastes* than did the other two cormorants, although differences were rarely statistically significant (Appendix 3.4). The same was true for cottids. The greatest differences, by far, occurred during the atypical year 1976. Between the Double-crested Cormorant and Brandt's Cormorant, which has a longer but more slender bill than the former, results were mixed. In four annual comparisons in which more than ten otoliths (and fish sizes) were available, Brandt's Cormorant ate the larger fish in three.

Although results from other researchers indicate a direct correlation between predator size and prey size in some seabirds (Ashmole 1968, Bédard 1969a, M. P. Harris 1970), the significance of these findings for cormorants is not so clear. Double-crested Cormorants were obviously exploiting a different prey population—that of estuaries—where in the case of cottids the large *Leptocottus armatus* dominated the diet. Brandt's Cormorants also fed mainly on this species of cottid, whereas the Pelagic fed on the wide variety of cottids that inhabit rocky substrates. Among the two species of fish that Harris compared in the diets of three auks, mean sizes differed by 23 to 50%, which is a much more dramatic result than our findings, i.e., differences less than 20%. Pearson (1968), too, found little size difference in the prey eaten by the several seabirds he investigated.

We were able to compare the sizes of certain prey captured by three auks, for which we had sufficient observations for nine summers, 1974–82, and for three prey species, rockfish, sanddabs, and anchovies. Only rockfish were common to all three predators' diets, and again results were mixed (Appendix 3.7). The longer-billed murre ate larger *Sebastes* than did the smaller puffin in only two of eight years, and compared to the still smaller guillemot it ate larger *Sebastes* in two of nine years. So far, results are not consistent with expectations. The larger puffin took larger *Sebastes* than did the smaller guillemot in only three of eight years. Within years, differences in the size of *Sebastes* taken by these three predators were actually very small (virtually all in the 10% range).

Expectations were assaulted still further in comparisons for the other two fish species. The larger murre took larger sanddabs than did the smaller guillemot in only three of seven years, and the larger murre took smaller anchovies than did the smaller puffin in all years. Again, actual differences in the sanddab comparison were quite small. In the anchovy comparison, differences were large. As with the cormorants' pattern of cottid consumption, the difference in anchovy size was likely related to differences in foraging habitat. Larger anchovies occur over deeper waters (Chapter 2), and puffins, unlike murres, restrict their foraging to the continental slope or to deeper waters. A. W. Diamond (1983) thought that the size differences he observed in the prey of tropical seabirds could well have been due to differing foraging areas, as did Volkman, Presler, and Trivelpiece (1980) in assessing size differences in the prey of three penguin species in the Antarctic.

Although we used different techniques to estimate prey length for cormorants (regressions from otolith diameters) and alcids (comparison to bill length), because of the extreme range in size among the six species (Figure 1.1), differences in prey size should be evident. Results for rockfish, anchovies, and cottids, however, were remarkably close; they all seemed to take fish of the same size, with the exception of the large anchovies taken by the puffin and Brandt's Cormorant. The sanddabs taken by the larger cormorants were much larger than those taken by the alcids, and are the most clear-cut results yet that relate to expectations based on predator size. At the lengths of these larger sanddabs, fish shape should also have dissuaded any birds but cormorants, which can open their mouths very wide, from taking them (Swennen and Duiven 1977).

SUMMARY

It is clear that the summertime natural history of breeding seabirds, and even visitors such as the Sooty Shearwater, in the Gulf of the Farallones is based on a "juvenile rockfish economy." When rockfish are available, foraging habitats, behaviors, and diets of many of the seabirds overlap extensively. When young rockfish are not available, which usually is the case during years of anomalously warm sea temperatures, avian predators diverge in their foraging habitats (except for some particularly social species), and diets diverge as well. In so doing, the birds switch to alternate prey, usually anchovies, squid, or other species. In all years, as the summer draws to a close, seabirds also switch to these other prey. This is true not only for those species that feed heavily on rockfish, but also for planktivorous species like Cassin's Auklet.

It is not clear whether the foraging of Sooty Shearwaters plays any role in affecting the foraging patterns of the resident species. When rockfish were abundant, shearwaters occurred in great numbers and foraged in the same areas as Farallon species. When rockfish were not abundant, few shearwaters were present, and those that were fed in closer association with some of the other species. Perhaps in the latter situation, the shearwaters had greater impacts on the foraging of some species, playing their role as "suppressors" in mixed species flocks (Hoffman, Heinemann, and Wiens 1981). The importance of that role, however, may be diminished when food is superabundant; it may also be diminished when food is much more scarce because so few shearwaters are present then.

The obvious switching from an economy based on one species to another economy that is much more diversified is an important finding. On the one hand, it means that the summertime diet is likely not representative of the entire year's, something that seabird researchers have sometimes assumed when summarizing or reviewing foraging patterns for large numbers of species (e.g., Ashmole 1971, Harper, Croxall, and Cooper 1985). Because the diet data available are almost always confined to the height of the chick period, generalizing further must be done with extreme caution. On the other hand, the switching of diet suggests that investigations during the nonbreeding period for a number of years will likely reveal some of the biological factors that affect timing and success of reproduction of Farallon seabirds, as well as of other populations. We will return to the subject of feeding ecology and relationships to breeding productivity in the various species' chapters, as well as in Chapter 12, where we will discuss overlaps in feeding niches.

4

Leach's Storm-Petrel and Ashy Storm-Petrel

David G. Ainley, R. Philip Henderson, and Craig S. Strong

Storm-petrels (family Oceanitidae) are among the least-known marine birds because they are among the most pelagic, breed primarily on distant and inaccessible islands, and are active there for little time, usually only at night. Furthermore, they nest out of sight in deep cavities or burrows, and desert their nests readily when disturbed. At sea, where storm-petrels spend most of their time, their small size and erratic flight render them inconspicuous and difficult to observe at length. For several seabird species in this book, much comparative information exists from other studies. This is not so for the two species of storm-petrels that nest at the Farallones. Thus, along with presenting observations on these species' biology, we will review aspects of the life-history patterns of other storm-petrels to learn more about the Farallon species through contrasts and similarities.

The two Farallon species are quite similar in morphology. They are nearly identical in the general shape of their bodies, tails, wings, feet, and bills, and are also nearly identical in size, although statistically distinct differences can be seen with sufficiently large samples (Figures 1.1 and 1.2). It has been known for some time that both species nest at about the same time of year on the Farallones (Bent 1922) and that their nest sites are interspersed on the same rocky talus slopes of the 8-ha southern quarter of the island. A study of the Ashy conducted by James-Veitch (1970) revealed its nesting habits to be almost indistinguishable from those of the several other storm-petrels studied to that point, including Leach's (as noted above). The most pronounced difference between the two species is in color: the Ashy is always entirely dark (except for the occasional albinistic feather; Baptista 1966), whereas almost all Leach's are dark with a white rump patch. A few of the Leach's on the Farallones are also entirely dark, as are almost all individuals in some of the populations nesting farther south in the eastern Pacific (Ainley 1980). Thus, because of the many similarities between the two species, a logical question arises: How do they differ biologically in ways that foster coexistence on the Farallones? Only rarely do more than two storm-petrels nest in the same vicinity, and information from the Farallones may help to explain why this is so.

The world over, there are 20 species of storm-petrels, but for 12 information on their biology and ecology is meager or nonexistent. Information on the remainder indicates many interspecific similarities in breeding biology; some of these characteristics and the sources of this information are summarized in Tables 4.1 and 4.2 (see also Palmer 1962). During a prebreeding occupation of nesting burrows lasting one to four months, the male visits the burrow more than the female, reaching a maximum visitation rate of 50 to 60% of nights. During the last 11 days or so prior to egg laying the female visits hardly at all. Presumably she is using that time to eat and to acquire the nutrient resources needed to form the egg. She then lays a single white egg usually having a faint ring of brown or reddish spots around the fat end. The egg is large, being about 21 to 29% of the female's body weight. If the egg is lost a second will be laid in fewer than 20% of the cases, and after about two weeks.

When the egg is laid, the male incubates it first, but both sexes eventually incubate an equal amount of time. The usual turn lasts about three days but may be as long as five days or as short as one day. The egg can withstand some chilling, and it is not unusual for it to be left unattended for at least one day and sometimes for several days at some time during the incubation period (all at once or intermittently). The male is usually not present when the egg is laid, and, since the female then leaves, the egg is often not incubated for the first day or so (see Boersma and Wheelwright 1979, Boersma, Nerini, and Wheelwright 1980).

Depending on how much it has been neglected, the egg hatches in 38 to 68 days (with no neglect, within 38 to 42 days). The newly hatched chick is semiprecocial; though it cannot move about much, it is covered by a thick thatch of down and its eyes open within a couple of days. A parent is almost always present when it hatches. Parents brood the chick for its first two to seven days of life, after which it is able to maintain its body temperature to some degree. In a few weeks it is entirely able to maintain its body temperature near the adult level. Before then, if it is not fed for three days or so, the chick enters

TABLE 4.1

Breeding chronology of storm-petrels

Species	Area	Colony occupation[a] (months)	Pre-egg period[b] (months)	Laying period[c] (months)	Incubation period[d] (days)	Nestling period[d] (days)	Source
Fregetta tropica	Signy I., Antarctica	5.5	1.5	1.0	41	(65–71)	Beck & Brown 1971
Hydrobates pelagicus	Skokholm, Wales	11.0	2.5	3.0	41 (38–50)	68 (56–86)	Davis 1957, D. A. Scott 1970
Oceanites oceanicus	Signy I., Antarctica	5.0	1.0	1.5	— (38–54)	60 (54–69)	Beck & Brown 1972
	Argentine I., Antarctica	5.5	1.0	1.5	43 (39–48)	—	Roberts 1940
	Pt. Géologie, Antarctica	4.5	0.8	—	41 (38–46)	54	Mougin 1968
Oceanodroma castro	Galápagos Is., Ecuador	?	2.5	3.5	42 (39–51)	70, 78e (60–72, 66–107)[e]	M. P. Harris 1969a
	Ascension I., South Atlantic	9.5	2.5	3.0	38	64 (59–74)	Allan 1962
Oceanodroma furcata	E. Amatuli I., Alaska	6.0	—	1.5	50 (37–68)	— (61–66)	Boersma, Nerini & Wheelwright 1980
	Little River I., California	12.0	3.0	4	— (40–42)	— (70–75)	S. W. Harris 1974
Oceanodroma homochroa	Farallon Is., California	12.0	3.5	4.5	44 (42–59)	84 (66–119)	James-Veitch 1970; this study
Oceanodroma leucorhoa	Kent I., New Brunswick	5.5	0.5	3.0	42 (38–50)	66 (63–70)	Gross 1935, Huntington 1963, Ricklefs, White & Cullen 1980
	Little River I., California	7.0	3.0	2.0	—	—	S. W. Harris 1974
	Farallon Is., California	7.0	2.5	2.0	43 (42–48)	61	This study
Oceanodroma tethys	Galápagos Is., Ecuador	10–12	2–3?	3?	—	66–86	M. P. Harris 1969a
Pelagodroma marina	Whero I., New Zealand	6.0	2.0	1.5	45 (45–59)	57 (52–67)	Richdale 1965

[a] First arrivals to last fledglings.
[b] First arrivals to first eggs.
[c] First to last eggs laid.
[d] Range and mean if more than three observations.
[e] Hot-season and cold-season figures, respectively.

TABLE 4.2

Storm-petrel breeding characteristics and success

Species	Area	Egg weight[a] (percent)	Brood period[b] (days)	Incubation stint[b] (days)	Hatch success (percent)	Fledge success (percent)	Fledge weight[a] (percent)	Source
Fregetta tropica	Signy I., Antarctica	26%	1	3	62%	25%	120%	Beck & Brown 1971
Hydrobates pelagicus	Skokholm, Wales	25	6–7	2–3	62	41	120	Davis 1957
Oceanites oceanicus	Signy I., Antarctica	28	2	2	35	33	150	Beck & Brown 1972
	Argentine I., Antarctica	27	1–2	2	—	35	160	Roberts 1940
	Pt. Géologie, Antarctica	29	2	—	50	40	170	Mougin 1968
Oceanodroma castro	Galápagos Is., Ecuador	21	2–3	6	60	30	130, 110[c]	M. P. Harris 1969a
	Ascension I., South Atlantic	25	6–7	> 5	54	36	110	Allan 1962
Oceanodroma furcata	E. Amatuli I., Alaska	21	5	2–3	74	54	110	Boersma, Nerini & Wheelwright 1980
	Petrel I., Br. Columbia	—	6	—	71	58	115	Vermeer, Devito & Rankin 1988
Oceanodroma homochroa	Farallon Is., California	22	5	2.5	59	42	100	James-Veitch 1970
					78	69		This study
Oceanodroma leucorhoa	Kent I., New Brunswick	22	4–5	4–6	—	—	105	Huntington 1963, Montevecchi et al. 1983
	Petrel I., Br. Columbia	—	6	—	65	57	125	Vermeer, Devito & Rankin 1988
	Farallon Is., California	—	—	—	60	20	110	This study
Oceanodroma tethys	Galápagos Is., Ecuador	22	1–2	5–6	33	23	—	M. P. Harris 1969a
Pelagodroma marina	Whero I., New Zealand	25	2–3	3–5	—	—	115	Richdale 1965

[a] Percentage of adult weight.
[b] Approximate average or most frequent period.
[c] Hot-season and cold-season figures, respectively.

a state of semitorpor to conserve energy. It grows rapidly during its first few weeks, reaching a weight that exceeds the adult's weight, sometimes by a factor of almost two. The excess weight results from a reserve of lipid and water needed in the event that parents bring food only intermittently, a pattern that is not uncommon in petrels. Most of the "real" weight increase occurs in the flight muscles, which by their bulk help the chick to maintain its body temperature. During its last few weeks as a nestling the chick grows its contour and flight feathers. Its weight then fluctuates greatly in response to feedings that are irregular in amount and timing; the general trend of body weight, however, is gradually downward. The chick is fed by its parents until it fledges. The entire nestling period lasts eight to 11 weeks (as long as 13 weeks for some individuals). The chick fledges weighing 100 to 150% of the weight of adults and learns to forage and avoid predators on its own. The excess weight presumably provides a buffer during the chick's first several days at sea (see Ricklefs 1983, Ricklefs, White, and Cullen 1980, Ricklefs et al. 1985).

Storm-petrels first begin to breed when four or five years of age. They make their first visits to their natal rookeries for a brief time during their second summer. In the next year they learn about the island, and in their next they may acquire a burrow and a prospective mate. Once they establish a pair bond, the members of the pair stay together for a number of years and use the same burrow. Their reproductive success is higher if they are familiar with their burrow and their mate than if either is new. If they lose their mate or burrow, a year or two will pass before they become reestablished. Storm-petrels, once they survive the rigors of their first few years, likely live on the order of 20 or more years. Highest mortality probably occurs during their initial years of prospecting for burrow and mate. A major cause of mortality is capture by avian predators (owls, gulls, skuas, and frigatebirds) at the nesting ground.

METHODS

Every five days, from April 1972 until April 1973, we attempted to capture 30 individuals of each species. In total we captured 321 Leach's and 981 Ashy storm-petrels. We captured the birds at night by playing tape-recorded flight calls of *Oceanodroma leucorhoa* (from Maine) to attract both species to mist nets placed at four locations on the talus slope. After each capture session, we banded the birds, recorded data, and released them one by one while it was still dark. We weighed each bird to the nearest 0.1 gm on an electric balance, measured its wing chord, and recorded its molt. We characterized the incubation patch on a scale of 0 to 5: downy, partially downy, bare, vascularized, refeathering, and refeathered but with evidence of recent vascularization (scaly skin).

We monitored burrow occupation by placing treadles and microswitches in the entrances to eight nest boxes during the years 1972–79 (Figure 4.1). When a switch was tripped by a bird either entering or departing, a record was made on an Esterline-Angus event recorder. Nest-box entrances were too small to allow entry by an auklet. We analyzed the resulting strip charts to determine visitation patterns and compared patterns on overcast (> 6/10 cloud cover) and clear nights (≤ 6/10 cover), as well as by moon phase. For the latter, months were divided three ways, by the five days on either side of the full moon, half moons, and new moon.

Another major task was to record the nesting success and chro-

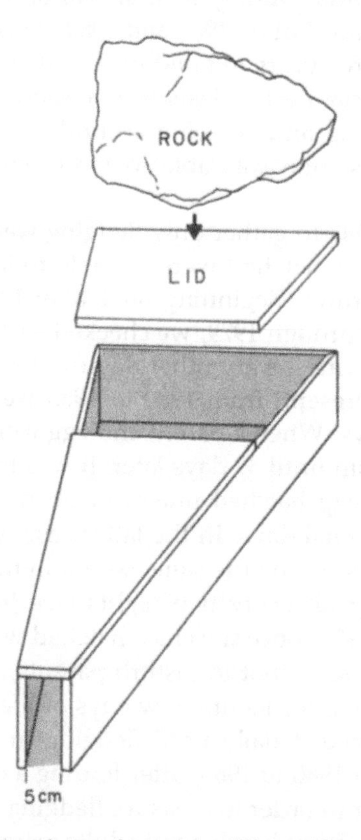

Figure 4.1. Schematic of nest boxes used in studies of Ashy Storm-Petrel breeding biology.

nology of storm-petrels breeding in marked nests. In this study, we were extremely careful to keep disturbance of nesting adults to a minimum because storm-petrels are known to be particularly sensitive to the activities of research (for example, see Boersma and Wheelwright 1979). In 1977, however, disturbance was higher than usual and nest desertions were rather common. In our subsequent analyses, we sometimes did not use results from 1977.

About one-quarter of the nests followed were natural cavities, but most were artificial nest boxes, first shown by James-Veitch (1970) to be effective as nest cavities (Figure 4.1). Use of these boxes helped to reduce disturbance. During the fall, after a pair had finished nesting in a natural cavity, we inserted a box in that place. In this way, as opposed to mere placement of the boxes at random but seemingly suitable sites, the storm-petrels accepted the boxes readily. Baird et al. (1983) report setting boxes randomly and subsequently found a low use rate by the birds. During the first year of our study, 1971, we followed about 25 nests, but by 1982 and 1983 we were checking over 100. It took about three years to build up an adequate sample of petrels nesting in artificial nests. Nests were easiest to find during the fall; by quietly walking and crawling around in suitable habitat on quiet, calm evenings, one was able to locate chicks by their soft peeping.

In 1971 we were able to gather only fledging weights of chicks because only after chicks hatched were we able to locate a number of easily accessible burrows. Beginning on 1 May 1972, however, and each year thereafter through 1979, we checked each of the numbered boxes and natural cavities every other day until a storm-petrel incubating an egg was present; from 1980 to 1983 we checked for eggs once every seven days. When a parent and egg were found, that nest was not checked again until 39 days later; it was then checked every other day until the egg hatched unless incubation continued more than about six additional days. In the latter case we skipped yet another day to avoid disturbing the same parent in the likely event that parents were trading off every two nights (see James-Veitch 1970). From 1972 through 1979, once the chick hatched we checked the nest again after five days so as not to disturb parents, who characteristically brood their young for its first few days of life. We then banded the chick and weighed it daily until fledging in order to calculate growth curves. From 1980 to 1983, after finding a chick, we began to weigh it 68 days later in order to measure fledging success and fledging weight. All chicks were banded but adults were not removed from nests for banding.

FARALLON POPULATIONS

Ainley and Lewis (1974) estimated the Farallon breeding populations of the Leach's and Ashy storm-petrels at 1,600 and 4,000 individuals, respectively, on the basis of a mark/recapture study. These likely were underestimates, particularly for the Ashy, because among the various sites on Southeast Farallon where birds were captured by mist nets, there was little interchange of individuals. Thus, it is probable that portions of the populations were not sampled adequately (a supposition supported by the observations of Furness and Baillie 1981). Indeed, more recent counts of Ashy Storm-Petrels each fall in Monterey Bay, 120 km to the south, consistently estimate on the order of 7,000 individuals (reports in *American Birds* summarized in Ainley 1976). Additional work (summarized by Sowls et al. 1980; Table 4.3) indicates that Ashy Storm-Petrels are known to breed at only five localities and few breed elsewhere than at the Farallones. The Farallon population of this species may number about 7,000 individuals, including subadults, individuals counted in Monterey Bay, and those still involved with nesting at the islands during September and October. About 85% of the world population of this species nests on the Farallones.

Ainley and Lewis' estimate of the population of Leach's Storm-Petrel on the Farallones is probably much more realistic, given the much higher rate of dispersal of individuals among the various capture sites. The population on the Farallones is quite small in comparison to others elsewhere in the eastern North Pacific (Table 4.3), accounting for less than 1% of the population on the Pacific coast of North America.

TABLE 4.3

Estimated numbers of Leach's and Ashy storm-petrels nesting off the North American Pacific coast, 1965–81

Area	Leach's	Ashy	Source
Alaska	4,000,000	—	Sowls, Hatch & Lensink 1978
British Columbia	219,000	—	Vermeer et al. 1983
Washington	7,700	—	Varoujean 1979
Oregon	1,011,800	—	Varoujean 1979
California	18,300[a]	8,200[b]	Sowls et al. 1980
Mexico	57,100+	50+	DeLong & Crossin ms
Total	5,313,900	8,300	

NOTE: Most recent data.
[a] Farallon population (1,600 birds) constitutes 9% of this figure (but see text).
[b] Farallon population (4,000 birds) constitutes 49% of this figure (but see text).

OCCURRENCE PATTERNS

Distribution at sea

Leach's Storm-Petrel is a bird of the open ocean and rarely occurs close to the coast (Ainley, Morrell, and Lewis 1974). During the period 1955–73, accounts in *Audubon Field Notes* and *American Birds* reported fewer than 25 Leach's Storm-Petrels within 80 km of the California coast between Point Conception (34.5° N) and Cape Mendocino (40.5° N), a north–south distance of about 800 km (information summarized by Ainley 1976). The information collected more recently by Briggs et al. (1987) confirms this offshore distribution (see also Chapter 3). Studies in the central portion of the North and South Pacific (Ainley and Boekelheide 1983, Crossin 1974, King 1970, Wahl, Benedict, Ainley, and DeGange, unpubl. data) report this species as one of the most abundant birds of the ocean between Hawaii and the Americas from September to May and uncommon, but present, during other months.

The Ashy Storm-Petrel has a very different oceanic distribution. It is confined to waters just seaward from the edge of the continental shelf and essentially to the waters of the California Current from central California south to northern Baja California (Ainley, Morrell, and Lewis 1974). The largest concentration occurs at the steep shelf break in Monterey Bay. Although a few Ashy Storm-Petrels can be found throughout the year in Monterey Bay, their numbers are highest in summer and fall. On many occasions, numerous birds have been observed within 50 km to the north and west of the Farallones, but Ashy Storm-Petrels have rarely been reported in mid-ocean (Chapter 3).

A comparison of changes in body weight within the pre-egg and nesting periods produced information that supports the pelagic occurrence patterns mentioned above. Both species weighed their maximum during the pre-egg stage when they were finishing or had just finished molt (Ainley, Lewis, and Morrell 1976). During these early weeks of island visitation, Leach's averaged 41.9 gm, compared to 39.6 gm for the Ashy. By fall and the end of the nesting season (periods 40–46), when chicks were almost fully grown, Leach's had dropped to 38.7 gm, almost the same weight as the Ashy (38.1 gm). Thus, although both species gradually lost weight, Leach's Storm-Petrel did so at a faster rate. The general trend, in which maximum weight is attained before egg laying and is then lost during nesting, has been described for the British Storm-Petrel, *Hydrobates pelagicus* (D. A. Scott 1970). The trend does not appear to be the same, however, in all species. In the Wilson's Storm-Petrel, *Oceanites oceanicus*, at Signy Island, there is only the slightest hint that weight is lost during the nesting season, and, as in the Ashy, weight begins to increase at the end of the nestling period (Beck and Brown 1972). In the Madeiran Storm-Petrel, *Oceanodroma castro*, of the Galápagos, body weight reaches a peak after the egg is laid and begins to decline only after parents begin to feed the chick, but in the Galápagos Storm-Petrel, *O. tethys*, at the same locality, no seasonal change in weight is evident (M. P. Harris 1969a). These differences in the trends of weight change indicate that patterns are likely the result of factors affecting different populations rather than being species-specific adaptations of weight fluctuation. With respect to the two populations on the Farallones, Ainley, Morrell, and Lewis (1974) hypothesized that the more rapid weight loss in Leach's over the course of the nesting season resulted from the greater energetic costs of longer travel between nesting and feeding areas. A second morphological comparison in the study by Ainley, Morrell, and Lewis (1974) revealed that Leach's Storm-Petrels carried less weight per unit wing area than the Ashy. This might allow Leach's less energy expenditure per unit distance traveled.

Seasonal attendance

Leach's Storm-Petrel visited Southeast Farallon for about 27 weeks (195 days) beginning the last week of February and ending the first week of September (Figure 4.2), or perhaps by mid-October on the basis of hypothetical fledging dates estimated from known laying dates (see below). The size of the adult population reached its peak about mid-March. Included were breeding birds and probably physiologically mature nonbreeders as well, to judge from studies of other storm-petrels (Beck and Brown 1972, M. P. Harris 1969a, D. A. Scott 1970). When chicks began to fledge during the middle of August, the number of adults visiting the colony declined sharply. At a site in northern California, S. W. Harris (1974) observed essentially the same timing and span of colony occupation, i.e., from late February to late September; in the summer-breeding population of this species on Guadalupe Island, Mexico, the same period of colony occupation is indicated (Ainley 1983). Baird et al. (1983) suggest a slightly longer occupation by Leach's nesting at sites in southeast Alaska: seven months, from late March or early April to late October. In New Brunswick and Maine, Leach's Storm-Petrels visit their colonies for a span of time similar in length to that in California, but arrival is in April, two months later (Gross 1935). On Guadalupe, the winter-breeding population likely occupies the colony for a more protracted period than do the other populations studied (Ainley 1983).

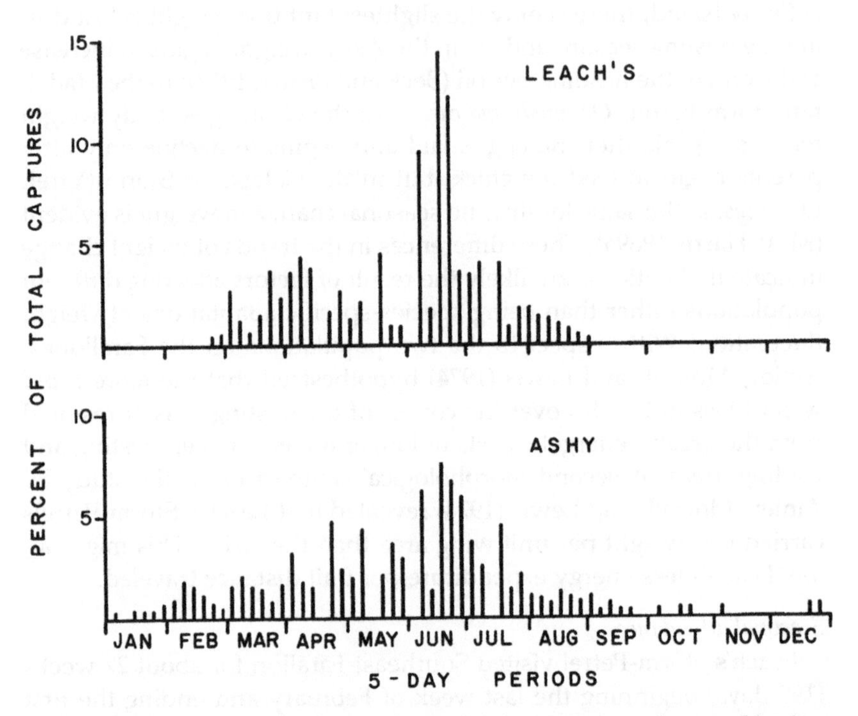

Figure 4.2. Rates of capture of Leach's and Ashy storm-petrels netted by five-day periods, 1972–73 (after Ainley, Morrell, and Lewis 1974, by permission).

During the first three weeks of June, large numbers of immature Leach's visited the island (forming the large peak in Figure 4.2). Immature individuals are distinguished from adults by their completely downy incubation patches and their browner, more worn and faded plumage. Late-season visits by immatures have been described for albatrosses (Fisher and Fisher 1969, Richdale 1950), shearwaters (M. P. Harris 1966, Richdale 1963), and the British Storm-Petrel (D. A. Scott 1970), but in other tubenoses, including at least two storm-petrels, immatures visit early in the breeding season (i.e., in the egg-laying period of adults), as in Leach's (Beck and Brown 1972, M. P. Harris 1969a).

Ashy Storm-Petrels visited the island over a more extended annual period of at least 325 days. Birds were captured throughout the year, except for a period of several weeks during late November and early December (Figure 4.2). Even during that hiatus a few adults must have visited, because we banded recently fledged chicks during December and January. Thoresen (1960), too, found adults feeding young in January. The fledgling reported by Bryant (1888) in early June and the freshly laid eggs reported by James-Veitch (1970) in March are additional evidence for the exceedingly long period of occupation by the Ashy Storm-Petrel population. Following the first visits of the new cycle during late December, the adult population increased to its peak by the beginning of February and then remained at that level for several months. The first chicks began to fledge during late August and the majority fledged by mid-November. Throughout the fledging period, the number of visiting adults slowly declined.

Our data from the monitored burrows confirm the year-round occupation of nests by Ashy Storm-Petrels (Figure 4.3). In fact, the pattern described by these data is virtually the same as that from the netting data. Particularly during the prebreeding months, December to April, but generally during all months, significantly fewer Ashies visited burrows on nights of a full moon than visited on nights of a new moon. This behavior is likely a response to predation by gulls or owls.

A long period of colony occupation is also exhibited by the British Storm-Petrel on Skokholm (Davis 1957, D. A. Scott 1970), the Fork-tailed Storm-Petrel, *Oceanodroma furcata*, in northern Calfornia (S. W. Harris 1974), and possibly the Madeiran Storm-Petrel in the islands of the Madeira group (M. P. Harris 1969a). Leach's Storm-Petrels on Guadalupe and Madeiran Storm-Petrels in the Galápagos are present

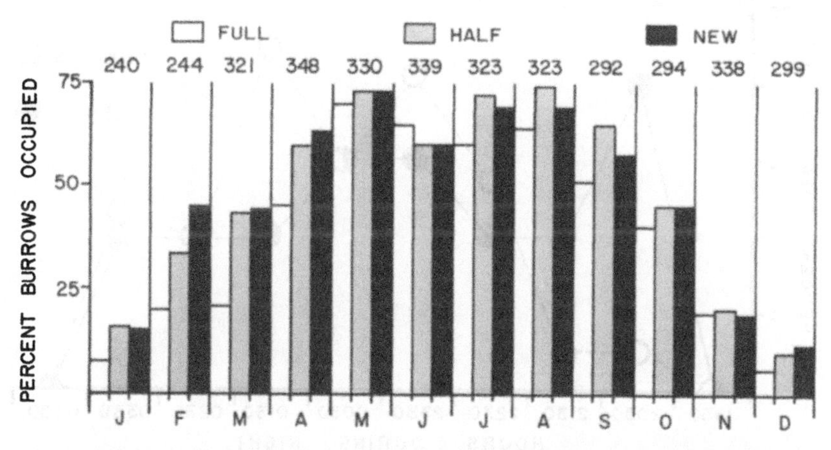

Figure 4.3. Percentage of eight electronically monitored nests that were occupied each month; data divided by phase of the moon, Ashy Storm-Petrel, 1972–79. Sample sizes at top (burrows × nights equipment operable).

in their colonies the year round, but in these two cases separate summer and winter populations are involved (Ainley 1983, M. P. Harris 1969a).

Immature Ashy Storm-Petrels, also recognizable by their downy incubation patches and their more faded coloration, began to visit during early April, much earlier than immature Leach's. They continued to visit until early July, although the low recapture rate of banded individuals suggests that few immatures made repeated visits during this period. During June 1972, we captured two one-year-olds banded as chicks the previous year. Thus, even immature Ashy Storm-Petrels visited over a more extended period than their Leach's counterparts, something like 100 vs. 20 days, respectively. Beck and Brown (1972) and M. P. Harris (1969a) hypothesized that early-season visits by immatures of the storm-petrel species they studied were an anti-predator strategy. For Leach's and Ashy storm-petrels, which on the Farallones would seemingly be subject to similar predation pressures, the markedly different patterns of island visitation exhibited by the immatures of the two species suggest that predation is not the only factor involved (see below).

Daily attendance

The two species also differed in their patterns of nightly visitation (Figure 4.4). Although both species arrived, flew about, visited their burrows, and departed the island only during darkness, Ashy Storm-

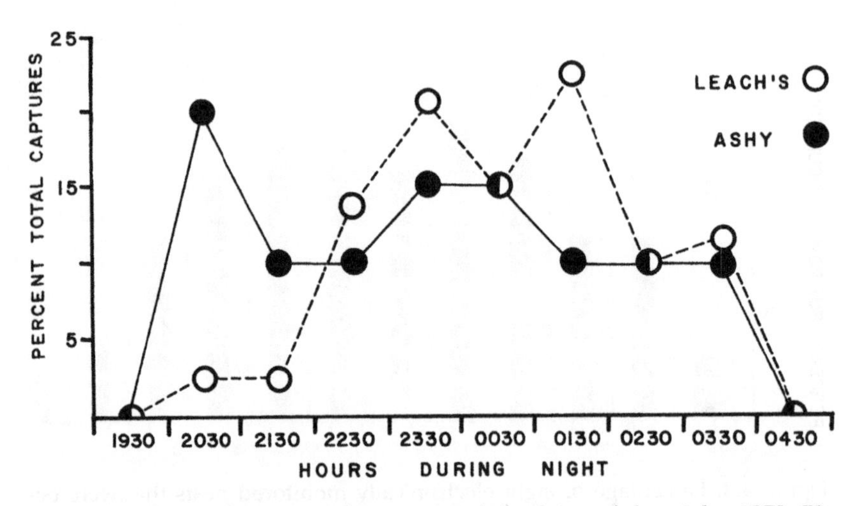

Figure 4.4. Rates of capture of storm-petrels by hour of the night, 1972–73. Net sample is the same as in Figure 4.2.

Petrels arrived in large numbers as soon as the sky was completely dark. Most birds captured then, and for the next hour or so, had just arrived, as indicated by their full stomachs. After the initial peak, the number of birds arriving or flying about declined to a lower level that continued throughout the remaining hours of darkness. During the middle hours of the night, we captured mostly immatures with downy incubation patches. Adults must have been in their nests during the middle of the night. In fact, not until about 2300 (an average 40 min earlier on nights of the new moon), two hours after darkness fell, did breeding individuals enter their burrows. Beginning about 0130 but certainly before morning twilight at about 0400, all birds either had departed or had entered nests for the day. Data from the monitored burrows indicate that departing adults actually began to leave their burrows about two hours before the last ones were captured in nets.

Leach's Storm-Petrels did not begin to arrive in really large numbers until an hour after darkness fell, or about 2200, and peak numbers did not arrive until midnight. Captures continued at a high rate for about two hours, then declined. As with its annual patterns of visitation, *O. leucorhoa* also apparently spent less time at the island each night than did *O. homochroa*. During those initial hours of darkness, Leach's Storm-Petrels were likely flying from their more distant feeding areas. At a site in Scotland, Waters (1964) and Furness and Baillie (1981), and, at a site in northern California, S. W. Harris (1974) found patterns analogous to those at the Farallones, i.e., the Leach's Storm-Petrel arriving much later and leaving earlier each night than the other storm-petrel species present, in the first case the British Storm-Petrel and in the second the Fork-tailed Storm-Petrel.

The amount of light from the moon had a marked effect on the activity of storm-petrels, although whether light level affected the two species differently is not known. During brightly moonlit nights few storm-petrels could be heard calling, few could be seen flying about, and it was futile to try to catch any in mist nets: it was apparent that they could see and avoid the net. The data from monitored burrows show that Ashies spend slightly more time in their burrows on clear nights, especially those during full moon (4.2 to 4.4 hours per night), than on overcast nights (3.5 hours per night). Similar patterns in response to bright moonlight have been observed in other studies. Most other researchers have concluded, however, that while moonlight may dampen activity it does not inhibit the visits of individuals except nonbreeders without nesting cavities. At the Farallones, on

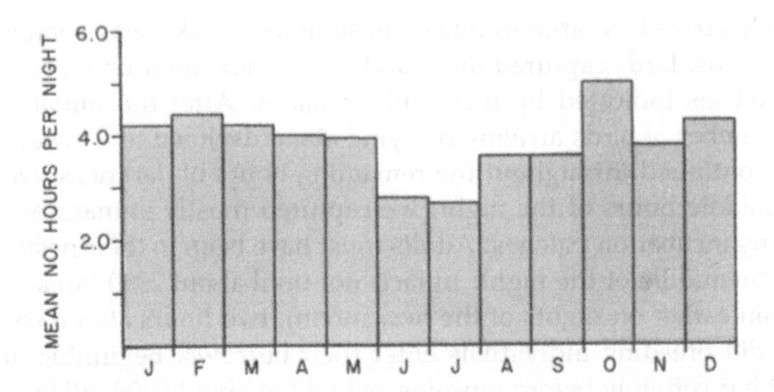

Figure 4.5. Mean number of hours between arrival and departure at nest boxes by Ashy Storm-Petrels, 1972–79. Sample is the same as in Figure 4.3.

the other hand, it did appear that moonlight affected the visitation of breeders (Figure 4.3).

The effect of moonlight was also evident in the growth and fledging patterns of chicks. For one thing, chicks were more likely to be fed by parents during the dark of the moon, a pattern manifested in the date on which they reached peak weight. During the period 1972–79, among the 185 Ashy Storm-Petrel chicks that survived to fledging and for which we had information on growth, 68.1% reached their peak weight during the dark half of the moon's monthly cycle (second quarter to the end of the third quarter), compared to 31.9% that reached peak weight during the light half (the years analyzed in a 2×8 table, $G = 14.914$, $p < .05$). Another indication that moonlight was important to storm-petrel activity was that chicks tended to fledge during the dark half of the moon's cycle (1972–83, $G = 24.954$, $p < .01$), and of those that fledged during the light half the majority (58.3%) left on overcast nights ($G = 27.396$, $p < .005$).

On the basis of data from monitored burrows (Figure 4.5), Ashies spend about the same amount of time in their nests each night regardless of the time of year, except for June and July. The lesser amount of time during those months could be a function of two factors. First, these are the months when most adults are incubating eggs and the pattern results from adults departing soon after being relieved of their shift at incubating (which may have lasted several days). Second, nonincubating breeders depart burrows sooner to display aerially, in response to the large influx of young nonbreeders that occurs then (see above).

REPRODUCTION IN LEACH'S STORM-PETREL

Leach's Storm-Petrels nest almost exclusively on the upper slopes of Lighthouse Hill. They also tend to nest deep within the talus and, as evidenced by the presence of dirt on the bills of several individuals caught in mist nets, they excavate their burrows as well. These characteristics reduced our access to this species and, as a consequence, we were unable to gather much direct information on its breeding biology. Because this species has probably been the subject of more research projects than any other storm-petrel, it is somewhat surprising that our knowledge of its basic breeding biology is less complete than that for several other species. Obviously, it is everywhere an elusive species.

Egg laying and incubation

Since 1971 we have been able to establish the laying dates for only 14 eggs, including three laid by various birds in the hands of banders. The mean Julian date for this sample was 154.8 ± 18.0 (4 June); the range was from 10 May to 14 July (Figure 4.6). Baird et al. (1983), using an equally small sample of eggs laid at sites in southeast Alaska,

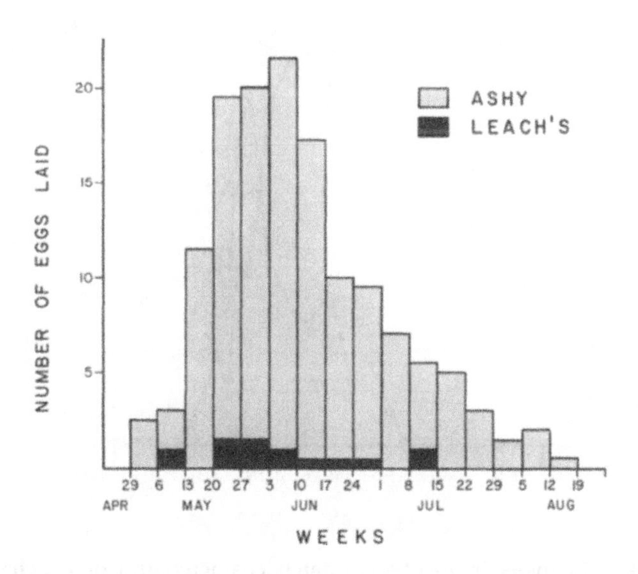

Figure 4.6. Frequency distributions of Leach's and Ashy storm-petrel laying dates, all years combined, 1972–83.

estimated a similar period, i.e., late April to mid-July. In the Bay of Fundy, egg laying by this species ranges from about the same time in May to mid-August (Wilbur 1969), about one month longer.

The laying period of Leach's Storm-Petrels on the Farallones averages about a week earlier and is more contracted than that of the Ashy (whose egg laying is discussed more fully below). Although this can be only tentatively suggested from the data in Figure 4.6, information on changes in the incubation patches of birds caught in the mist nets confirms the trend (Figure 4.7). Passerines and birds of many other taxa lose the down on the abdomen within a few days of egg laying; with the onset of incubation, the skin of this brood patch becomes heavily vascularized. The close timing of these changes provides criteria for determining when eggs are laid. This pattern, however, is not evident in storm-petrels (as noted by most observers) or in penguins (Ainley 1975), in which breeding birds lose the down from their abdomens up to a month or more before egg laying (even phys-

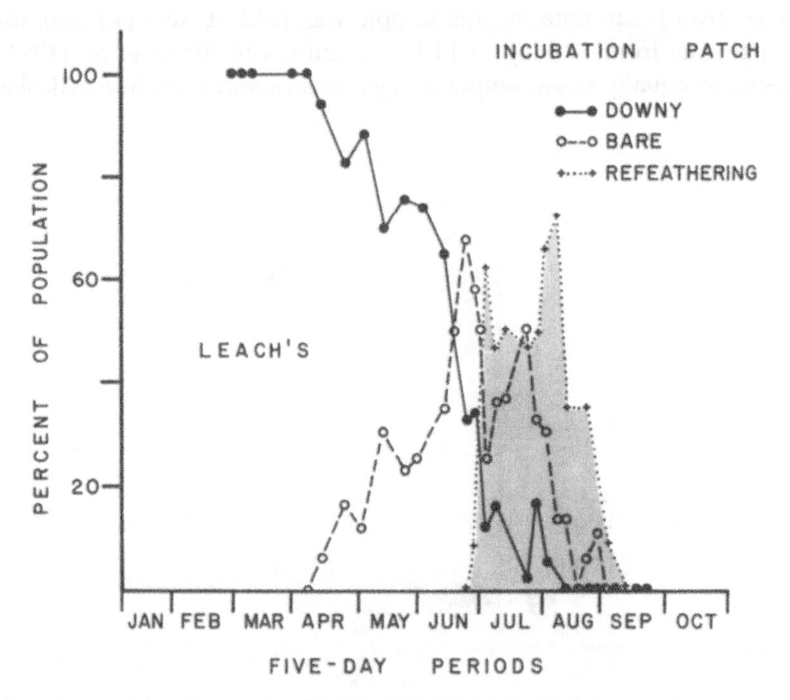

Figure 4.7. Changes in incubation-patch characteristics of Leach's Storm-Petrels, by time of year, 1972 (after Ainley, Morrell, and Lewis 1974, by permission). Phenology is best indicated by refeathering patches.

iologically mature and immature nonbreeders that visit the colony lose abdominal down). Furthermore, both our experience and that of other researchers show that it is difficult to determine whether the skin of the incubation patch is vascularized even in a petrel removed from its nest while incubating. We could, however, correlate certain characteristics of the patch to hatching. Between two and ten days after the egg hatches, down feathers begin to grow on the incubation patch, and within three to four weeks it becomes completely refeathered with down (n = 15 birds). The initiation of refeathering within a week after the chick hatches has also been observed in the Madeiran and the British storm-petrels (M. P. Harris 1969a, D. A. Scott 1970). Presumably the pattern is similar in Leach's. By correlating refeathering to egg hatching and by extrapolating we could approximate the egg-laying period of the storm-petrels on the Farallones. This procedure is especially valuable in the case of Leach's Storm-Petrel, for which we have so few known egg-laying dates.

During the net study in 1972, the first refeathering patch found in *O. leucorhoa* (Figure 4.7) appeared on 21 June (five-day period number 35), and the last was found on 29 August (period 49). In some individuals, the patch should have begun to refeather within two days of their chick's hatching, so the first egg probably hatched around 19 June. Counting back 42 days from this date, which is approximately the minimum incubation period in this species (Table 4.1), we arrive at 8 May as the probable laying date. Though the egg could have been incubated for longer than 42 days, that date is within a week of the earliest known laying date. If we next count back 21 days (the minimum time needed for refeathering of a patch after egg laying) from the date on which the last patch showed evidence of refeathering, we arrive at 8 August as the latest hatching date (latest known hatching date is 29 August). Extrapolating from this, 29 June becomes the latest laying date, about two weeks earlier than the latest known laying date. Thus, even though incubation-patch data were gathered in but one year and egg dates were gathered over nine years, reasonable agreement exists between the two. The incubation-patch data seem to confirm the pattern of egg laying drawn from the meager sample of known laying dates. Similarly, the incubation-patch data for the Ashy Storm-Petrel also agree with the pattern drawn from egg dates (Figure 4.8).

Both methods of analysis indicate a relatively contracted synchronous laying period for *O. leucorhoa* on the Farallones, more contracted than that for the Ashy and apparently more contracted than those for

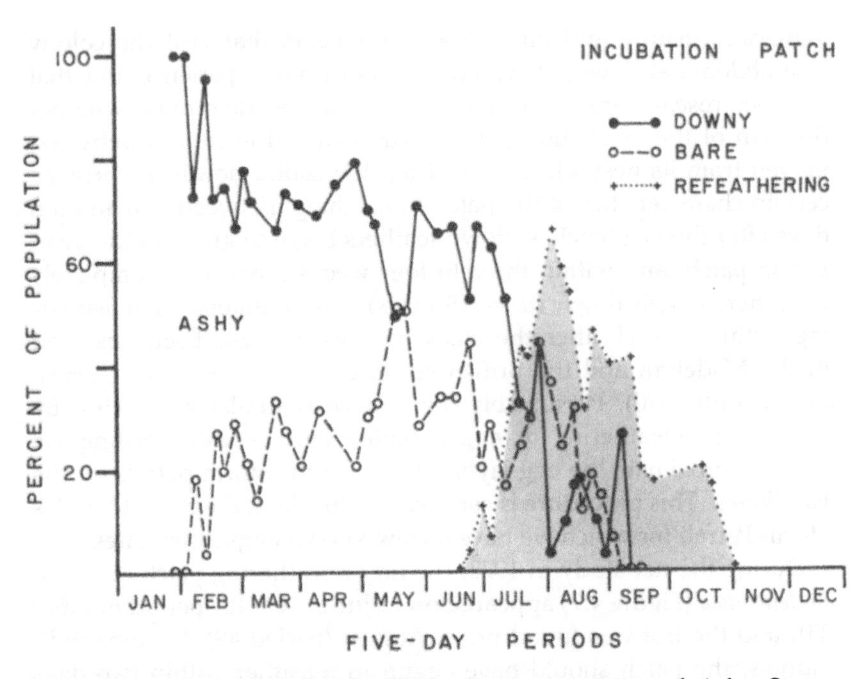

Figure 4.8. Changes in incubation-patch characteristics of Ashy Storm-Petrels, by time of year, 1972 (after Ainley, Morrell, and Lewis 1974, by permission). Phenology is best indicated by refeathering patches.

Leach's Storm-Petrels breeding in the Bay of Fundy (Gross 1935, Wilbur 1969; Table 4.1) and during the winter on Guadalupe Island, Mexico (Ainley 1983). The nesting period of this species on the Farallones appears to be similar in timing and duration to that of the populations also nesting during the summer on Guadalupe (Ainley 1983), in northern California (S. W. Harris 1974), and in southeast Alaska (Baird et al. 1983).

Breeding success

We were able to determine the fate of ten eggs for which we knew the laying date: six of them hatched (60.0%), and two chicks ultimately fledged (20.0% nesting success). At least one of the four chicks that died was apparently killed and eaten by a feral house mouse. Among the four eggs whose hatching date was known to within a day, the incubation period was 43.0 ± 1.2 days. The figure for hatching success is comparable to that calculated in most other studies of storm-petrels. Unfortunately, this appears to be the extent of direct information available on nesting success in this species (Table 4.2).

Wilbur (1969) presented a figure for hatching success of 98% in a sample of nests on Kent Island. This, however, is likely an overestimate, because his method did not allow for eggs that disappeared before his check (he collected data only from nests having eggs or chicks during a one-week period well past peak hatching), and he assumed that all of the 25 eggs still being incubated would hatch. As discussed below, it appears that the disappearance of eggs or their failure to hatch is an important factor lowering reproductive success in most of the other species of storm-petrels studied. The overall success rate of 20% in the Farallon sample is likely to be rather low. Possibly the relatively low hatching rate of the ten eggs on the Farallones, if not a function of small sample size, was related to the quality of the birds involved. It is possible they were of poor quality, if their choice of suboptimal, near-surface nesting habitat is any indication. The "prime" breeders in this species do not use nest cavities that are so accessible and near the surface; otherwise we would know more about Leach's Storm-Petrel.

Egg laying completely overlapped in the two Farallon storm-petrel populations, a function of both species' beginning to lay at about the same time. This contradicts Cody's (1973) hypothesis that two sympatric species of storm-petrels should have non-overlapping egg-laying periods. More discussion of the factors that contribute to the timing of reproduction in these two species on the Farallones is presented below.

REPRODUCTION IN THE ASHY STORM-PETREL

In marked contrast to our study of reproduction in the Leach's Storm-Petrel, we were able to gather a great deal of information on the Ashy. There was no evidence to indicate that Ashies excavated burrows, and from our crawling around on the talus slopes at night and listening for their calls from within the hillside, it was obvious that many nested quite close to the surface. Fortunately, the species seemed to accept our nest boxes readily. Among the 43 nests studied in 1971 and 1972, most were natural cavities, and many were used for only one or two seasons once we began to study them. Thereafter, however, as we increased our sample of nest boxes, the attrition rate was lower and rather constant. Almost all nests used for more than four seasons were in nest boxes. Thus our study benefited greatly from the use of artificial burrows. Before we switched to boxes, we probably caused undue disturbance to petrels in study burrows be-

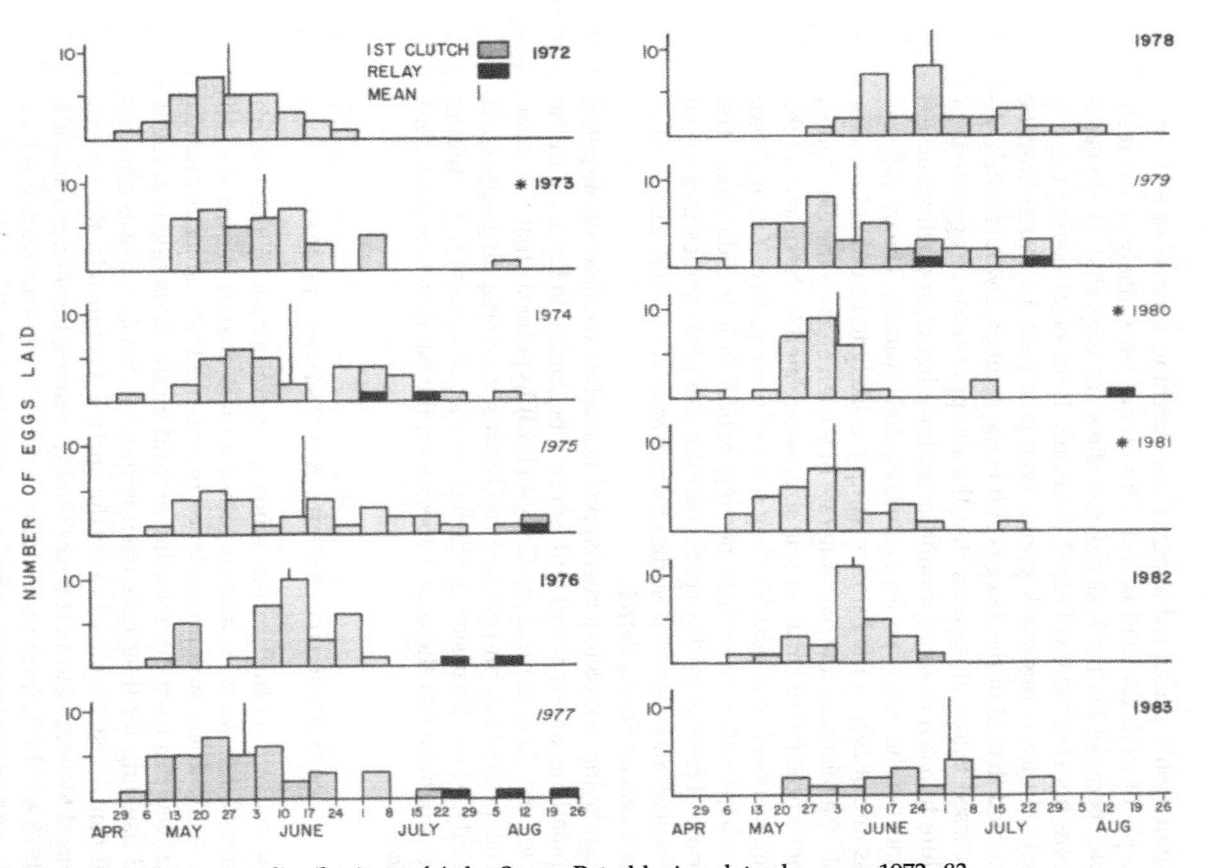

Figure 4.9. Frequency distributions of Ashy Storm-Petrel laying dates by year, 1972–83; those distributions marked by an asterisk (*) were significantly skewed and peaked (t test, $p < .05$).

cause we had to move rocks. In addition, our boxes were less likely to be usurped by Cassin's Auklets.

Egg laying and incubation

The proportion of occupied cavities (i.e., adult storm-petrel observed in residence during at least one nest check) in which eggs were laid remained more or less constant throughout the study (Appendix 4.1). The range from 82 to 100% of burrows was higher than D. A. Scott (1970) observed for the British Storm-Petrel, of whose occupied burrows only 70 to 85% contained eggs. The smaller non-breeding portion of the Ashy Storm-Petrel population occupying nest sites could indicate more intense competition for sites. This is consistent with a population that does not excavate its own nests but must depend on the availability of natural cavities and must contend for nest sites with a larger, more vigorous competitor, in the Ashy's case, the Cassin's Auklet (see below). The gradual decline over the years in the number of study nests actually used by petrels, as indicated in Appendix 4.1, is also partly the result of interference by the auklets (again, see below).

We were able to record laying dates to within one day for 274 first eggs during the period 1972–79, and to within six days for 107 eggs during the period 1980–83 (Figure 4.9). Because we checked burrows every seven days during the latter four years, the distribution of these laying dates is presented graphically for all years by one-week intervals, replacement eggs included. The average laying date for all 381 eggs was 9 June ± 20 days, with a range of 1 May to 16 August. Second eggs were laid at least as late as 26 August. Some eggs were laid perhaps as much as a month later than this, however, because we captured recently fledged young as late as January (see also Thoresen 1960, James-Veitch 1970). Although the date of the first egg varied between 1 and 30 May, the date of the last first-laid eggs extended twice that range, from 30 June to 16 August.

The pattern of egg dates deviated significantly from a normal distribution during only three of the 12 years: 1973, 1980, and 1981. In those years, laying was skewed somewhat toward the earlier part of the laying range, and the peak was more pronounced. The only years in which egg laying was negatively (but not significantly) skewed, i.e., toward late in the laying period, were 1976, 1982, and 1983, two of the four ENSO years (see Chapter 2; 1973 was also such a year). A comparison of the mean laying dates also indicated little significant interannual variability. If the years of lowest and highest variability in egg laying, 1972 and 1975, respectively, are excluded, variability among all other years was the same (F_{max} variance ratio = 2.94, $p > .05$). Egg laying in 1972 and 1977 was significantly earlier than in the five years of latest egg laying, 1974–76, 1978, and 1983, but was not appreciably earlier than in the other years. Laying in 1978 and 1983, however, was significantly later than in all other years (SNK test, $p < .05$). If we recall from the discussion in Chapter 2 that 1974–75 was the period of cold water and intense drought and that 1978 and 1983 were the two most intense warm-water years, it is apparent that the pattern of laying was atypical in one or another way during years of anomalous conditions. We will return to this subject in a later chapter.

The mean number of days between laying and hatching for a sample of 184 eggs gathered during the years 1972 to 1979 was 44.8 ± 5.2. The number of days ranged from 42 to 59 (Figure 4.10); 42 days is

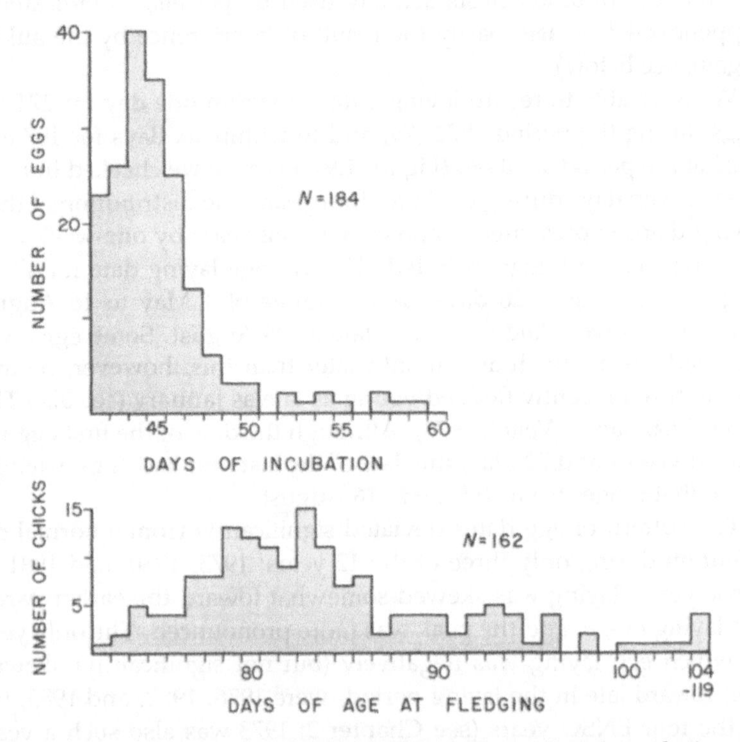

Figure 4.10. Frequency distributions of Ashy Storm-Petrel incubation and nestling periods; all nests and eggs combined, 1972–79.

probably much closer to the hatching period of Ashy Storm-Petrel eggs incubated continuously. Boersma and Wheelwright (1979) found the number of days exceeding the minimum "incubation period" to be a direct function of the number of days a parent did not incubate the egg. They found for the Fork-tailed Storm-Petrel in southeast Alaska that an egg could be neglected for as many as 31 days (i.e., the "incubation period" would be about 73 days) and still hatch, and that about half of their sample was not incubated for 11 or more days. Our data for the Ashy indicate a maximum egg neglect of only 17 days (median 3 days).

Boersma and Wheelwright (1979) suggested that the ability to leave eggs unattended, which is a function of nesting in a cavity safe from predators and having embryos that are resistant to chilling, allowed storm-petrels to cope better with a variable environment, particularly with storms and other factors that might affect their ability to find food. To test this hypothesis we looked at within-year and between-year variability in the amount of egg neglect. First, we divided egg-laying dates equally among early, middle, and late periods (1972–79) and compared the amount of egg neglect (days of incubation in excess of 42) within years. The proportion of eggs neglected more than three days was much higher among those laid during the first third of egg laying (46.5%, $n = 71$) than among those laid during either the middle (26.0%, $n = 71$) or the last third (23.7%, $n = 38$) (if the first and middle thirds are compared after arcsin transformation, $t = 2.616$, $p < .01$). This could be a function of wind strength inversely affecting food availability or directly affecting the amount of time required to fly to and from the island. As indicated in Chapter 2 (Figure 2.1), during May (first third of egg laying) winds exceeded 14 knots on more than half of all days, but during June and July (middle and late thirds) winds exceeded that level on only a third of all days. In contrast, between years we could detect no difference in either the mean number of days of neglect or the proportion of eggs neglected more than three days. The mean number of days of neglect ranged from 2.2 ± 1.9 (in 1972) to 4.2 ± 3.3 (in 1974), and the proportions ranged from 23% (in 1973) to 46% (in 1974); thus, there was not much interannual variability. Neglect was highest, however, in 1974 (46%) and 1975 (39%; compared to 23–33% for the other years), the two years of most persistent, strong northwesterly winds during mid- to late summer (Chapter 2). Thus, it does appear that wind conditions may affect egg neglect. Both Richdale (1965) and D. A. Scott

(1970) felt that extreme irregularity in the return of storm-petrels after foraging was a function of wind inhibiting flight rather than food gathering.

Over the entire span of the study, an average 77.7 ± 9.6% of eggs hatched, with a range of 67% in 1983 to 94% in 1973 (Appendix 4.1). Ignored here is the figure for 1977, when undue disturbance was an element of the study (see Methods). Hatching success was not significantly different among years (G = 10.050, DF = 10, $p > .05$). Within years, however, there was a significant trend in hatching success. Combining data from 1972 through 1979, we separated eggs by whether they were laid during the first, second, or last third of each year's laying period. Among eggs laid during the first period, 84.9% (n = 93) hatched, compared to 73.9% (n = 111) and 75.0% (n = 52) for the two later periods (G = 117.434, DF = 2, $p < .05$). This indicates that neglect, which was highest among eggs laid early, did not reduce hatching success in this species. Indeed, the percentage of chicks ultimately fledged relative to eggs laid was the same for eggs neglected six or more days (89.5%, n = 36) as for eggs neglected less (88.9%, n = 229). On the assumption that the less-experienced and first-time breeders lay later, as is the case in the British Storm-Petrel (D. A. Scott 1970) and in other seabirds (Ainley, LeResche, and Sladen 1983), the high hatching success of the earliest-laying pairs is likely related to the greater breeding proficiency of these birds (as Scott surmised for his sample).

Chick development

Chicks remained in the nest an average 84.4 ± 6.5 days (n = 162, 1972–79). The chick period ranged in length from 72 to 119 days and was much more variable in length than the incubation period (Figure 4.10); the coefficient of variation for the chick period was, in fact, 65% greater than for the incubation period.

At hatching, the Ashy Storm-Petrel chick is covered in down but is helpless. The parents brood it on average for its first 3.7 ± 2.5 days (range 0–11; calculated from James-Veitch's [1970] table XIX), with 80% brooding it for the first five days. The development of the chick, including feather growth and changes in the size of various appendages, is described in detail by James-Veitch (1970).

We used a logistic model to analyze the Ashy Storm-Petrel's growth curve. Hatching weight, asymptotic weight, and the growth constant (K) averaged 9.2 gm, 48.8 gm, and 0.1079 per day, respectively (Table 4.4). James-Veitch (1970) calculated an average hatching weight of

TABLE 4.4

Mean growth parameters for Ashy Storm-Petrel chicks, 1971–83

	Logistic growth model, ANOVA					
Year	n	Hatching weight (gm)	Asymptotic weight (gm)	Growth constant (K; per day)	Fledging weight n	Fledging weight gm[a]
1971					18	42.3 ± 5.2
1972	10	12.1	52.1	0.0776	24	42.5 ± 3.6
1973	10	10.8	53.2	0.0952	19	44.0 ± 4.2
1974	5	8.6	43.0	0.1147	22	41.5 ± 4.1
1975	20	9.7	49.6	0.0966	22	41.3 ± 3.4
1976	26	8.7	47.3	0.1237	26	40.3 ± 4.2
1977	15	7.7	46.7	0.1202	16	40.7 ± 3.7
1978	19	8.6	48.4	0.1085	19	42.0 ± 4.6
1979	20	9.1	49.5	0.1088	21	42.4 ± 5.4
1980					12	40.2 ± 3.5
1981					15	41.8 ± 3.8
1982					14	40.0 ± 4.0
1983					6	40.8 ± 5.9
Total/average	125	9.2	48.8	0.1079	234	41.5 ± 4.3

[a] Mean ± SD.

6.7 ± 1.9 gm (n = 44), which is not that different from our estimate. The growth curve itself was typical of that of other storm-petrels (compare Baird et al. 1983 and Figure 4.11); once chicks reached an asymptote exceeding adult weight, their weight fluctuated depending on the frequency and size of meals.

Hatching weights showed no significant year-to-year variability, but asymptotic weights and the growth constant did (Table 4.5; $F > 4.37$, DF = 7,116, $p < .001$). There appeared to be no connection between the variability and warm-water or cold-water years. Years with lower growth rates had higher asymptotic weights ($r_S = -.740$, $p < .05$). This result is possibly an artifact of time spent in the nest, which, as discussed above, is a function of when the dark of the moon occurs relative to when chicks are ready to fledge.

Asymptotic weight showed a tendency to decrease for late-fledging chicks within a season, but no statistically significant seasonal trends were apparent in the growth parameters (ANOVA, DF = 116; hatching weight, $p > .5$, $t = 3.31$; asymptotic weight, $p < .10$, $t = 12.04$; growth constant, $p > .4$, $t = 3.67$). The relationship for asymptotic weight might have been significant statistically were it not for several outlying values. Even these, however, equaled or exceeded adult weights.

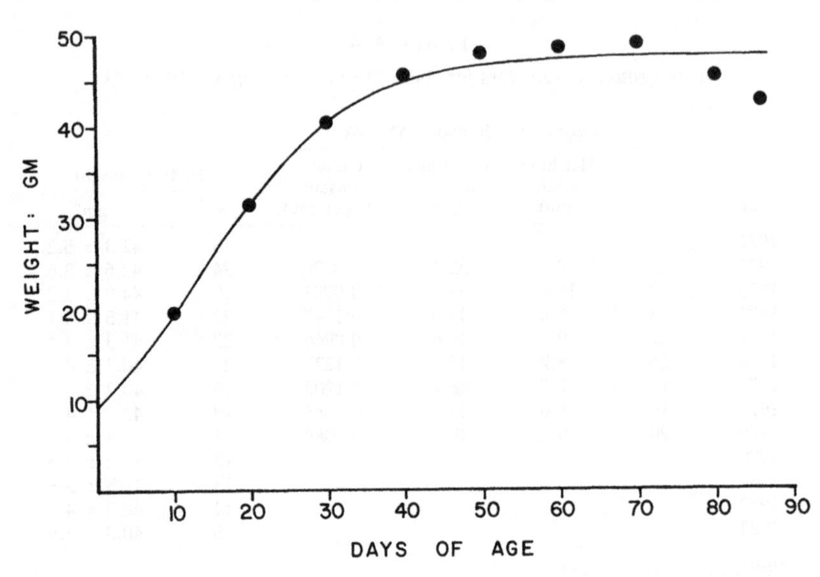

Figure 4.11. Logistic growth curve of Ashy Storm-Petrel chicks (ANOVA; n = 125). Dots are mean values for respective ages; all years combined, 1972–79.

At fledging, chicks averaged 41.5 ± 4.3 gm (n = 234), or not quite 10% greater than adult weight (Table 4.4). Chicks, therefore, lost weight after reaching the asymptote, the typical pattern in storm-petrels. James-Veitch (1970) determined a mean fledging weight of 37.2 gm (n = 20), but to judge from some of the individual weights he listed (down to 20 gm), either his scale was not reading accurately or some of the birds he measured were dying and failed to fledge. His calculation of hatching weight was also about 3 gm less than ours.

TABLE 4.5

Mean growth parameters in the Ashy Storm-Petrel, 1972–79

Asymptotic weight (gm)							
1974	*1977*	**1976**	**1978**	*1979*	*1975*	**1972**	**1973**
43.0	46.7	47.3	48.4	49.5	49.6	52.1	53.2
Growth constant (K; per day)							
1972	**1973**	*1975*	**1978**	*1979*	1974	*1977*	**1976**
0.08	0.10	0.10	0.11	0.11	0.11	0.12	0.12

NOTE: Lines connect similar means; Tukey-Kramen Studentized Range test; p > .05.

Between years in the present study, the average weight at fledging ranged from 40.0 ± 4.0 to 44.0 ± 4.2 gm, but during no year was mean fledging weight significantly higher or lower than in the others. Mean fledging weight was correlated with asymptotic weight ($r_S = .912$, $p < .05$). Within years, chicks that fledged during the last third of the fledging period averaged 1.8 gm lighter than those that fledged earlier, a trend consistent with the weak trend observed for asymptotic weight. In a stepwise multiple regression analysis, relating fledging weight to laying date, hatching date, fledging date, incubation period, nestling period, and total egg–nestling period, the only significant correlation was with fledging date ($r = -.2198$, $F = 3.858$, DF = 1,76, $p < .05$). In another analysis, this one relating fledging date to laying date, hatching date, incubation period, nestling period, and total egg–nestling period, the most important variable affecting fledging date was the length of the chick period ($r = .975$, $F = 2889.5$, DF = 1,73, $p < .01$). This result is consistent with the hypothesis mentioned above that the inverse rankings of growth rate and asymptotic weight were the result of differences in the lengths of nestling periods.

Although chicks hatched late in a season tended to fledge at lighter weights than did earlier ones, "late" chicks were *significantly* lighter only during 1975, 1976, and 1983, when they were 3.6 to 8.8 gm lighter. The only other year in which late chicks averaged more than 2 gm lighter was 1978 (2.2 gm). Three of these four years, 1976, 1978, and 1983, were years of significant ENSO or warm-water anomaly (Chapter 2). During three years, 1973, 1974, and 1979, the average weight of late chicks was actually greater than that of early chicks (0.7 to 2.5 gm). D. A. Scott (1970) also found little difference between the weights of early- and late-fledging British Storm-Petrels. The lack of much within- and between-year variation in weight-related growth parameters of Ashy Storm-Petrels is consistent with the findings of Ricklefs et al. (1985), who concluded that weight gain in Leach's Storm-Petrel nestlings at Kent Island (Canada) was a function of chick demand rather than the provisioning abilities of parents. In other words, the demands of chicks did not push parental provisioning capabilities to the limit.

Breeding success

If all years except 1977 are considered, chicks eventually fledged from 69.4% of all eggs laid and 0.69 fledglings were produced per breeding pair (Appendix 4.1). As a check against whether our interference may have significantly affected nesting success, for nests occupied at least the median number of seasons (i.e., the occupants were seemingly able to cope with our intrusions), we calculated the percentages and the number of chicks fledged and found them consistent with the overall figures (Table 4.6). Failure of eggs to hatch and the death of chicks were the most important factors depressing nesting success. Failure of eggs to hatch has been observed to be an important factor in the breeding biology of most other storm-petrels (Table 4.2). Among chicks that died, most succumbed during the first week of life.

Although the percentage of pairs raising young varied between 61.1 and 81.1%, there were no significant interannual differences ($G = 5.962$, $DF = 10$, $p > .05$). The number of chicks fledged per breeding pair did not correlate significantly with hatching success ($r_S = .207$, $t = 0.635$), but it did correlate with nestling survival (i.e., the number of chicks fledged relative to the number that hatched; $r_S = .602$, $t = 2.262$, $p < .05$). Within seasons there was no significant difference in nesting success relative to the time of egg laying. If the laying period is divided into the early, middle, and last third and the results for the years 1972–79 are combined, chicks fledged from 72% of eggs ($n = 78$) laid early, compared to 70% ($n = 105$) and 67% ($n = 48$) of eggs laid midway and late in the year.

Several factors likely affected chick survival. One factor was pre-

TABLE 4.6

Nesting record for 25 Ashy Storm-Petrel burrows, 1972–83

Variable	*n* (%)
A. Total seasons	301
B. Total eggs laid	274 (91.0% A)
Failed to hatch	26 (9.5% B)
Abandoned	9 (3.3% B)
Disappeared	8 (2.9% B)
Broken	2 (0.7% B)
C. Total eggs hatched	229 (76.1% A) (83.6% B)
Chicks died	16 (7.0% C)
Chicks disappeared	28 (12.2% C)
Chicks eaten by mice	1 (0.4% C)
D. Total chicks fledged	184 (61.1% A) (67.2% B) (80.3% C)
E. Chicks per breeding pair	0.67

NOTE: Burrows recorded were occupied for at least five seasons.

dation of chicks by the feral house mouse (Table 4.6), which in turn could be a function of the amount of rain and resulting seed crop. However, no data are available to support or negate a relationship to seed availability. Another likely factor was interference from Cassin's Auklets, and this appeared to be important. For instance, among storm-petrel nests that were used fewer than five seasons, 23.5% were usurped by auklets, but among nests used five or more seasons, only 13.5% were usurped. In many cases the auklets took over the nest while the storm-petrel chicks were present, although they did not lay eggs themselves in that site until the following season. This might explain the slightly lower chick survival during the period 1972–74, which was when we were establishing our nesting boxes (Appendix 4.1). In later years, with boxes in place, auklets had a harder time gaining access to our study nests. Nevertheless, some auklets did enter boxes, and as a result we had progressively fewer nest sites occupied by storm-petrels (some were also taken over by mice and some disintegrated; Appendix 4.1). Survival of Ashy Storm-Petrel chicks was highest in 1976, 1982, and 1983, warm-water years when few if any auklets nested on the Farallones (see Chapter 10). This finding adds support to the idea that auklet disturbance significantly depresses the storm-petrels' reproductive success. In an analogous situation, Richdale (1965) observed that Narrow-billed Prions, *Pachyptila belcheri*, usurped the nests or reduced the breeding success of the White-faced Storm-Petrel, *Pelagodroma marina*, on Whero Island, New Zealand.

PREDATION BY WESTERN GULLS

Predation of adult storm-petrels by Western Gulls and also rarely by owls that occasionally find their way to the island (DeSante and Ainley 1980) is another factor that can affect chick survival. In a few cases in which we suspected one parent to have died, the other could not successfully rear its chick. Thus, predation appears to be a significant factor, although immature nonbreeding storm-petrels receive the brunt of the predation pressure. Over the years, we have found the remains of more than 100 storm-petrels in gull pellets. Of the 2,326 storm-petrels banded during 1971 and 1972, 15 (1%) have since been found in the pellets. On some small islands near the California–Oregon border, where Leach's Storm-Petrels also nest within Western Gull colonies, Osborne (1972) detected an even higher mortality rate. He found the remains of 220 storm-petrels in gull pellets during one year, which, out of an estimated population of 7,800 petrels there, represents more than three times the predation rate at the Farallones. Capture by avian predators at the breeding ground is also an important factor for the British (D. A. Scott 1970), Madeiran (M. P. Harris 1969a), Leach's (Gross 1935, Huntington 1963), and Wilson's storm-petrels (Beck and Brown 1972).

Predation by Western Gulls, in other ways, may have affected the reproductive biology of both the Ashy and Leach's storm-petrels at the Farallones. Nesting in cavities and nocturnal activity by storm-petrels are generally recognized by ornithologists as strategies for avoiding large predatory birds (see Lack 1968). At the Farallones, such predation also probably determines where and when the storm-petrels nest. The storm-petrels choose sites mostly on the south-facing talus slope of Lighthouse Hill and on the flat area abutting the base of the hill to the south. This is the only area where a lot of talus and nest cavities exist, as well as being the only area that until 1976 had few nesting gulls. Thus, by nesting in this area, the storm-petrels have reduced their contact with gulls. This seems to be a positive response to low density of gulls rather than a coincidence, because suitable terrain for petrel nesting occurs elsewhere on the island, but in those areas gulls have been established for many decades.

Gull predation may also modify the timing of breeding by storm-petrels on the Farallones and may be the main factor responsible for the complete overlap in egg laying between the two species. The primary factor affecting the timing of breeding, as in most seabird populations, is likely the availability of food for chicks and fledglings (Lack 1954, Perrins 1966). In addition to the two storm-petrels, five other species of Farallon birds also feed mostly on fish and squid (see Chapter 3). The eggs of these other species hatch at the same time of year, usually about three weeks before the hatching of storm-petrel eggs. This disparity suggests that another factor modifies the primary one for storm-petrels, and that factor may well be gull predation. Young storm-petrels begin to fledge after almost all other species have left the island, the gulls being among the last to leave. In fact, there is little overlap between storm-petrel fledging and gull occupation (see Figure 12.1).

At the breeding grounds, storm-petrels are no match for gulls, except under the protective cloak of darkness. This is even more true for fledglings, which are clumsy, disoriented, and naive. As an example, during walks on autumn nights, we have picked up more than 25 fledglings found sitting on the ground. Unless they are injured, adults seldom sit quietly in the open above ground. By delaying egg laying a few weeks, and by delaying fledging until the dark of the moon or until nights are overcast (thus adding even more time to the chicks' already long growth period), both storm-petrel species on the Farallones time their breeding and their fledging so that the young depart when fewest gulls are present and when detection by gulls is least likely. In three ways, a potentially important factor in fledgling mortality is minimized.

Furthermore, the advantage gained by storm-petrels in avoiding gull predation may well outweigh any pressures from food supply for earlier fledging. In central California, surface waters are most productive from April to September (Chapter 2). The young of most species must grow and fledge before food becomes limiting. Because storm-petrels and other tubenoses have slow growth rates (reviewed by Lack 1968, Ricklefs 1968; see above), however, daily food demand per individual may not be as high as in other birds. In addition, storm-petrel eggs can withstand chilling and chicks can go for several days without being fed (as noted above). Through these and other adaptations, storm-petrels do not forfeit a significant degree of breeding success by delaying nesting; in fact, there does not appear to be much difference in the nesting success of early and late nesters, as noted earlier.

THE STORM-PETREL BREEDING STRATEGY

In the few situations where two storm-petrel species have been studied at or near the same site, it appears that the greatest ecological separation occurs in foraging distance from the colony, as well as perhaps in nesting habitat preference. At the Farallones, Little River Island, and Skokholm, Leach's Storm-Petrel forages at a much greater distance than the Ashy, Fork-tailed, and British storm-petrels, respectively. Among the Galápagos Islands, the Madeiran apparently ranges much less widely than does the Galápagos Storm-Petrel. In the South Orkneys, the situation is similar, respectively, between the Black-bellied and Wilson's storm-petrels, but in this case, there is an added difference in water-type preference; the Black-bellied forages in waters generally warmer than where the Wilson's forages (Ainley and Fraser, unpubl. data)—the same, of course, could be said for the water-type preferences of Leach's and Ashy on the Farallon Islands (Chapter 3). The difference in foraging range manifests itself in such phenomena as (1) the length of the breeding cycle—the species feeding closer to the nesting area tend to frequent the colony over a longer period (except in the high-latitude Antarctic example; Table 4.1)—and (2) the nightly visitation pattern: the distant-feeding species tend to arrive later and leave earlier. For the more "inshore," sedentary storm-petrels, lengthening of the breeding cycle occurs through a longer pre-egg period, less synchronous egg laying, and the nestling's longer growth period. These patterns are affected strongly by local conditions; examples are the different nestling periods observed for three separate populations of *Oceanodroma castro* and the extremely short nestling periods of populations breeding under especially cold, rigorous conditions, i.e., the two Signy Island species and *O. leucorhoa* in Maine. In each situation, the laying period of one species is completely overlapped by that of the other more asynchronous species (again, the partial exception is in the high-latitude Antarctic, where the two laying periods are practically identical). This contradicts Cody's (1973, 1974) hypothesis that coexisting storm-petrels should have laying periods that do not overlap.

There may also be species-specific adaptations involved in the lengths of breeding cycles; for example, the nestling period seems to be generally longest in the genus *Oceanodroma* among the four genera so far studied. One could surmise that the longer colony occupation and the longer nestling period in the more sedentary, "inshore" species could be a response to increased competition for nesting sites and food (and therefore reduced availability and increased procurement costs), respectively, but at the moment such speculation may be premature. The storm-petrels in the populations mentioned above do not support A. W. Diamond's (1978) hypothesis that the more offshore species should be much more abundant (because of lessened competition for food). Only at Little River Island is the more offshore species (Leach's) also the more abundant. On the other hand, relative to other seabirds, *all* storm-petrels are "offshore" feeders. The group, therefore, may not provide a good test of the hypothesis.

Other patterns are also apparent among the several storm-petrels now studied extensively (Table 4.2). Egg size may be to some degree a function of taxonomic relationship, as it appears that species in the genus *Oceanodroma* lay the smallest egg relative to body size. Interestingly, the longest developmental periods, as noted above (Table 4.1), and the lightest fledging weights occur in this genus as well (Table 4.2). Perhaps these three factors are interrelated (see Ricklefs 1983, and studies cited therein). Finally, the frequency with which mates switch incubation turns could be a function of food availability. If one accepts some usual assumptions about food availability in different faunal zones, incubation stints are shortest in the high-latitude Antarctic, where food is readily available, and longest in tropical waters, where food is less available.

SENSITIVITY OF STORM-PETRELS TO DISTURBANCE

It appears that the measurement of hatching and reproductive success in storm-petrels has been strongly affected by the research. In several studies, hatching success has been 60% or below and breeding success has been 40% or below (Table 4.2). In the case of the British Storm-Petrel and the Wilson's Storm-Petrel, D. A. Scott (1970) and Beck and Brown (1972), respectively, noted that nesting success was not sufficient to maintain the population and postulated that immigration was necessary for population stability. Exceptions to the generally low figures for nesting success are apparent in the Fork-tailed Storm-Petrels studied by Boersma, Nerini, and Wheelwright (1980) and the Ashy Storm-Petrels in the present study. In both cases, nests were checked only four or fewer times between laying and hatching of the egg. In the other studies, researchers checked nests repeatedly, even daily, to determine, among other things, the incubation schedule of mated pairs. Most investigators, except the latter, also handled burrow occupants to band or mark them as a basis for a study of mate and nest fidelity. These procedures likely caused disturbance to a degree that affected some results. In one season, in fact, Boersma, Nerini, and Wheelwright (1980) measured a hatching success of 58% in burrows checked daily, compared to 78% in burrows checked only four times while eggs or chicks were present. At the Farallones, James-Veitch (1970), over a three-year period of daily nest checks, measured a hatching success of 59% and a breeding success of 42% in the Ashy Storm-Petrel, compared to our results of 78 and 69%, respectively, over 12 seasons. In 1977 at the Farallones, when researcher disturbance was great, our results were 62 and 49%, respectively. In situations similar to those experienced by M. P. Harris (1969a) in the Galápagos, where there is intense competition for nest sites and much interference among competing birds, storm-petrels could be even more sensitive to disturbance.

In conclusion, it appears that in studies of storm-petrels, which are extremely sensitive to human disturbance at their nesting grounds, researchers should assess nesting success in samples independent of those in which they assess other aspects of reproduction. In situations where the sample of nests is small (i.e., about 35 per year in which eggs are laid), if information on reproductive success is desired, then the study of other nesting variables should be sacrificed.

Much of the Farallon Islands consists of precipitous cliffs, a habitat that even gulls use for nesting because of the intense competition for space. (Photo: W. Parsons)

On boats such as this one under the boom at East Landing, the Farallon studies have depended for logistics, including surveys of seabird feeding habitat. (Photo: T. Penniman)

The Shubrick Point blind, where biologists spent many hundreds of hours collecting much of the data presented in this volume. (Photo: J. Penniman)

Bob Boekelheide observing murres and Teya Penniman observing gulls and cormorants from the blind atop Shubrick Point. (Photo: J. Penniman)

Southeast Farallon Island was among the first sites to be included in the National Wildlife Refuge system, a factor that has contributed to the large numbers of Western Gulls and other seabird species that breed there. (Photo: W. Parsons)

The marine terrace on Southeast Farallon provides habitat for surface-nesting gulls and for humans, as well as for burrow-nesting auklets. The rain-catchment apron from which gull pellets were collected is visible in the background. (Photo: W. Parsons)

Pinnipeds, such as California and Northern sea lions (below; *Zalophus californianus* and *Eumetopias jubatus*) and elephant seals (above; *Mirounga angustirostris*), perhaps once occupied much more space on the Farallon marine terrace than they do at present—a possibility that, if true, would have had great impact on the amount of space available for nesting birds. (Photos: R. Boekelheide, B. Heneman)

A series of specimens of Leach's Storm-Petrel, showing the gradation in the number of white feathers of the rump (upper tail coverts). This entire range in color patterning occurs within the Farallon population, although the coloration of the third, fourth, and fifth specimens (from the left) is similar to that of about 70 percent of Farallon individuals. (Reprinted from Ainley 1980)

An Ashy Storm-Petrel chick about 20 days old and weighing no more than 20 grams. (Photo: I. Tait)

An adult Ashy Storm-Petrel, the smallest of the Farallon-nesting seabirds. (Photo: I. Tait)

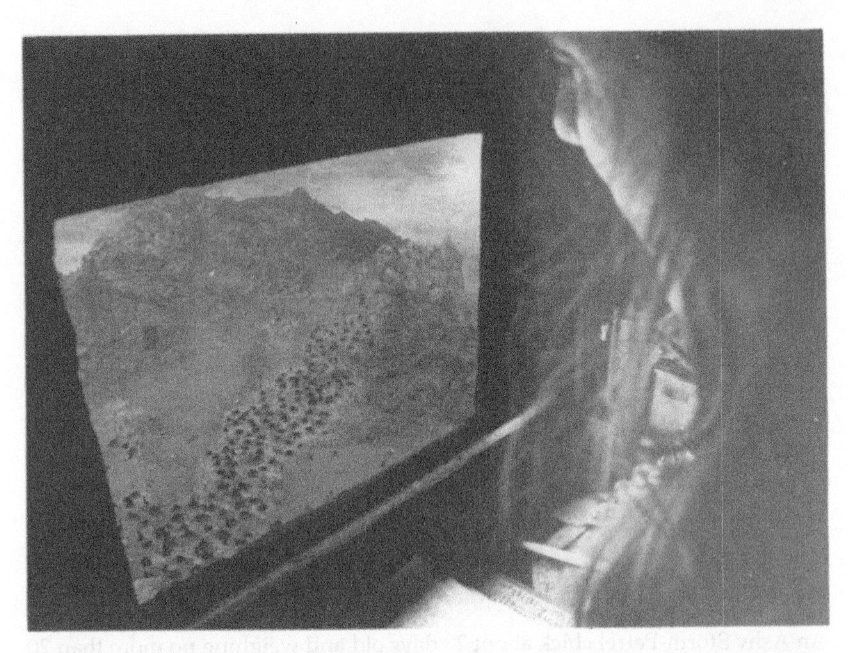

The view from the Sea Lion Cove blind, overlooking a colony of Brandt's Cormorants. (Photo: L. Romero, San Francisco *Examiner*)

A nesting colony of Brandt's Cormorants near Sea Lion Cove. (Photo: J. Penniman)

The extensive nesting colony of Brandt's Cormorants and Common Murres on Tower Point, overlooking Fisherman's Bay, ca. 1975. The North Farallones are visible in the distance. (Photo: PRBO files)

An adult Brandt's Cormorant with its two 20-day-old chicks at a marked nest site. (Photo: B. Heneman)

A pair of Pelagic Cormorants at their nest, which is stuck to a cliff face 30 meters above the sea. (Photo: I. Tait)

An adult Pelagic Cormorant perched on a rock at Southeast Farallon.
(Photo: W. Parsons)

An adult Double-crested Cormorant at its nest site. (Photo: I. Tait)

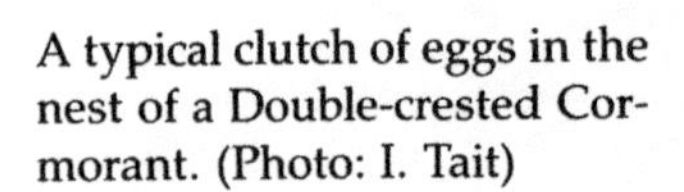

A typical clutch of eggs in the nest of a Double-crested Cormorant. (Photo: I. Tait)

A banded adult Western Gull with its chicks on a marked territory. (Photo: R. Boekelheide)

An adult Western Gull regurgitating food to its two chicks. (Photo: L. Spear)

An adult Western Gull rides an updraft near the summit of Lighthouse Hill. (Photo: W. Parsons)

Common Murres at their territories just before egg laying. (Photo: R. Boekelheide)

A pair of Common Murres with their 12-day-old chick. (Photo: L. Spear)

An adult Common Murre. (Photo: W. Parsons)

The view from the blind on Shubrick Point, showing murres and Brandt's Cormorants nesting in June 1981. (Photo: L. Romero, San Francisco *Examiner*)

A pair of adult Pigeon Guillemots perched near their nesting cavity. (Photo: PRBO files)

The nesting cavity of Pigeon Guillemots, with the maximum two-egg clutch. (Photo: I. Tait)

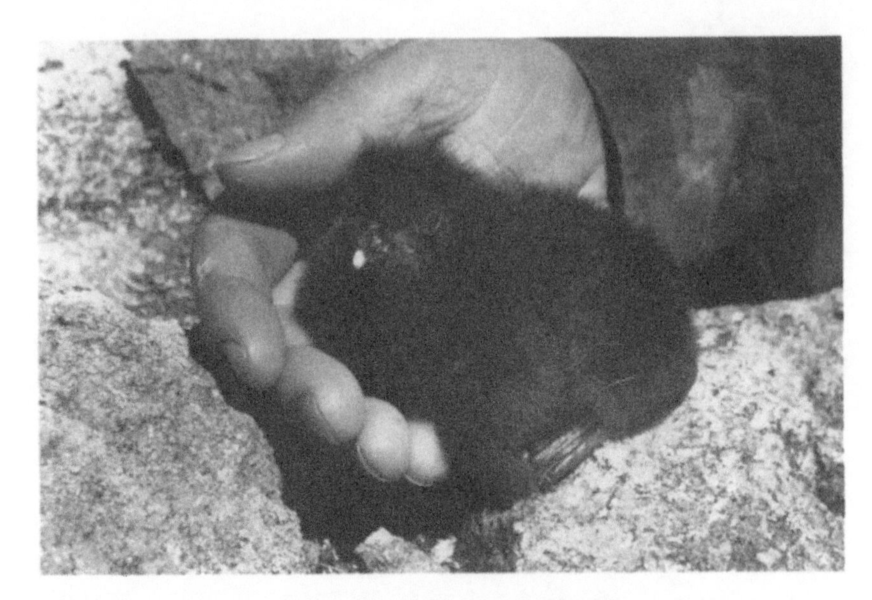

A ten-day-old Pigeon Guillemot chick. (Photo: PRBO files)

A Cassin's Auklet chick about 20 days of age. (Photo: B. Heneman)

An adult Cassin's Auklet weighs about 165 grams. This species is active on the island only at night. (Photo: W. Parsons)

A Rhinoceros Auklet about to enter its nesting cavity on Southeast Farallon. (Photo: J. Foote)

A pair of Rhinoceros Auklets on a favorite perch by their nesting cavity on Southeast Farallon. (Photo: J. Foote)

Adult Tufted Puffins spend much time standing by the entrances to their burrows, perhaps in display of their gaudy plumage. (Photo: W. Parsons)

An adult Tufted Puffin with a juvenile rockfish in its beak. (Photo: PRBO files)

5

Brandt's Cormorant

Robert J. Boekelheide, David G. Ainley, Stephen H. Morrell, and T. James Lewis

Cormorants constitute a group of about 30 species found worldwide from tropical to polar regions in both marine and freshwater habitats. They are heavy-bodied birds with long necks, slender hooked bills, and large totipalmate feet, which they use to propel themselves underwater (Chapter 3). Unlike many seabirds, cormorants are not ocean wanderers; they are never found farther than a few hours' flight from land. They return to shore daily to roost, preen, and dry their feathers, which apparently lack the waterproof qualities of other, more pelagic seabirds' (Rijke 1967, Clark 1969, Curry-Lindahl 1970, Elowson 1984). As discussed by Owre (1967), their wettable plumage is in fact beneficial, because it reduces buoyancy and permits greater underwater maneuverability. They are reputed to be voracious consumers of fish and squid, as they ingest between 10 and 27% of their body weight daily (Chapter 3; Wetmore 1927, Van Dobben 1952, du Plessis 1957, Bowmaker 1963, Duncan 1968, Junor 1972, Berry 1976). They have frequently been condemned as competitors with fishermen, but most evidence shows that they largely consume species with little market value (Lewis 1929, Serventy 1938, Robertson 1974, Batchelor 1978, Ainley, Anderson, and Kelly 1981). In some areas where they do consume commercially valuable fish, such as anchovetas, *Engraulis ringens*, off Peru and pilchards, *Sardinops ocellata*, off southern Africa, they atone for this sin by producing huge guano deposits, probably the finest phosphate sources on the planet.

The cormorants' basic color scheme, particularly in adult birds, is either solid black, as in most Northern Hemisphere species, or else countershaded, with black dorsally and white ventrally, as in many Southern Hemisphere cormorants. Their black feathers shine with iridescent green, violet, or bronze, and their eyes are exotic blue, green, red, or aquamarine. The dark feathering may aid in the absorption of solar radiation, whereas the light ventral surface may conceal birds from shoaling fish near the surface (Simmons 1972, Siegfried et al. 1975). During breeding, however, cormorants are not conservative dressers. Many species develop white plumes or feathered patches on the head, neck, back, and flanks, plus gaudily colored patches of skin on the face, gular pouch, and eye-ring.

Among Farallon species, a range of color patterns exists, although all three are basically dark. In the Double-crested the iridescence is bronze, in the Brandt's it is green, and in the Pelagic it varies from purple on the head and neck to green on the remainder of the body. Double-cresteds have aquamarine eyes and an orange throat, Brandt's have green eyes and a cobalt blue throat, and Pelagics have green eyes and a red throat and face. During breeding, the two crests of a Double-crested are plumes that look like horns above each eye; they range from white to dark in color. Brandt's produce white "whiskers" on the sides of their head and long white plumes on the back, and Pelagics produce large white patches on their flanks. In all three species these breeding plumes disappear during the nestling stage of reproduction.

When breeding, cormorants choose a colony site inaccessible to mammalian predators, usually on an island, on a steep cliff, in trees, or in some combination of the above. They do not defend their nests from large predators, and their eggs and chicks are not concealed with cryptic coloration. Their chicks, being altricial and nidicolous, are entirely helpless for several weeks after hatching.

On flat ground their nests are usually spaced within colonies only as far apart as neighboring birds can reach with their beaks while sitting on the nest; this spacing limits access within the colony by avian predators, such as gulls, skuas, and ravens (Siegel-Causey and Hunt 1981).

Cormorants have a higher reproductive rate than other marine birds. Clutch sizes are larger and more variable; they usually average three or four eggs but range from one to seven eggs. Each egg is small, amounting to about 3% of the female's body weight (Snow 1960, Berry 1976). As discussed by Lack (1968) and A. J. Williams and Burger (1979), brood reduction is an important component of the cormorant nesting scheme and helps to maximize the number of chicks fledged each year. In such a scheme, species that regularly experience annual variation in food supplies should always lay a clutch large enough to take advantage of years of abundant food. Through a combination of adaptations such as asynchronous hatching and rapid chick growth with deferred fat storage, the younger, smaller chicks are at a disadvantage relative to their older siblings when feeding conditions are poor. Consequently, in poor years the youngest chicks die at an early age and only the older, healthier chicks survive; in better years all chicks may survive. The younger chicks also serve as insurance in case the older chicks die at a young age. Decreasing egg size by laying date within clutches, an adaptation that is often a component of brood reduction, does not occur in the Shag, *Phalacrocorax aristotelis* (Coulson, Potts, and Horobin 1969), the only cormorant species in which within-clutch variation of egg size has been investigated.

Before our study, little was known about the breeding biology of the Brandt's Cormorant. This is perhaps due to the inaccessibility of nesting colonies and to the birds' sensitivity to disturbance. Although early visitors to the Farallones frequently mentioned this species, few attempted systematic observations other than population estimates and cursory notes on breeding activities (Bryant 1888, Loomis 1896; see Ainley and Lewis 1974 for historical information). At Point Lobos, near Monterey, California, L. Williams (1942) detailed the behavioral displays of this species, including characteristics of nest sites and nest construction. A cursory study of breeding biology was accomplished by Hunt et al. (1981) in the Channel Islands of southern California.

METHODS

From 1971 to 1985, we observed nesting Brandt's Cormorants on Southeast Farallon Island at two colonies. The first, Colony I, is located on a rocky peninsula near Sea Lion Cove on the northwestern shore of the island (Figure 1.3). Colony I is of medium size and contains more than 500 nests during better breeding years. It is composed exclusively of Brandt's Cormorant nests, other than a small pocket of Common Murres, which began to nest on its southeast edge in 1981. The second site, Colony II, is located on Shubrick Point on the island's extreme eastern corner, where a similar number of Brandt's Cormorants and several thousand Common Murres nest among one another. Western Gulls also nest around the perimeters of both of these colonies. Beginning in 1986, we observed cormorants only in Colony I.

Each colony was viewed from a small wooden blind perched on an overlooking ridge—Colony I from the "Cormorant Blind" and Colony II from the "Murre Blind." We obtained access to the blinds by climbing the slope out of sight from the cormorants, so the birds were not aware of our presence. During each breeding season, we observed samples of 25 and 40 nests in each colony, respectively. Information recorded daily or every other day included date, time of observation, number of adults present, number of eggs and/or chicks in each nest, and condition of the nest.

In general, we observed the same nest sites throughout the entire 13 years, although different sites were occasionally added to the sample if older sites were not occupied for two or three years. This was especially true at Colony II, where during the 1970's Common Murres occupied more and more cormorant sites as the murre population expanded (Chapter 8). We selected nests on the basis of whether their contents were easily visible from the observation blinds when the birds stood up. This meant that virtually all contiguous nests within the "best" field of view were included in the sample.

From the nest observations, we determined the chronology of nest formation, egg laying and hatching, and chick fledging, plus the fate of all eggs laid and chicks hatched. It is possible that we missed an occasional egg if it was lost on the first day of incubation prior to our observations. We believe, however, that this occurred infrequently.

Disturbance of cormorant colonies was strictly controlled, with the exception of infrequent helicopter overflights in the early 1970's. After 1974, the U.S. Coast Guard, USFWS, and PRBO developed a helicopter flight procedure that corrected this problem. Nests abandoned by cormorants during helicopter overflights were not included in the analyses.

For several analyses, we compared parameters among early, middle, and late nest sites. Early and late sites for a particular year are here defined as those with clutch-initiation dates either before or after one-half standard deviation from the mean clutch-initiation date for that year. Sites classified in mid-season are those with clutch-initiation dates within one-half standard deviation from the mean. This system more or less separated each year's nests into three groups of approximately equal size.

We determined seasonal patterns of colony attendance by counting the total number of individual cormorants holding sites and pairs at nest sites in three subcolonies of Colony I. Counts were made daily at 0930. By subtracting the number of pairs from the total number of individuals we obtained a relative comparison of the number of occupied sites in different years. These data were grouped by five-day means to display seasonal attendance patterns.

We used three methods to estimate numbers of birds attempting breeding. In 1971 and 1972, counts were made of developed nest sites in September after chicks had fledged (Ainley and Lewis 1974). From 1980 to 1986 we counted occupied sites during the peak of the cormorants' breeding period, choosing one day in mid- to late June with favorable weather and visibility. Nest sites not visible from accessible parts of the island were counted by boat. We used binoculars and spotting scopes to count distant colonies. From both these methods the total number of birds was estimated by doubling the number of occupied nest sites.

For the years 1973–79 and 1983, we estimated the breeding population by using the counts taken daily in Colony I. After grouping the daily counts into five-day means, we used the highest mean from each year to estimate the total population of Brandt's Cormorants. This method assumes that the daily colony attendance counts for the three subcolonies in Colony I reflected the overall population present on the South Farallones. Using the 1980 to 1982 data, for which we have both total population counts and daily colony counts, we calculated a factor by dividing the total counts by the highest five-day mean of the daily colony counts (1980–82 mean factor = 53.7 ± 1.7). This factor was subsequently multiplied by the highest five-day means from 1973 to 1979 to yield estimates of numbers of Brandt's Cormorants attempting breeding in those years.

From 1970 to 1983 PRBO banded 8,156 Brandt's Cormorants at the Farallon Islands, all as chicks. Banding was performed at night to prevent predation by Western Gulls on eggs and small chicks. Each bird was given two bands, a USFWS numbered band on one leg and a colored wrap-around plastic (polyvinyl chloride) band on the opposite leg. The color and leg position of bands were year-specific. Age-related breeding patterns are still under investigation.

We tallied recoveries and sightings reported to us by the USFWS Bird Banding Laboratory. Dates were grouped by season to correspond to the birds' annual cycle: (1) breeding season, March to August; (2) fall dispersal, September to November; and (3) winter period, December to February. Locations were grouped by geographical sections of the mainland coast: (1) south of Point Conception, (2) from Point Conception to Cape Mendocino, and (3) north of Cape Mendocino. Age categories included hatching year, second year, and after second year. We assume that the pattern of band recoveries reflects the overall distribution of the population, although no systematic observations have been made to confirm this.

FARALLON POPULATION

The most abundant cormorant at the Farallon Islands is the Brandt's, which is an exclusively marine species inhabiting coastal waters of western North America from southeastern Alaska to the Sea of Cortez (Palmer 1962). It is characteristic of coastal upwelling areas of the California Current System (Chapters 1 and 2), filling a niche somewhat similar to that of the Cape Cormorant, *P. capensis*, of the Benguela Current (Rand 1960, 1963, Berry 1976, Crawford and Shelton 1978) and the Guanay, *P. bougainvillii*, of the Peru Current (Murphy 1936, Jordán 1967, Duffy 1983c). Although not as abundant as these species, the Brandt's is very gregarious. It usually feeds and roosts in large concentrations, sometimes in the company of gulls, pelicans, and other cormorants (Chapter 3).

Brandt's Cormorant is numerically concentrated in central and northern California (Table 5.1), where about 75% of the world's population nests (Sowls et al. 1980). The largest breeding colony in the world is at the Farallones, but at least 11 sites in central and northern California also harbor greater than 1,000 pairs (Sowls et al. 1980). In addition, at least three colonies of that size occur in both southern California and Baja California (Sowls et al. 1980, DeLong and Crossin ms). Breeding concentrations north of California are small, and the species

TABLE 5.1

Estimated numbers of Brandt's Cormorants nesting in North America, 1965–79

Area	Breeding pairs	Source
Alaska	< 100	Sowls, Hatch & Lensink 1978
British Columbia	370	Manuwal & Campbell 1979
Washington	280	Varoujean 1979
Oregon	79	Varoujean 1979
California[a]		
N of Pt. Conception	58,431	Sowls et al. 1980
S of Pt. Conception	5,779	Sowls et al. 1980
Mexico, Pacific coast	10,600	DeLong & Crossin ms
Total	75,639	

NOTE: Most recent data.

[a] At its peak, the Farallon population (11,900 pairs) constitutes 19% of the California total.

was not confirmed as breeding in either Canada or Alaska until the 1960's (Stirling and Buffam 1966, Isleib and Kessel 1973).

Prior to human habitation of the Farallones, the size of the population of Brandt's Cormorant may have been limited by that of the Common Murre. These two species use similar nesting habitat, and, if the murre population once numbered 400,000 birds (Ainley and Lewis 1974), it is possible that these species competed for nesting space (see Chapter 12). Despite their smaller size, murres are able to usurp space from cormorants because they return to the colony earlier in the year, their nesting density is very high, and the general commotion of their breeding activities discourages cormorant nesting. We observed this particularly in Colony II, where the percentage of sample sites occupied by cormorants declined steeply over time (Appendix 5.1). Correspondingly, the number of murres nesting there among the cormorants increased (Chapter 12).

The cormorants' sensitivity to human disturbance undoubtedly affected population sizes during the egging period (1850's to 1890's) and throughout the period when families, pets, and military personnel occupied the island (Chapter 1). By the early twentieth century, population estimates indicated that fewer than 5,000 Brandt's Cormorants used Southeast Farallon (Ainley and Lewis 1974). By at least 1959, the population had increased to an estimated 18,000 to 20,000 birds.

Numbers of Brandt's Cormorants attempting to breed varied annually from a low of 2,000 in 1983 to a high of 23,800 in 1974 (Figure 5.1). The overall mean was 16,049 ± 5,673 birds for the entire 16 years. Virtually all warm-water years—1973, 1976, 1978, 1983, and 1986 (see Chapter 2)—saw dips in the number of breeding cormorants. Interestingly, numbers of cormorants were high in 1974, 1977, and 1979, immediately after most warm-water years. Exceptional, however, was 1983. This pattern suggests that unfavorable breeding conditions were sometimes, but not necessarily, accompanied by catastrophic mortality of adults. Apparently cormorants regularly skip breeding if insufficient food is available. However, other evidence suggested that elevated levels of mortality did occur in warm-water years, especially 1983 (Stenzel et al. 1988).

A second measure of the size of the breeding population is the percentage of occupied sites at which eggs were eventually laid (Appendix 5.1). Every year some sites were occupied, yet the birds never laid eggs. This may have been due to such factors as immaturity, failure to find a committed mate, or to nest abandonment or destruction

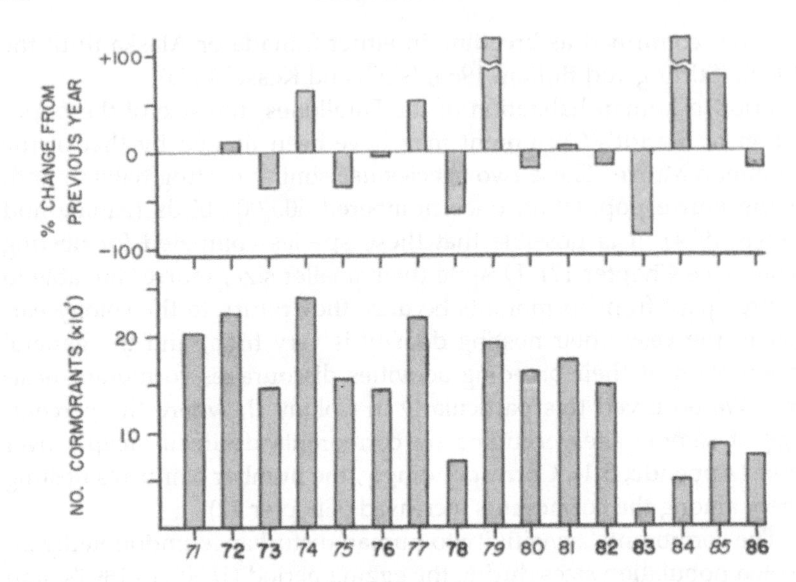

Figure 5.1. Estimated breeding populations of Brandt's Cormorants, 1971–86.

prior to egg laying. At both study colonies the warm-water years again stand out as having fewest breeding attempts among occupied sites. In the early 1970's (1972 to 1975), the percentage of sites at which eggs were laid was higher than in later years. This is consistent with the lower frequency of warm-water years during that period (Chapter 2).

Extreme annual fluctuations in breeding numbers are also evident among the Cape Cormorant and Guanay, two cormorants that also exploit the food webs of upwelling zones affected by ENSO (i.e., warm-water years, Chapter 2; Rand 1960, 1963, Jordán 1967, Crawford and Shelton 1978, Tovar 1978). Major declines in guano production in both Peru and South Africa correlate well with ENSO events. Like the Brandt's Cormorant, these birds do not attempt breeding in some years and will abandon nest sites en masse if food supplies decline in the middle of the season (Rand 1960, 1963, Jordán 1967). As discussed by Penhale (1972) and here in Chapter 6, Pelagic Cormorant breeding populations in California also vary significantly from year to year.

On the Farallones, Brandt's Cormorant nests occur on flat areas, rocks, and boulders, and are cemented in place by guano. Most Brandt's nesting areas of significant size are located on gentle slopes facing from northwest to northeast, into the prevailing spring and summer winds (Figure 5.2). A similar orientation of colonies is evident in the Channel Islands (Hunt et al. 1981). The largest Farallon colonies occur on the northeastern side of Lighthouse Hill and within Pelican Bowl on West End (Figure 1.3), where the topography and terrain are especially favorable. A significant number of birds breed as well in smaller outlying colonies, each containing fewer than 500 nests.

Colonies vary from year to year in their size and location. Cormorants occupied several "new" colonies during this study, always during years when feeding conditions were favorable and many new breeders paid visits (see Chapter 3). We noticed new areas especially in 1974, 1977, and 1979, all years with larger-than-average breeding populations (Figure 5.1).

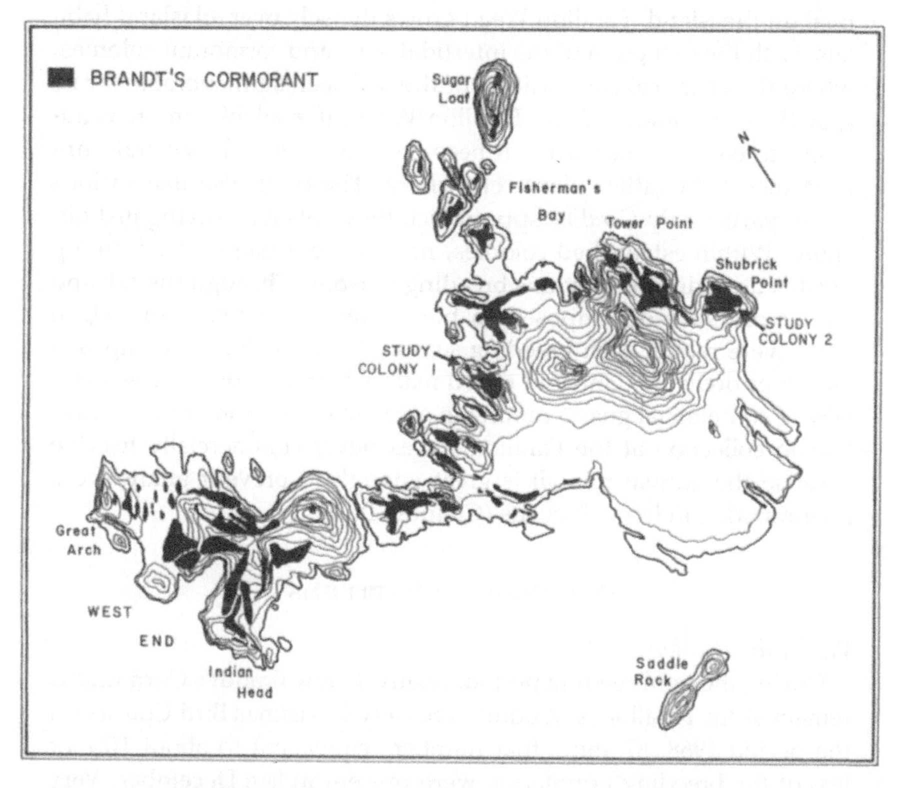

Figure 5.2. Brandt's Cormorant nesting areas on the South Farallon Islands; shaded areas are colonies used during at least one year between 1979 and 1983.

All new colonies developed where large numbers of males actively gathered plants for nesting material (see below). Occasionally, males attracted to a collecting melee stayed in the area, occupied sites, and began displaying for mates. Although most displaying males did not remain at these sites for long, females were occasionally attracted to the activity and paired with the males. If just one pair began to reside in suitable habitat, a colony quickly formed as more and more birds chose sites. Because of the inaccessibility of the new colonies we could not accurately determine breeding success in them, but it appeared that the rate of nest abandonment and failure was higher than in more established colonies. Some new colonies existed for only a few days to a few weeks, not long enough to produce chicks.

The nesting material used by Brandt's Cormorants consists primarily of Farallon Weed, *Lasthenia minor maritima*, the most abundant annual on the island. Farallon Weed grows densely over all island habitats, with the exception of the intertidal zone and cormorant colonies, where the chemical composition of the soil seems unfavorable for its growth. Cormorants gather Farallon Weed, if available, in areas adjacent to colonies, but when it becomes scarce they fly several hundred meters to gather plants elsewhere. The birds also use various algae, particularly *Coralina* spp., which they obtain by diving just offshore. Within established colonies, nests were usually placed on top of sites used during previous breeding seasons. Through the fall and winter old sites decomposed to horizontal platforms, upon which nests were rebuilt each breeding season. Such platforms, composed almost entirely of guano and rotted material, are, of course, the mainstay of commercial guano mining in areas like Peru and South Africa. Guano collection at the Farallones was never commercially feasible because the annual rainfall is great enough to prevent guano from accumulating in large deposits (Chapter 2).

OCCURRENCE PATTERNS

Winter distribution

During the nonbreeding period, relatively few Brandt's Cormorants remain at the Farallones. Audubon Society Christmas Bird Counts for the period 1968–82 show that numbers equivalent to about 10% or less of the breeding population were present in late December. Very rarely, these birds occupied nesting areas; usually they roosted on offshore islets and headlands as do individuals roosting along the mainland shore.

TABLE 5.2

Recoveries of Farallon-banded Brandt's Cormorants, 1970–82

Year	Number banded	Died at SE Farallon[a]		Died in year 1		Died in year 2		Died after year 2[b]	
		n	%	*n*	%	*n*	%	*n*	%
1970	100	NA		4	4.0%	1	1.0%	0.2	0.2%
1971	2,048	NA		93	4.5	4	0.2	0.1	< 0.1
1972	1,000	NA		70	7.0	4	0.4	0.9	< 0.1
1973	424	NA		27	6.4	0	0.0	0.7	0.2
1974	720	NA		23	3.2	6	0.8	2.8	0.4
1975	406	27	6.7%	13	3.2	6	1.5	0.2	< 0.1
1976	325	35	10.8	28	8.6	0	0.0	0.8	0.2
1977	592	15	2.5	28	4.7	4	0.7	0	0.0
1978	241	6	2.5	2	0.8	2	0.8	2.5	1.0
1979	410	NA		20	4.9	0	0.0	0	0.0
1980	690	43	6.2	43	6.2	5	0.7	NA	
1981	605	65	10.7	25	4.1	NA		NA	
1982	593	24	4.0	24	4.0	NA		NA	
Total/ average[c]	8,154	215	6.2%	400	4.9%	32	0.5%	0.75	< 0.1%

NOTE: All birds were banded as chicks at Southeast Farallon. *NA*, data not available.
[a] I.e., did not fledge.
[b] Average per year.
[c] Totals and averages for years with available data only.

On average, almost 5% of banded Brandt's Cormorants were recovered at sites other than the Farallones during their first year (until the following 31 July; Table 5.2). This compares with the fewer than 3% reported by Hatler, Campbell, and Dorst (1978) for Brandt's Cormorant chicks banded at Great Bear Rocks, British Columbia. We assume that most of the birds recovered were found dead on beaches; the causes of deaths were rarely known. This discussion of dispersal will only include those recoveries reported to us as "found dead."

From the time juvenile Brandt's Cormorants first departed the Farallones in August until the following November, most banded juveniles recovered were found north of Cape Mendocino (Figure 5.3), primarily in Humboldt Bay and along the Oregon coast. Few recoveries were logged in southern California during fall, despite the fact that many more people visit beaches in southern California. This pattern indicates that young birds disperse north after the breeding season. Despite fewer recoveries during winter, it appears that most juveniles remain in the north through that season. Even though the percentage of recoveries in southern California increased during winter, recoveries there were still fewer than in more northerly areas.

Once cormorants reached their second year, occurrence shifted to

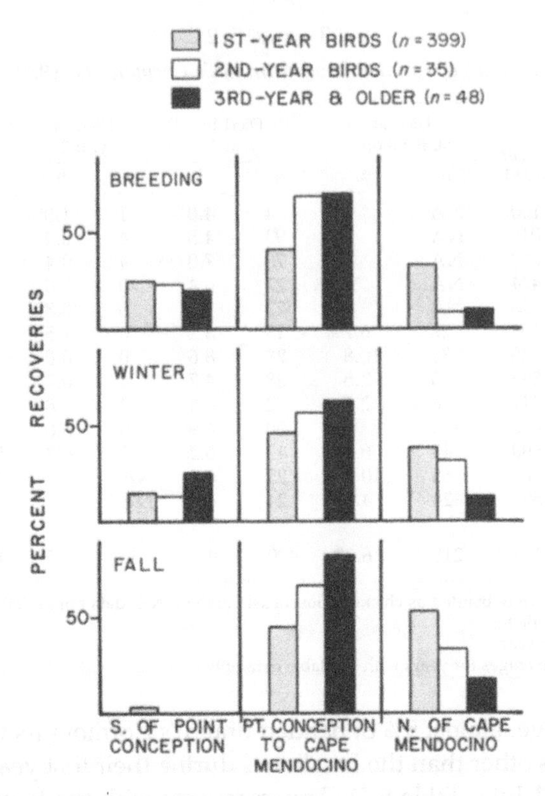

Figure 5.3. Recoveries of Brandt's Cormorants banded at the Farallon Islands, 1971–82; recoveries separated by time of year, age, and section of coastline for fall (August to November), winter (December to March), and the breeding season (April to July). Samples indicate number recovered.

central California (broadly defined as Point Conception to Cape Mendocino; Figure 5.3). In the fall, no second-year or older birds were recovered in southern California. The higher percentage of recoveries north of Cape Mendocino in the fall, however, suggests a northward movement by adults after the breeding season, a pattern similar to that of first-year birds. Several other Pacific coast seabirds, such as Brown Pelicans, *Pelecanus occidentalis*, Heermann's Gulls, *Larus heermanni* (Anderson and Anderson 1976), and Farallon species such as Western Gulls (Spear 1988a, Coulter 1975), also move north in fall. Through the winter it appears that adult cormorants then move southward (Figure 5.3), and second-year and adult birds are recovered then in southern California.

Christmas Counts for other localities, as recorded in *American Birds*, also show a general concentration of Brandt's Cormorants in central California during winter. During ten of 13 years from 1970 to 1982, the highest numbers were counted in central California, from Monterey Bay to the Gulf of the Farallones. High counts in 1978 and 1979 were recorded at the Pender Islands, British Columbia, and in 1982 at the San Juan Islands, Washington. Munro and Cowan (1947), Vermeer (1977), and Hatler, Campbell, and Dorst (1978) also recorded large concentrations of Brandt's Cormorants near Vancouver Island during winter.

Recoveries of banded cormorants have been analyzed in greatest detail in the British Isles, where chicks of both *P. aristotelis* and *P. carbo* have been banded for several decades (Coulson 1961, Balfour, Anderson, and Dunnet 1967, Coulson and Brazendale 1968, Potts 1969, Elkins and Williams 1974, Swann and Ramsey 1979). The recovery rate in these studies was much higher (17 to 23%) than in ours with Brandt's Cormorants, but the percentage of birds killed by man, especially by shooting, was also much higher. Apparently bounties were paid in several areas of Great Britain where cormorants are considered competitors with fishermen. The general dispersal patterns revealed by these studies were that (1) young birds dispersed farther than adults, (2) dispersal direction did not differ with age, (3) birds from different geographic areas of breeding had different dispersal patterns, and (4) irregular "irruptions," apparently due to food shortages, occurred in the exclusively marine species, *P. aristotelis*.

Seasonal attendance

Brandt's Cormorants usually began to occupy nesting areas on the Farallones during mid- to late March, although this varied from early March to late April in different years (Figure 5.4). Thus, colony occupation may be a month or two later than in southern California, although southern California data are available only for 1976 (Hunt et al. 1981). An extremely late season at the Farallones was that of the strong ENSO year 1983, when most nesting areas were not occupied until the first half of May. On the basis of observations of banded individuals, the first birds to return were "old" males, usually nine years of age or older, that occupied sites used by them in previous breeding seasons. These early males occupied sites sporadically at first, accumulating nesting material and roosting at their sites from late afternoon to early morning, but remaining absent during the middle of the day. Sites occupied by early males were usually centrally located in the breeding colonies, providing a nucleus for future growth as new birds arrived.

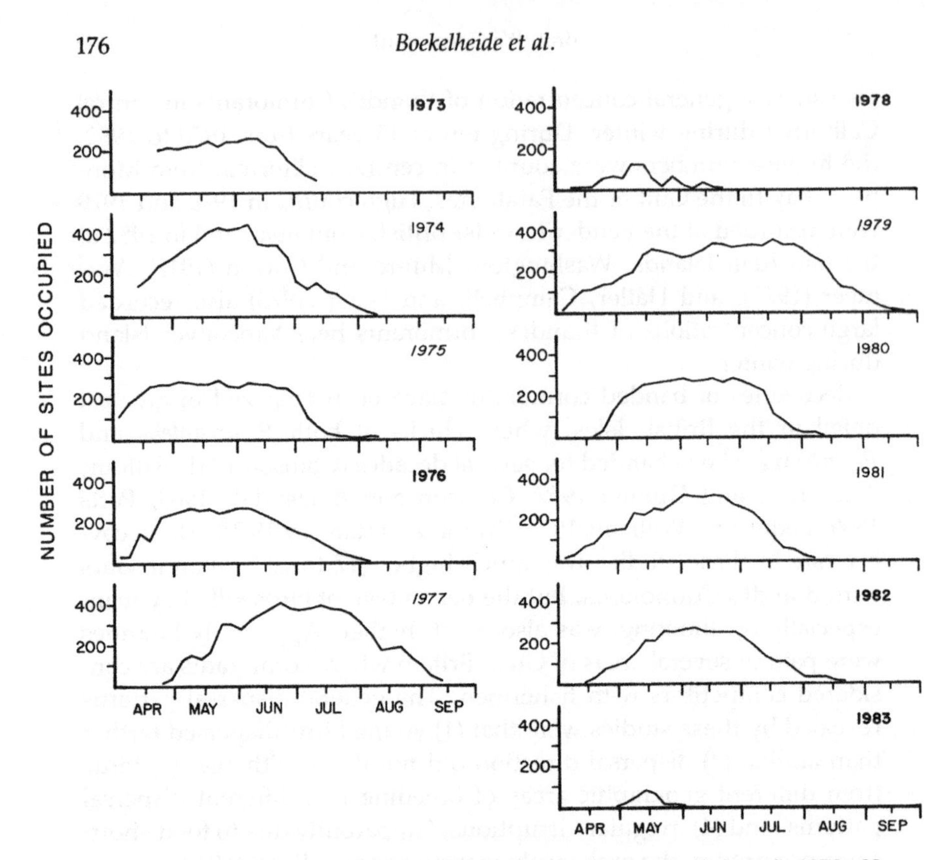

Figure 5.4. Attendance patterns of Brandt's Cormorants in Colony I, 1973–83; censuses made daily at 0930 PDT.

Prior to pairing, male cormorants gather as much nesting material as possible, over which they stand while displaying for passing females. After pairing, the female remains at the site and constructs the nest while the male makes repeated trips for more nesting material. One bird must always remain at the site to protect the nest material from thieving neighbors. By the time of egg laying, nest sizes vary from mounds 15–20 cm high to small incompletely built semicircles.

Brandt's Cormorants compete for nesting material both among themselves and with Western Gulls. They typically undergo a flurry of nest-building activity in April, May, and, in years of late breeding, June and July, after gulls have built nests and laid eggs. In the excitement of gathering nesting material, male cormorants form "raiding parties," repeatedly flying back and forth from nest sites to patches of available Farallon Weed. Entire large areas are denuded by their activities. Gull nests in the path of these raiding parties are frequently stolen, even though the gulls may protest loudly and remain seated on their eggs. Cassin's Auklet burrows may also be trampled by the cormorants' vigorous weed pulling (Manuwal 1972).

During years of early upwelling, older females began to arrive and pair with available males in late March and early April. In contrast, during years of latest breeding, 1978 and 1983, we did not observe pairs until mid-May. Like males, early females were somewhat sporadic in attendance, and remained for only short periods during early visits. After becoming firmly paired, at least one member of each pair was always present at the nest.

Throughout the period when birds were initiating their breeding activities, numbers of birds fluctuated daily. A significant factor affecting numbers of birds present appeared to be weather, especially wind speed. When strong northwesterly gales blew, few birds arrived, and many site holders, especially unpaired males, temporarily deserted their territories. In contrast, on days when winds were relatively light, the largest increases in colony attendance occurred. Thus, colony size actually increased in a stepwise fashion, with steps corresponding to mild days.

Brandt's Cormorants continued to establish territories through April and May, and colony size peaked by late May and early June during most years (Figure 5.4). Site occupation then stabilized for two to eight weeks, after which numbers declined as failed breeders departed and successful breeders left their chicks alone at the nest. Two years, 1977 and 1979, were noteworthy for the duration and persistence of high numbers of birds at the island. During both these years prey were available near the Farallones well into the fall, as large flocks fed daily through September and October (Chapter 3). In extreme contrast were the warm years 1978 and 1983, when the food web failed to develop and few cormorants attempted breeding. Evident in Figure 5.4 is a lack of breeding birds during 1978 and 1983; few birds occupied nesting areas and all deserted early in the season.

EGG LAYING

On average, egg laying began during late April and early May, although extreme dates of first clutch initiation ranged from 12 April to 15 May in Colony II and from 18 April to 22 May in Colony I (Figure 5.5). For the entire study period, cormorants in Colony II began lay-

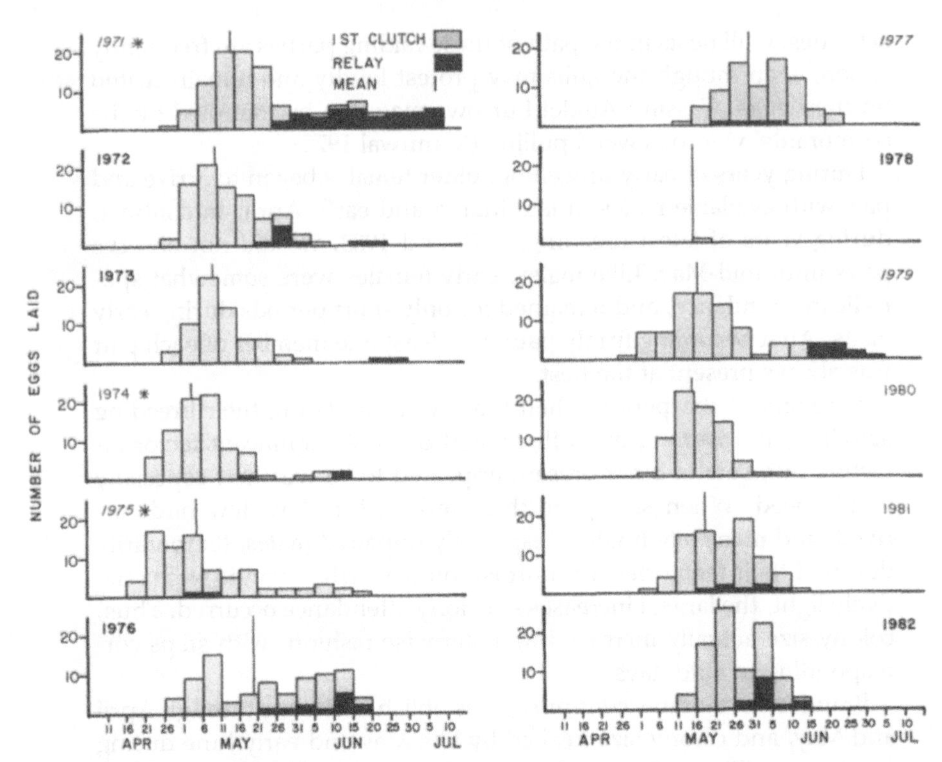

Figure 5.5. Frequency distributions of Brandt's Cormorant egg-laying dates in Colony I, 1971–82; distributions marked with an asterisk (*) are significantly peaked and skewed (t test, $p < .05$).

ing significantly earlier than those in Colony I ($t = 5.14$, $p < .001$), and rankings of mean dates of clutch initiation for 1972 to 1982 in the two study colonies were strongly correlated ($r_S = .806$, $p < .001$). Thus, even though laying dates differed consistently, the colonies showed similar year-to-year variation in timing. Years of cooler ocean temperatures in the early spring, particularly the four from 1972 to 1975, generally stand out as having the earliest laying dates at both colonies (see Chapter 2). Years of delayed upwelling, such as 1977, 1978, 1982, and 1983, were among the years with latest laying dates (Table 5.3).

Within years, clutch-initiation dates at Colony I were positively skewed toward the early season in only three years—1971, 1974, and 1975 (Figure 5.5). Similarly, egg laying showed a strong peak around the mean clutch-initiation date in 1971 and 1974. At Colony II, clutch

TABLE 5.3

Mean clutch-initiation dates, Brandt's Cormorants, 1971–83

Colony I

1974	*1975*	**1973**	**1972**	1980	*1971*	*1979*	1981	**1976**	**1982**	*1977*
5/3	5/4	5/9	5/10	5/15	5/15	5/17	5/18	5/20	5/22	5/31

Colony II

1974	*1975*	**1972**	**1973**	**1976**	1981	*1979*	**1982**	1980	*1977*	**1978**
4/28	4/29	4/30	5/7	5/7	5/8	5/9	5/15	5/18	5/21	6/6

NOTE: 1983 not included at either colony, and 1978 not included at Colony I, owing to the small number of nests with eggs. Colony II not observed in 1971. Lines connect similar means; SNK test, $p > .05$.

initiation was strongly peaked and skewed toward the early season in 1973, 1974, 1976, and 1977. After 1977, laying distributions were approximately normal in all years at both colonies.

Brandt's Cormorants laid their eggs at night or very early in the morning, as do Shags in Great Britain (Snow 1960). If most eggs were laid at approximately the same time of day, then the intervals between successive eggs within clutches were consistent, regardless of year, clutch size, or order of eggs within a clutch (Figure 5.6). Except for

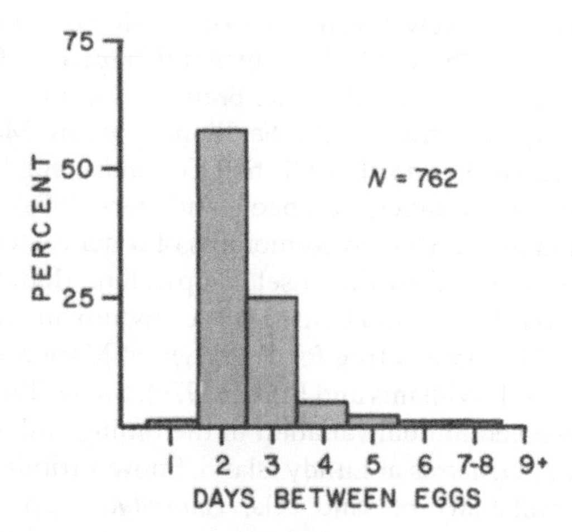

Figure 5.6. Number of days between successive eggs laid by Brandt's Cormorants; all clutch sizes and years combined, 1971–79.

rare fifth or sixth eggs, two days predominated as the interval between eggs, outnumbering three days by about two to one in frequency. The interval betwen first and second, second and third, third and fourth, and fourth and fifth eggs of a clutch averaged 2.6 ± 1.0, 2.5 ± 0.9, 2.7 ± 0.9, and 2.7 ± 0.6 days, respectively. During some years, such as 1971 and 1974, there was a tendency toward longer periods between third and fourth eggs, but the total differences were not statistically significant. Among other cormorants, intervals between eggs are usually three days in *P. aristotelis* (Snow 1960), one day in *P. auritus* (Mitchell 1977), two days in *P. olivaceus* (Morrison, Shanley, and Slack 1979), and between one and three days in *P. capensis* (Berry 1976). Like Snow (1960), we infrequently observed much longer intervals between eggs, occasionally over ten days, regardless of order within a clutch.

The timing of breeding by other Brandt's Cormorant populations is not well documented, but there appears to be a latitudinal trend. Of greatest interest, Michael (1935) observed some Brandt's Cormorants at La Jolla, in southern California, building nests and laying eggs in December 1933 and fledging young in April 1934; most of the birds at La Jolla, however, began nesting activities in March and April. At Santa Barbara Island in southern California, laying commenced in late February 1976; in 1977, laying began in early March (Hunt et al. 1981). Brandt's Cormorants in Baja California, Mexico, often nest that early as well (Ainley pers. obs.). These earlier dates relative to phenology at the Farallones are likely a response to the earlier onset of upwelling at lower latitudes in the California Current (Chapter 2). Correspondingly, in extreme northern California, Brandt's Cormorants begin laying eggs slightly later than at the Farallones, in late May and June (Osborne 1972). Farther north, in British Columbia, egg laying occurs in late June and July (Hatler, Campbell, and Dorst 1978).

The initiation of breeding in cormorants of other eastern boundary currents is also affected by the onset of upwelling (Rand 1960, 1963, Berry 1976, Berry, Millar, and Louw 1979, Crawford and Shelton 1978, Duffy 1983c). The same is true for *P. atriceps* at Marion Island in the Indian Ocean (A. J. Williams and Burger 1979). Snow (1960) and Coulson (1973) observed annual variation in the timing and synchrony of egg laying by *P. aristotelis* at Lundy Island; Snow attributed this variation to the availability of sand eels, *Ammodytes* spp., in different years.

A fairly complete description of the physiological changes experienced by cormorants in response to environmental events has been contributed by Berry, Millar, and Louw (1979). They note that *P. capensis* initiates breeding in response to seasonal changes in photoperiod as well as in wind speed and direction, which lead to increased pituitary activity. As food becomes more available in spring, the birds' circulating levels of triglycerides and sex hormones increase, leading to a rise in body and gonadal weights. Cape Cormorants reach their annual maximum in body weight during the prebreeding period, increasing in weight an average of 18% over winter levels. Thus, according to Berry, Millar, and Louw, nutritional levels are the ultimate factor controlling the onset of breeding, even though the initial stimuli are photoperiod and weather.

Clutch size

Clutch sizes averaged slightly greater than three eggs at both study colonies, with clutches at Colony I slightly but not significantly larger ($t = 1.81$, $.1 > p > .05$). If the years in which no eggs were laid (1978 and 1983) are excluded, clutches at Colony I in different years averaged 2.4 to 3.8 eggs, whereas clutches in Colony II averaged 2.7 to 3.3 eggs (1983, with its small sample size, excluded). Despite a pattern of lower clutch sizes in 1976, 1977, and 1978 at both colonies, no correlation existed between the colonies in the rankings of different years ($r_S = .457$, $t = 1.54$, $p > .1$). At Colony II, clutch sizes did not vary significantly among years, but during 1976 and 1977 at Colony I they were significantly smaller than in all other years except 1975 (Table 5.4). Recall that laying was significantly later at this colony in 1976 and 1977 than in many other years. What is indicated is an association between delayed breeding and reduced clutch size, possibly due to poorer body condition of females prior to egg laying. The delay in breeding and reduction of clutch size were extreme during the years of warm water and late upwelling, 1978 and 1983, when nearly all females failed to lay eggs.

TABLE 5.4

Mean clutch sizes, Brandt's Cormorants, Colony I, 1971–83

1977	1976	1975	1974	1972	1980	1981	1979	1973	1971	1982
2.4	2.6	2.9	3.2	3.2	3.2	3.2	3.2	3.3	3.6	3.8

NOTE: Colony I does not include 1978 or 1983, when nests were abandoned soon after clutch initiation. Lines connect similar means; SNK test, $p > .05$. All means were similar for Colony II.

Clutch sizes ranged from one to six eggs, but about 81% contained either three or four eggs, and over 97% contained two to four eggs. Five of the six one-egg clutches occurred during 1976 and 1977, years of late breeding. At Colony I, two years, 1971 and 1982, stand out in particular because one-half or more of clutches contained at least four eggs; these two years also lead with the highest number of five-egg clutches. In all, among 483 clutches (1971–83), 1.2% consisted of one, 16.6% of two, 57.1% of three, 22.7% of four, and 2.4% of five eggs.

At Santa Barbara Island, clutches averaged 2.4 eggs in 1976 and 2.3 eggs in 1977 (Hunt et al. 1981). Among other cormorants studied, clutches average from a low of 1.8 eggs in *P. carunculatus* (A. Nelson 1971) to a high of 3.8 eggs in *P. auritus* (Mitchell 1977), although clutches generally range between 2.5 and 3.5 eggs (Batchelor 1978, A. J. Williams and Burger 1979, Norman 1974, Derenne, Mary, and Mougin 1976, Lock and Ross 1973, Morrison, Shanley, and Slack 1979, Erskine 1972, Snow 1960, Berry 1976, Cline and Dornfeld 1968, Vermeer 1969a,b). Few of these studies lasted longer than one to three years, however, so it is difficult to say whether the large amount of between-year variation in the clutch size of Brandt's Cormorants is unusual.

Among Brandt's Cormorants at the Farallones, despite much annual variation, differences in clutch sizes within breeding seasons were not clearly obvious (Appendix 5.2), a pattern similar to that reported by Coulson, Potts, and Horobin (1969) and Coulson (1973) for *P. aristotelis* but opposite that reported by Snow (1960). In Colony I, for the total 13 years, clutches laid mid-season were significantly larger than late clutches, but early clutches were not different from either middle or late clutches. Within years, however, clutch sizes decreased through the season, especially in 1972, 1976, and 1979. In contrast, clutch size increased through the laying season in 1977 and 1981. During the three years in which mean clutch size ranked highest at Colony I, 1971, 1973, and 1982, clutches were largest at mid-season, although never significantly so. A comparison between extremely early and extremely late layers (those earlier than or later than one standard deviation away from the mean laying date) also showed no significant difference in clutch sizes ($t = 1.39$, DF $= 65$, $p > .05$).

Incubation period

The incubation period of all eggs in first clutches combined was 29.9 ± 1.1 days ($n = 170$). There was a significant decrease, however, in incubation period between first and second and between second and third eggs (Figure 5.7; t tests, $p < .05$). A decrease is also

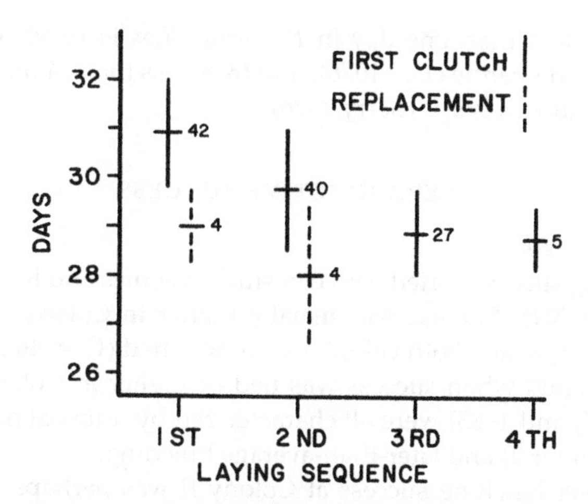

Figure 5.7. Incubation periods of Brandt's Cormorant eggs, 1972–79, in relation to laying order. Means are indicated by horizontal bars; standard deviations, by vertical bars; numbers are clutches observed.

apparent between first and second eggs in replacement clutches, although sample sizes were too small for statistical comparison. A pattern seems clear, especially with the first three eggs in a clutch, namely, that with each later egg the incubation period decreases by about one day. Berry (1976) found that incubation periods between eggs in the Cape Cormorant decreased by three days, with first eggs hatching in 28 days, second eggs in 25 days, and third eggs in 22 to 23 days. For *P. atriceps*, which lays a smaller clutch (mean clutch size 2.6 eggs), the hatching of first eggs required 30.5 days, second eggs 28.8 days, and third eggs 28.7 days (A. J. Williams and Burger 1979). Similarly, in *P. auritus* at Mandarte Island, British Columbia, first eggs hatched in 29.9 days, second eggs in 28.4 days, and third eggs in 27.9 days (Drent et al. 1964).

The mechanism behind the decrease in incubation period among successive eggs of a clutch appears to be incomplete incubation of earliest-laid eggs (Palmer 1962, Berry 1976). Adults are less settled on their first eggs, lowering egg temperature and slowing embryo development (Drent 1973). Recall that cormorant eggs are laid, on average, two days apart, so if the incubation period is decreased by one day with each successive egg, chicks hatch only one day apart. This system provides a method for decreasing the early start given first chicks within a clutch, possibly to decrease size differences among siblings. This, however, works against a "brood reduction" type of nesting scheme (see below). Among other cormorants, recorded hatching intervals are less than one day in *P. auritus* (Lewis 1929), 1.6 days in *P. albiventer* (Derenne et al. 1976), and from less than 24 hours to up to three days in *P. capensis* (Berry 1976).

REPRODUCTIVE SUCCESS

Hatching success

Hatching success varied between study colonies and between years (Appendix 5.1). Success was usually higher in Colony I for the 12 years during which both colonies were watched ($G = 46.3$, $p < .05$). The four years when success was tied or higher in Colony II (1976, 1978, 1982, and 1983) were all characterized by unusual ocean conditions (Chapter 2) and later-than-average breeding.

The lower hatching success at Colony II was perhaps due to two factors. First, as mentioned above, Colony II is a mixed colony of Brandt's Cormorants and Common Murres. In this colony, nearly all cormorant nests are partially or totally encircled by murres, leading to many more social interactions than in Colony I, a colony composed exclusively of Brandt's Cormorants. The additional commotion and fighting for space leads to both murres and cormorants occasionally losing eggs. Second, individual Western Gulls nesting near Colony II feed to a great extent on eggs (L. Spear pers. comm.), primarily those of neighboring Common Murres. These gulls seem especially vigilant in their search for eggs left exposed by nesting birds, and take cormorant eggs when possible.

When years are compared, 1983 stands out because complete nest abandonment occurred at both study colonies; abandonment also occurred in Colony I during the warm-water year 1978. In Colony I, hatching success differed significantly only between the three years with highest (1979, 1980, and 1981) and the three with lowest success (1976, 1978, and 1983) (t tests following arcsin transformation, $p < .05$). At Colony II, where hatching success was less than 50% in all but three years (1980, 1981, and 1982), the only differences were those between 1981 and 1974 and between 1982 and 1974, 1978, and 1973.

Annual variation in hatching success was apparently due to differences in the abandonment of nest sites and the disappearance of eggs during incubation rather than to differences in numbers of unfertilized eggs or embryo deaths. Among the 920 and 635 eggs observed in Colonies I and II, respectively, during the 1971–83 period, 11.3 and

13.4% failed to hatch although incubated to term. Some birds continued to incubate dead eggs much longer than the normal incubation period, even up to 90 days, a phenomenon similar to that observed by Mitchell (1977) in *P. auritus*. Again, the higher disappearance rate, Colony II (44.6%) compared to Colony I (29.1%), is consistent with more disturbance and commotion due to murres in Colony II.

If nests are compared on the basis of whether eggs were laid early, midway, or late in the laying period, hatching success did not differ over the period of study (*G* test, $p > .05$; Appendix 5.3). There were distinct differences within years, however. Hatching success was substantially better among early than among later breeders in 1972 and 1975, but the converse was true in 1971 and 1977. Birds laying midseason had much better success than early and late individuals in 1973 and 1976. In two years, 1971 and 1977, hatching success improved as the season progressed, with late layers having substantially better success than earlier ones. In brief, each year seemed to have its own pattern, without a consistent relationship to whether it was a warm- or cool-water year.

Breeding success

Fledging success varied substantially, both between years and between study colonies (Appendix 5.4). The two colonies differed in overall fledging success (*G* test, $p < .05$), although when ranked among different years, success correlated between the two colonies ($r_S = .616$, $t = 2.35$, $p < .05$).

Through the entire study period, cormorants fledged about three of every four chicks hatched. At both colonies, chick mortality was extremely low in 1977 and 1979, although rates were significantly different from only 1976 at Colony II and from 1973, 1981, and 1982 at Colony I (*t* tests following arcsin transformation, $p < .05$). During 1977 and 1979, prey seemed very abundant near the Farallones when chicks were being fed, as large feeding flocks occurred daily late into the breeding period (Chapter 3). Even most small, late chicks survived to fledging during these two years.

After data for the entire 13 years were combined, no differences in fledging success among early, middle, and late nesters were apparent (Appendix 5.3). In only two years, 1973 and 1982, did fledging success distinctly decline as the season progressed. Only in 1975 did early- or middle-nesting birds have obviously poorer fledging success than those nesting later.

The number of chicks fledged per egg-laying site, like the other variables on which overall success depends, ranged widely both between and within years (Appendixes 5.3 and 5.4). Because of the large within-year variability, between-year variability was not high, although fluctuations at both study colonies were correlated closely ($r_S = .661$, $t = 2.645$, $p < .05$). At Colony II, overall success was similar during all years except 1982, when the highest number of chicks fledged per site. Success during that year was significantly higher than during 1973, 1974, 1976, and 1978 (Table 5.5). At Colony I, cormorants were extremely successful during 1979 and 1980 and were least successful during 1975, 1976, 1978, and 1983. Success during 1975 and 1976, however, was significantly poorer than during only the three best years, 1979, 1980, and 1981. The production of chicks was strongly correlated with the proportion of juvenile rockfish in the diet, 1973 to 1977 (Colony I, $r_S = .625$, $p < .1$; Colony II, $r_S = .975$, $p < .01$).

To determine the relationship among important breeding variables, we calculated correlations for the period 1972–83 at both study colonies (Table 5.6). The only variable that consistently and highly correlated with the number of chicks fledged per nest was hatching success. In poorer years, nest abandonment during incubation is an obvious limitation to breeding success, but it seems that in all but the very best years the loss of eggs, either through abandonment or incidental loss, has the greatest influence on overall success.

In a comparison of clutches of different sizes, the number of chicks fledged increased with clutch size, although only two-egg clutches

TABLE 5.5

Mean number of Brandt's Cormorant chicks fledged per nest, 1971–83

Colony I										
1976	1975	1973	1982	1977	1971	1972	1974	1981	1980	1979
0.63	1.11	1.11	1.36	1.37	1.46	1.52	1.74	1.92	2.18	2.56

Colony II										
1978	1976	1974	1973	1975	1977	1972	1979	1980	1981	1982
0.46	0.47	0.67	0.68	1.05	1.11	1.36	1.38	1.40	1.50	1.77

NOTE: No chicks fledged at Colony I in 1978 or 1983, or at Colony II in 1983. Lines connect similar means; SNK test, $p > .05$.

TABLE 5.6
Spearman correlation coefficients, Brandt's Cormorant breeding, 1972–83

Breeding variable	Row number				
	1	2	3	4	5
Colony I					
1. Percent occupied sites with eggs	—				
2. Timing of laying	.4388				
3. Clutch size	.0455	.2203			
4. Hatching success	.1678	.3147	.7220[a]		
5. Fledging success	.6643[b]	.1399	.0962	.3916	
6. Chicks fledged per laying site	.2850	.2080	.6066[b]	.9196[c]	.6101[b]
Colony II					
1. Percent occupied sites with eggs	—				
2. Timing of laying	.7797[a]				
3. Clutch size	.7308[a]	.7308[a]			
4. Hatching success	.0227	−.1154	−.0507		
5. Fledging success	.2850	.0997	.2605	.5944[b]	
6. Chicks fledged per laying site	.0839	.0594	.2308	.9143[c]	.6801[b]

[a] $p < .01$. [b] $p < .05$. [c] $p < .001$.

fledged significantly fewer chicks than did other clutches (*t* test, $p < .05$; Figure 5.8). The number of chicks fledged per egg laid was highest for three-egg clutches. An apparent pattern is for birds laying larger clutches to fledge increasingly more chicks, but birds laying only three eggs seem more successful in rearing each egg to fledging. This is consistent with observations of *P. lucidus* (Oliver and Kuyper 1978) and *P. varius* (Norman 1974). Such a pattern could be complicated by the tendency for larger clutches to be laid only in years when food is easily available, which are also years when chick survival is greatest. This, however, is apparently not the case; rankings of clutch size did not correlate with those of fledging success at either study colony (Table 5.6).

One measure of breeding consistency in Brandt's Cormorants is the number of years in which each study site fledged at least one chick (Figure 5.9). From this tally it appears that Brandt's Cormorants are remarkably inconsistent in their breeding success, because at both study colonies most nests failed more than 50% of the time. The mean number of successful years per nest in Colony II was 4.3 ± 2.2 years ($n = 20$ nests) for the 12 years 1972–83. Colony I fared significantly better, however, with an average 6.1 ± 2.5 ($n = 20$) successful years per nest ($t = 2.40$, DF $= 38$, $p < .05$). The lower number of successful years per nest in Colony II is consistent with the pattern shown by other breeding variables at that colony, particularly hatching success.

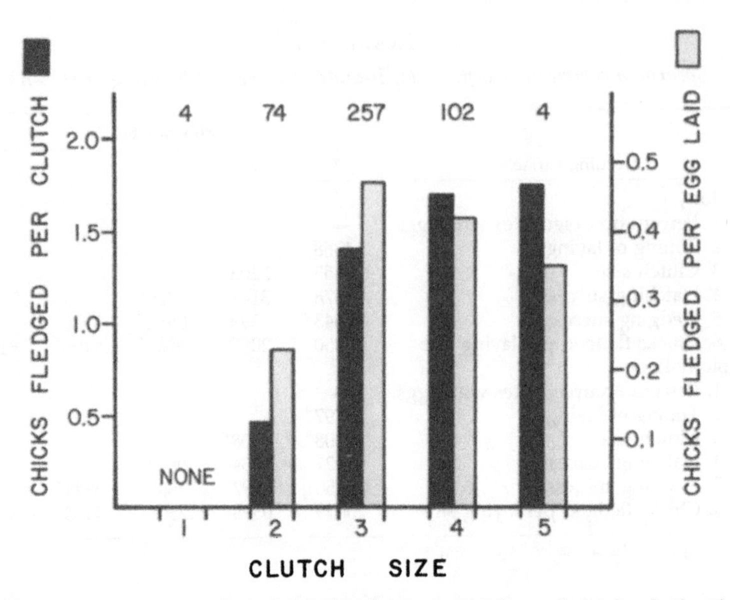

Figure 5.8. Relationship between clutch size, number of chicks fledged per clutch, and number of chicks fledged per egg laid, Brandt's Cormorants, 1972–82, study colonies I and II combined; numbers at top are clutches observed.

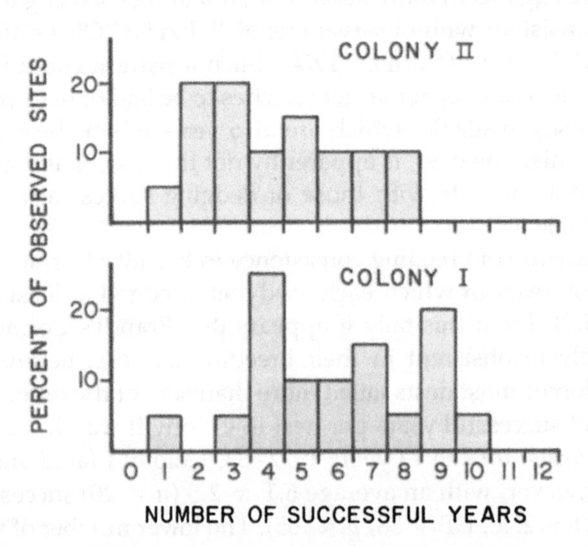

Figure 5.9. Number of years of successful breeding (i.e., at least one chick fledged) at individual nest sites, 1972–83; n = 20 sites at both colonies, including only Brandt's Cormorant sites followed throughout the entire 12-year period.

It is also consistent with the greater commotion brought by murres as well as their encroachment on cormorant nesting habitat.

As with Brandt's Cormorants at the Farallones, nesting success elsewhere and among other cormorant species is quite variable. At Santa Barbara Island, Brandt's Cormorants in one colony fledged 0.6 chick per pair in 1976, compared to 1.1 per nest in 1977 (Hunt et al. 1981). In most studies of other cormorant species, hatching success ranged from 40 to 70%, and fledging success ranged from 60 to 95% (A. J. Williams and Burger 1979, Norman 1974, Derenne et al. 1976, Morrison, Shanley, and Slack 1979, Snow 1960, Cline and Dornfeld 1968, Mitchell 1977, Lock and Ross 1973, Drent et al. 1964, McLeod and Bondar 1953, Kortlandt 1942, Batchelor 1978). This means that cormorants typically fledge chicks from fewer than one-half of the eggs they lay, and rarely do they fledge chicks from even two-thirds of their eggs (see Chapter 12). Many of these species may be similar to the Brandt's Cormorant in that excellent production during certain years may maintain the population, and entire year classes may be absent because of periodic nesting failure.

Abandonment of nesting attempts during years of scarce food, leading to decreased population sizes, has been documented for cormorants of other eastern boundary currents. Guanay populations fluctuate dramatically as a function of ENSO events in the Peru Current; during the severe ENSO in the late 1950's the Guanay population dropped from 28 million to 6 million breeding birds but recovered to 18 million by 1963 (Jordán 1967). Similarly, failure of the food supply in the Benguela Current may lead to mass desertions of nest sites and failure to breed by the Cape Cormorant (Rand 1960, Crawford and Shelton 1978).

Replacement clutches

After losing their first clutch, Brandt's Cormorants lay replacement eggs if conditions permit. Throughout the entire study, second clutches were laid in 57% ($n = 108$) of nests from which first clutches were lost. In six cases, we observed three clutches attempted, and once, in 1980, we observed four clutches attempted at a site in Colony I. Since Brandt's Cormorants sometimes switch mates after a nest failure (PRBO, unpubl. data), however, we do not know whether the same birds were involved in each clutch attempt at each site. All replacements were laid at sites that had lost eggs, never at nests that had progressed to hatching and then lost chicks. Unlike Kortlandt (1942), Drent et al. (1964), A. Nelson (1971), and Deans (1972), we never observed pairs attempting second broods following successful fledging of the first brood.

Replacement clutches had relatively poor success, fledging chicks from only 26.3% of all eggs laid, compared to 41.1% for first clutches (Appendix 5.5). Additionally, clutch sizes were smaller and fewer chicks fledged per clutch attempt. During favorable years like 1977 and 1979, however, even late-laying individuals such as those attempting replacement clutches may successfully fledge most of their chicks.

Replacement clutches are common among other cormorants. Snow (1960) found that replacements by *P. aristotelis* were more successful than original clutches laid at the same time. She suggested that replacements were laid by older birds that had started their first clutches earlier in the breeding season. McLeod and Bondar (1953) reported that 30 to 50% of *P. auritus* pairs in an inland population renested after losing eggs. Morrison, Shanley, and Slack (1979) found that in *P. olivaceus* sizes of replacement clutches were no different from those of original clutches.

Chick development

Through observations from our blinds, we characterized development patterns and determined critical events in the prefledging life of Brandt's Cormorants. Cormorant chicks are completely dependent on their parents for protection and food. When chicks are small, one parent must remain at the nest to prevent predation by Western Gulls and harassment by other cormorants, and to provide warmth. Parents must maintain a regimen of feeding trips and nest reliefs throughout the nestling period, feeding the chicks while attending the nest. Discussed in Chapter 3 was the fact that duration of feeding trips differed among years. Years in which trips were long were those in which fledging success was lowest.

As discussed above, eggs are laid two days apart, but because of decreasing incubation period with respect to laying order, eggs hatch only one day apart. This pattern ensures that eggs hatch more synchronously than they are laid and that size differences between chicks are less pronounced. Several days may elapse, however, between hatching of the first and last chicks in a brood, decreasing the smallest chicks' abilities to compete for food with larger, older siblings. This, of course, is a basic tenet of brood reduction theory, as previously discussed.

Within the first five to ten days after hatching, Brandt's Cormorant chicks are completely brooded by parents except on extremely warm days, when parents instead stand to provide shade. The chicks rarely show themselves and lift their heads only when begging for food. After ten days, they become more coordinated and actively beg with snakelike neck movements. They may even occasionally sit up. It is during this time that chicks of the presumably similar Double-crested Cormorant become effective endotherms, able to maintain a stable body temperature within normal ambient temperatures (Dunn 1976). Down begins to cover their bodies, but their heads remain naked until the juvenal plumage grows in at five to six weeks of age. At ten to 20 days, chicks begin to grow rapidly, but they still remain in the nest, protected by a parent. At this stage, chicks regularly sit up in the nest to beg or to preen. Much of their time is spent sleeping, with long necks occasionally drooped over the edge of the nest bowl.

A parent cormorant carries food to its chicks in its esophagus and proventriculus; food is then passed to the chicks through regurgitation. The exact method of food transfer changes as the chicks grow. When chicks are very young, parents pass small morsels of food into the chicks' mouths. Within a week, chicks begin to insert their heads inside their parents' mouths, and larger amounts of food pass with each feeding. Eventually, as a chick becomes larger, it vigorously thrusts all of its head and much of its neck far inside its parent's mouth and throat.

Extremely critical in the prefledging life of cormorant chicks are the first few days they are left alone by their parents. If too young, chicks may be killed by gulls or by other cormorants. Immature or nonbreeding cormorants harass unaccompanied chicks in or out of nest sites, pecking them repeatedly as if to assert dominance or to usurp their space. Small chicks forced out of nests by this activity may be unable to return and eventually may die, or they may be pecked as they attempt to crawl into other nests with adults in attendance. The age at which parents left chicks alone at nest sites varied greatly among years (Table 5.7). At both study colonies, the two years of exceptional rockfish abundance, 1977 and 1979 (Chapter 3), stand out with parents staying at nests for long periods after hatching. At both colonies, parents stayed with chicks significantly longer in 1979 than in all other years and significantly longer in 1977 than in all other years except 1971 at Colony I and 1981 at Colony II (SNK test, $p < .05$). Recall that in 1977 and 1979 breeding seasons were very long and fledging success was highest at both colonies. In fact, rankings of fledging success and the number of days parents stayed with chicks

TABLE 5.7

Mean time Brandt's Cormorants remained at nest sites with chicks, Colonies I and II combined, 1971–82

(*Days*)

Year	Time with chick		
	Mean ± SD	n	Range
1971	43.0 ± 6.16	12	29–51
1972	29.9 ± 5.48	13	20–40
1973	24.5 ± 2.69	17	18–31
1974	29.1 ± 5.06	18	17–40
1975	34.5 ± 7.04	26	25–46
1976	29.6 ± 6.48	22	20–43
1977	44.8 ± 7.07	31	43–60
1979	58.8 ± 9.83	36	40–84
1980	33.2 ± 5.46	31	22–45
1981	29.8 ± 7.69	33	19–61
1982	29.8 ± 6.09	30	20–56
Total	37.0 ± 4.45	269	17–84

NOTE: Includes only sites that fledged chicks. Time with chick = number of days between hatching date of first hatched egg and date when parents first left chick or chicks alone at nest. No breeding in 1978 or 1983.

were highly correlated (Colony I, $r_S = .740$, $t = 3.48$, $p < .01$; Colony II, $r_S = .864$, $t = 4.85$, $p < .01$). Parents stayed with chicks for the shortest average period in 1973, especially at Colony II, where the interval was significantly shorter than in all years except 1976 (SNK test, $p < .05$). There is obviously an advantage for parents to stay at nest sites as long as possible, but this seems possible only in years when food is abundant near the island.

The ages at which chicks died (Appendix 5.6) also showed variability among years. During two warm-water years, 1973 and 1976, chicks suffered high mortality throughout the first 30 days of life, in contrast to years such as 1975, when many chicks died at a very young age but few did so later. Again, in the prolonged breeding seasons of 1977 and 1979 chick mortality was very low.

Brandt's Cormorants continue to feed their chicks for several weeks after they wander away from the nests, sometimes into late September and October. This was particularly true in years such as 1977 and 1979, when prey sources persisted near the island. After the last chicks wander from the nest site, parents usually remain in the colony only to feed their offspring. Otherwise they are harassed so vigorously by chicks wanting food that they must roost away from the colony. Although neither parents nor older chicks actually occupy nest sites during this time, the site becomes a place for chicks and parents to meet when the latter arrive with food. Hungry chicks closely observe each arriving parent in hopes that it might be theirs; if it is, they chase it, usually to the colony periphery, until the parent regurgitates a meal. When food is plentiful and chicks are well fed, they do not beg so vigorously, and parents may spend several minutes searching through the colony for their chicks. Large chicks also enter the water below the colony daily for bathing and diving, although they are still reliant on their parents for food.

During seven of the eight years between 1975 and 1982, we kept records of the number of banded chicks we found dead in Colony II after the completion of the breeding season (Table 5.2). We found 2.5 to 10.8% of the chicks we had banded, averaging 6.2% of the total banded for all years. Because chicks are fed by their parents for 20 to 40 days following their departure from nest sites, it appears that mortality of older chicks about to disperse from the island is lower than that of young chicks still in nests (cf. Appendix 5.6).

THE CORMORANT BREEDING STRATEGY

Several breeding adaptations displayed by cormorants set them apart from other seabirds, including, with some exceptions, other members of the Pelecaniformes: (1) clutch size is relatively large and variable, (2) investment in eggs is small, as measured by egg size relative to the female's body size, (3) associated with the low investment in eggs, hatching success drives overall nesting success, (4) chicks are extremely altricial, and (5) the first and last chicks to hatch in a brood are quite different in size and competitive ability as a result of asynchrony in hatching. Other life-history characteristics of the group include a relatively early age of first breeding and higher annual mortality of adults (Ashmole 1971, Boekelheide and Ainley 1989). In general, these characteristics result in extreme variability in annual breeding effort and success as a response to variation in food abundance. On the other hand, to a degree much greater than most other seabirds, cormorants are able to increase their reproductive output substantially when food is especially abundant.

Brandt's Cormorants at the Farallones illustrate the life-history characteristics of cormorants quite well. The size of their breeding population fluctuates significantly from year to year. Even though they may not attempt to breed during years when food is limited, they rapidly respond with increased numbers in favorable years. Their propensity to forgo breeding may have important survival value, permitting adults to attempt breeding later when conditions improve. Mean clutch sizes may vary by as much as one egg between years, an indication that cormorants can take advantage of years of exceptionally abundant food by increasing reproductive output. The timing of their breeding effort is particularly sensitive, but, as shown in 1977, even if laying is delayed they may respond quickly if food availability improves. They are steadfast in their incubation and chick-rearing duties when conditions permit, but they quickly abandon efforts when food is not available. Together, these adaptations are particularly suited to conditions in the Gulf of the Farallones, where annual variability in food supply is the norm rather than the exception. Some of these characteristics are discussed more fully in relation to other Farallon species in Chapter 12.

6

Pelagic Cormorant and Double-crested Cormorant

Robert J. Boekelheide, David G. Ainley, Harriet R. Huber, and T. James Lewis

The Pelagic and Double-crested cormorants are the other two members of the cormorant family that nest on the Farallones. With the Brandt's Cormorant, they occur along much of the North American West Coast, although their zoogeographic affinities differ. The Pelagic Cormorant's range centers along the cold, foggy, windswept coastlines of the subarctic Pacific Ocean, from northern Japan and eastern Siberia east, through the Aleutian Arc and Bering and Chukchi seas, south through the Gulf of Alaska to the California Current. It is an exclusively marine cormorant. The Double-crested, in contrast, is ubiquitous throughout much of North America. It breeds in marine, estuarine, and freshwater habitats along both coasts and in the interior, particularly around the Great Lakes, in the Mississippi–Missouri River drainages, and the northern plains of the United States and Canada (Palmer 1962). It is abundant in subtropical habitats in the Gulf of California and Florida.

Despite familial similarities, the Pelagic, Double-crested, and Brandt's cormorants differ in important ways from each other. The Pelagic is best known for its cliff-nesting habits; it cements its nest onto small, precarious ledges on vertical cliffs, sometimes in loose colonies, but more often in a scattered fashion depending on the availability of ledges (Siegel-Causey and Hunt 1981, 1986). The Double-crested, in contrast, nests in discrete colonies, sometimes in trees or mangroves, but most often on the ground on islands (Lewis 1929). The species seemingly prefers to build nests on the ground, but will use trees if no treeless islands exist nearby (Vermeer 1973). Along the Pacific coast, colonies typically occur on moderately steep rocky slopes of offshore islands. Where ranges overlap, these two cormorants generally do not overlap in nesting habitat, because the Pelagic restricts itself to steep cliff faces. The Double-crested is much more likely to nest alongside Brandt's Cormorant, although the Double-crested prefers the shoulders of hillsides, higher slopes, and summits of islands for its colony sites (Michael 1935, Drent et al. 1964, Siegel-Causey and Hunt 1981, 1986; this study).

The Pelagic Cormorant is a much smaller bird than its sympatric cousins. The subspecies that ranges from British Columbia to Baja California, *Phalacrocorax pelagicus resplendens*, is much smaller than the more northerly nominate race (Palmer 1962) and weighs between 0.8 and 2.2 kg (PRBO data; Figure 1.1).

Taxonomically, Double-crested Cormorants are divided among four subspecies. Through at least the 1931 edition of the *AOU Checklist of North American Birds*, the western subspecies of Double-crested Cormorant, *P. auritus albociliatus*, first described by Ridgway, was known as the Farallon Cormorant; its type locality is the Farallon Islands (American Ornithologists' Union 1957). This subspecies ranges along the coast from British Columbia to the Gulf of California and inland to Nevada and western Idaho (Palmer 1962). It is the largest subspecies of Double-crested Cormorant, weighing up to 2.7 kg (Palmer 1962; Figure 1.1).

Most studies of breeding Double-crested Cormorants have been carried out on islands in freshwater lakes and marshes (McLeod and Bondar 1953, Cline and Dornfeld 1968, Hanson 1968, Vermeer 1969a,b, 1970, 1973, Greichus and Greichus 1973, Mitchell 1977, Roney 1979). Some colonies along the eastern seaboard have been studied also (Lewis 1929, Mendall 1936, Lock and Ross 1973, Ross 1977, Ellison and Cleary 1978). Results of these studies indicate a clutch of three to four eggs and one to two chicks fledged per nest attempt. The timing of egg laying is variable, depending on a number of environmental factors such as temperature, ice cover, and habitat availability. The only detailed study of Double-crested Cormorant breeding ecology on the Pacific coast was conducted at Mandarte Island, British Columbia (Drent et al. 1964), where the species' nesting characteristics appear similar to those on the Farallones.

Detailed studies of the Pelagic Cormorant's breeding ecology are few, because the species' dispersed nesting habits and preference for cliffs limit accessibility and sample sizes. Scattered observations in Alaska arose from the Outer Continental Shelf Environmental Assessment Program (Baird et al. 1983). Among the Alaskan birds, breeding effort and success varied little; clutches were generally of three to four eggs, and most pairs fledged two chicks. Only at Middleton Island, in the Gulf of Alaska, did breeding success vary. Farther south, at Mandarte Island, Pelagics hatched about half their eggs and fledged about two chicks per pair (Drent et al. 1964). The two years of nesting failures reported by Penhale (1972) near Morro Bay, California, appear extreme (but see below).

Neither the Double-crested nor the Pelagic Cormorant is easily accessible as a study subject at the Farallon Islands. What little information we have from the last 15 years has been obtained opportunistically through the least amount of disturbance. This is especially true for the Double-crested Cormorant, whose nesting colony can be observed only by telescope from a distance of more than 0.5 km. Despite these handicaps, we have been able to construct a fairly complete picture of these birds' breeding biology. Comparisons with other cormorants are contained in Chapter 5.

METHODS

We observed Pelagic Cormorants at two main study areas on Southeast Farallon: (1) at Shubrick Point and (2) along the northwest shoreline adjacent to Maintop Bay (Figure 1.3). At Shubrick, study nests were situated on the point's northeastern face and at the entrance to Great Murre Cave. Both these areas were visible from the "Murre Blind" at the crest of Shubrick Point. The Maintop Bay sites were located on cliffs and promontories adjacent to small embayments known as Boiler Cove and Sea Lion Cove. We observed the Maintop Bay sites from both the "Cormorant Blind," which overlooks Boiler Cove and the southwest section of Sea Lion Cove, and from the "Pelagic Blind," which overlooks the remainder of Sea Lion Cove. From all these vantage points we observed between 16 and 41 nest sites, averaging 31 per year, 1971–83.

As in studies of other species, we generally observed the same nest sites of the Pelagic Cormorant year after year. These were chosen because they were visible from established observation points where we caused no disturbance to the cormorants (see Chapter 1). The island's greatest concentrations of Pelagic Cormorants, located on the north ridge of Lighthouse Hill and on the east face of Maintop, were not accessible for study.

From 1972 to 1979 we observed sites every day, noting nest occupancy, nest condition, and the number of eggs and chicks. In 1971 and from 1980 to 1983, we noted the same information but observed nests every other day. From these observations, we determined reproductive chronology and success. In this chapter, we also make comparisons among early, middle, and late nesting attempts, using definitions given for Brandt's Cormorants (Chapter 5).

Prior to 1976, we occasionally noted important breeding events in the Double-crested Cormorant colony in the *Farallon Journal*. From 1976 to 1983, we observed this colony daily, weather permitting, recording the number of adults present and the number of occupied nest sites. Other important milestones noted were the dates when well-built nests and chicks were first visible. These chicks were probably two to three weeks old. We also recorded the date each year when we first observed fledglings away from the Maintop colony, usually in Fisherman's Bay or at East Landing.

During the 1971 breeding season, at different times of day, we counted the number of Pelagic Cormorants visible on the north ridge of Lighthouse Hill. From this information, we obtained a measure of their diurnal attendance patterns. For analysis these counts were divided among three-hour periods within the prebreeding, incubation, and nestling stages of reproduction. During the 1976–86 breeding seasons, on one evening every month we counted all the individuals, pairs, and nest sites of Pelagic Cormorants visible on Southeast Farallon, West End, and the adjacent islets. Using the high count for each year, in combination with the proportion of study sites where birds laid eggs, we estimated the total breeding population (egg-laying pairs) for the entire island group. Estimates for the breeding population in 1972 were obtained from DeSante and Ainley (1980).

We banded 598 Double-crested Cormorants and 109 Pelagic Cormorants from 1967 to 1983, all as chicks prior to fledging. We banded Double-crested Cormorants at night to prevent predation by Western Gulls. Like other Farallon birds, each chick was usually given two bands, a USFWS metal band on one leg and a wrap-around plastic band, colored to denote year class, on the other leg. We tallied all recoveries reported by the USFWS Bird Banding Laboratory by the age of the bird and the location of recovery.

FARALLON POPULATIONS

The populations of Pelagic and Double-crested cormorants at the Farallones are small compared to those in the more central areas of the species' breeding ranges. During the 1979–80 count period used by Sowls et al. (1980), 12% of Pelagic Cormorants and 10% of Double-crested Cormorants along the California coast nested on the Farallones. Historically, Farallon populations may have been much larger, but cormorants have been affected by human disturbance and other factors (see Chapter 1).

Pelagic Cormorants are widely distributed at the South Farallones. Essentially, they nest on all suitable cliffs (Figure 6.1). They are not colonial in the strict sense, but instead form loose aggregations in favorable habitat. The largest aggregations occur on the north ridge of Lighthouse Hill and on the east face of Maintop (Figure 1.3). Each of these groups contained between 150 and 200 nest sites in 1982. Many smaller groups, usually containing fewer than 25 nests and, in some cases, as few as three, exist on other promontories and around the island perimeter, such as on cliffs overlooking surge channels and on offshore islets.

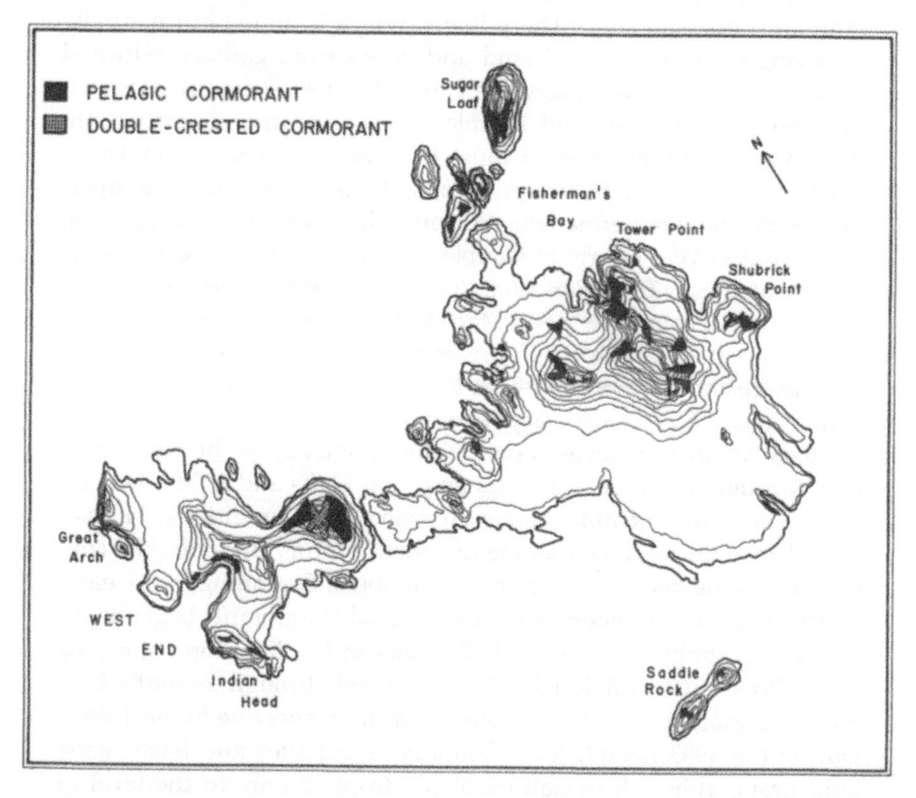

Figure 6.1. Nesting areas of Pelagic and Double-crested cormorants at the South Farallon Islands; shaded areas are colonies used during at least one year between 1979 and 1983.

The Double-crested Cormorant, in contrast, has the most restricted nesting distribution of all Farallon birds. Through the period of this study, and as far as we know through this entire century, all Double-cresteds nested on the upper slopes of Maintop (Figure 6.1). Most nested on Maintop's summit and northeast face, but through the 1970's the nesting area extended down the southwest side. Of the 229 well-built nests on Maintop in 1982, 163 occurred on the northeast slope and 66 occurred on the southwest.

Like Brandt's Cormorant, both the Pelagic and Double-crested construct nests of plant material that are solidified by compaction and feces. The Pelagic, in fact, cements its relatively small nest to cliff ledges by using its own excrement. Both species use annual terrestrial vegetation and marine algae to create their nests, but the Double-crested is unique in that its nests contain bones, large feathers, and even an occasional stick. These items, which include both naturally occurring bones from the island and bones from garbage returned by Western Gulls, are interwoven into the nest structure in much the same way as mainland Double-cresteds incorporate sticks into their nests. Nesting material must be somewhat limited for Farallon Double-cresteds, however, because their nests pale in comparison with the semipermanent "towers" that eventually result from the year-after-year efforts of Double-cresteds that build nests on the ground at sites with shrubs (e.g., San Miguel and Mandarte islands, G. Hunt pers. comm.; Isla San Martín, Ainley pers. obs.).

Through the twentieth century and until the 1970's, numbers of Double-crested Cormorants at the Farallones remained fewer than 50 pairs (Figure 6.2). Ainley and Lewis (1974) attributed the Double-cresteds' failure to recover in part to the decimation of the Pacific Sardine throughout its California range in the 1940's and 1950's by overfishing at a time of unfavorable environmental conditions. Human disturbance through most of the twentieth century undoubtedly contributed to the decline in cormorant numbers. Beginning in the early 1970's, however, numbers of Double-crested Cormorants began to increase perceptibly and reached 229 pairs in the Maintop colony by 1982. Occasional individuals also wandered through Brandt's Cormorant colonies on Southeast Farallon, as if prospecting for nest sites. During the 1983 ENSO, fewer birds occupied sites and fewer pairs built nests, although overall numbers dropped only to the level of 1980. Some mortality must have occurred during the 1982–83 ENSO, because populations after that event were lower than in 1982. Because

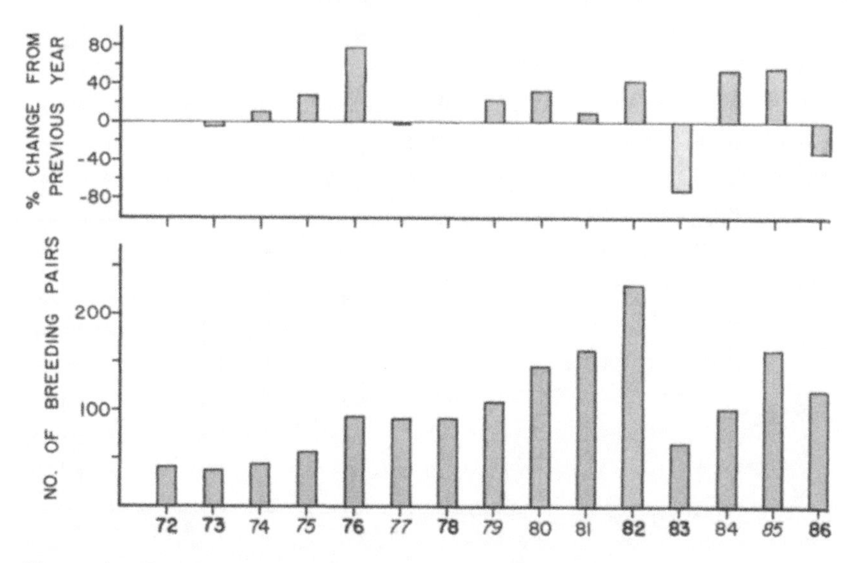

Figure 6.2. Population estimates for Double-crested Cormorants at the Maintop colony, 1972–86.

of the low level of disturbance now prevailing at the Farallones, this species seems to have the potential for future increase, provided that prey are available.

The Double-crested Cormorant has also increased recently in other areas of North America. It has increased throughout much of its Canadian range within the last decade, primarily because of a decrease in human disturbance of nesting colonies (Vermeer and Rankin 1984). It has also recovered along the New England coast (Drury 1973). Additionally, the use of organochlorines, which caused eggshell thinning in this species in southern California (Gress et al. 1973), has been controlled; affected populations have since begun to breed successfully.

Population trends for the Pelagic Cormorant are poorly documented, probably because the species is relatively noncolonial. Through the nineteenth and early twentieth centuries, the Pelagic was the least numerous cormorant at the Farallones (Ainley and Lewis 1974). As we witnessed between 1976 and 1983, however, the size of the species' breeding population is annually quite variable (Figure 6.3). During poor years, such as 1978 and 1983, most Pelagics were absent. In 1979, their numbers recovered rapidly when feeding conditions presumably improved following the poor year. Population levels be-

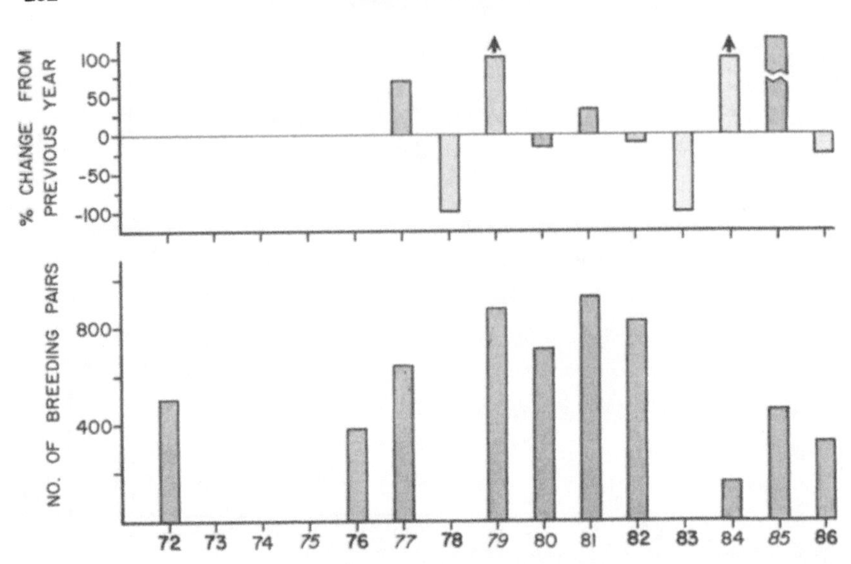

Figure 6.3. Population estimates for Pelagic Cormorants at the South Farallon Islands, 1972 and 1976–86 (1972 estimate from Ainley and Lewis 1974).

tween 1979 and 1982, in fact, may indicate the maximum number of potential breeders, i.e., approximately 800 to 900 pairs. In contrast to the rapid recovery following 1978, the population did not recover after the 1982–83 ENSO, but another factor, high mortality due to entanglement in gill nets, was likely at fault (see Chapters 8 and 12).

At other localities Pelagic and Double-crested cormorants may move between colonies or between islands if conditions become unfavorable (Drury 1973, Benz and Garrett 1978, Nysewander and Barbour 1979, Carter, Hobson, and Sealy 1984). Carter et al. suggested that Pelagic Cormorants move to new colony sites because of human disturbance, predation, unfavorable weather, or natural destruction of nesting ledges. None of these factors played a significant role at the Farallones during this study.

We do not know what the Pelagic Cormorants did during years such as 1978 and 1983, when nearly all deserted their nests early in the season and stayed away from the island. It is unlikely that they bred at other sites within the Gulf of the Farallones, for during these years ENSO-like conditions prevailed throughout California. During 1983, many probably died, as indicated by the lack of recovery in subsequent years (Figure 6.3). It would be of value to document movements of cormorants among colonies by means of banding.

OCCURRENCE PATTERNS

Nonbreeding distribution

Of the 109 Pelagic Cormorants banded between 1967 and 1983, only two were recovered at localities other than Southeast Farallon. Both birds were juveniles and were found dead in the Monterey Bay area, one at Moss Landing and one at Pacific Grove. One was found in September, the other in March. It is difficult on the basis of so few records to speculate on the dispersal of Pelagic Cormorants away from the Farallones, but it seems that they do not disperse very far. Sizable roosts of Pelagic and other cormorants occur along the mainland coast of the Gulf of the Farallones during the fall and winter. It is probable that many of these individuals are Farallon birds.

Recoveries of banded Double-crested Cormorants were also concentrated in central California, although, as for Brandt's Cormorants, they indicated a pattern of significant movement during the first year. We received reports of 35 banded birds at sites other than the Farallones. Of these 35, 71% were juveniles. This is consistent with Houston's (1971) observations in Saskatchewan, where 66% of recoveries were of birds less that six months old. Recoveries of first-year Farallon Double-cresteds ranged from San Diego, California, to Bamfield, British Columbia. There appeared to be a southerly trend to dispersal, however, because only two of the first-year birds were recovered north of Point Arena, California, whereas nine were recovered south of Monterey. The remaining first-year birds recovered were either in San Francisco Bay or from the California coast between Point Arena and Monterey. Five of seven recoveries of birds older than one year occurred either in San Francisco Bay or along the coast of the Gulf of the Farallones; the other two occurred in southern California.

If the localities of band recoveries accurately reflect the dispersal of Double-crested Cormorants away from the Farallones, they also reveal the habitat preferences of these birds when not breeding (Chapter 3). Seventeen of 35 recoveries were in shallow coastal bays, lagoons, and estuaries, another five were in San Francisco Bay, six were on coastal beaches, four were on rivers within 15 km of the coast, and only one was farther inland, near Los Baños in the San Joaquin Valley. The primary recovery sites, other than San Francisco Bay, were Tomales Bay and Morro Bay. Additionally, we regularly observed Farallon-banded Double-crested Cormorants throughout the year in Bolinas Lagoon and Drake's Estero, within 35 km of Southeast

Farallon Island. It is likely that Farallon birds mix at these coastal sites with individuals that move from interior breeding sites for the non-breeding period.

Colony attendance

After breeding, and usually by late September, the number of Pelagic Cormorants attending nest sites dropped to a level that was 25%, or lower, of their average peak breeding population. Numbers continued at the lower levels through the fall. In some years, Pelagics returned in appreciable numbers to occupy nest sites and to begin nest building as early as December or January, but in others, particularly the warm-water years such as 1978 and 1983, numbers remained at extremely low levels through the winter and did not begin to increase until April or May. Individual males were usually the first to occupy sites, but even during winter and early spring we often observed pairs together at nest sites. Even in favorable years, however, different colonies varied in winter occupancy—the outlying nesting sites often remained deserted for several weeks after major nesting areas were occupied (see Siegel-Causey and Hunt 1981, 1986).

We counted Pelagics annually at Christmas as part of the Audubon Society's Christmas Bird Count. Numbers ranged from a low of 18 in 1976 to 2,000 in 1971, averaging about 400. The winter population ranged from essentially 0 to 100% of the breeding population. This variation was probably related to prey availability near the island during winter.

Once nest building began, whether in mid-winter or mid-spring, nest occupancy became more persistent. One member of each pair remained at the site to protect nesting material from thievery by other cormorants, although this tenacity waned when conditions indicated a scarcity of food near the islands (Chapters 2 and 3). In years favorable for breeding many sites were occupied for several months before egg laying. The tendency for birds to return so early to ensure site ownership may indicate that sites suitable for this species are limited on the Farallones (see also Siegel-Causey and Hunt 1986).

Pelagic Cormorants reached peak numbers in May and June, when most pairs had eggs and chicks. This peak continued through the nestling period. Chick feeding continued through June and July, and the earliest chicks departed nest sites during the second half of July. During extended breeding seasons such as in 1977 and 1979, some chicks continued returning to the nest for feedings into late September and early October.

The diurnal attendance pattern of Pelagic Cormorants prior to lay-

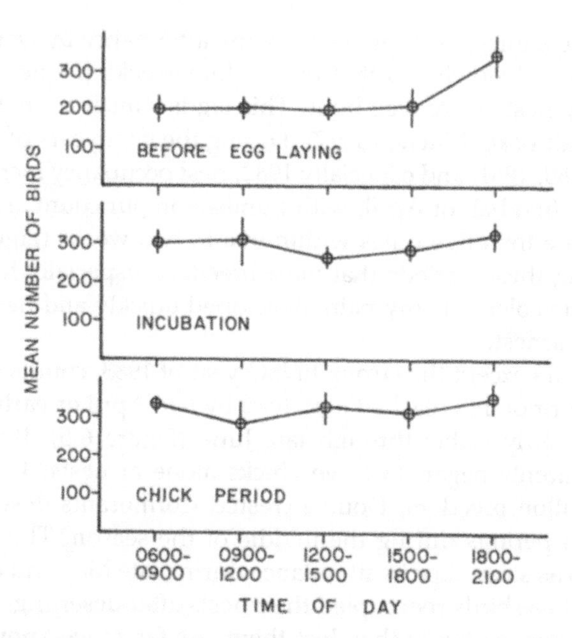

Figure 6.4. Diurnal attendance patterns of Pelagic Cormorants nesting on the north ridge of Lighthouse Hill during 1971; vertical lines show standard deviations.

ing showed consistent nest occupancy through the morning and afternoon and a distinct increase in attendance in the evening (Figure 6.4). The evening increase undoubtedly resulted from the return of mates at day's end. This pattern ceased once laying began. Then, evening numbers did not greatly differ from those of the morning and afternoon.

Double-crested Cormorants were relatively consistent in their winter attendance—essentially none were present from the time the last breeding birds departed in September until the first breeders arrived in March (Table 6.1, Figure 12.1). Occasional individuals or small groups visited during the winter, but they usually flew past without stopping. We observed Double-crested Cormorants on only four of the 17 Christmas counts between 1968 and 1985: one bird in 1971 and 1977, three in 1970, and four in 1973.

The earliest Double-crested Cormorants occupied nest sites in mid- to late March, although during 1978, an extremely late warm-water year, they did not arrive until the first week of April (Table 6.1, Figure 6.5). As with Brandt's Cormorants, the first Double-cresteds to arrive were males, who advertised vigorously at their nest sites.

After the earliest birds arrived, colony attendance by Double-cresteds developed slowly, so that one to four weeks frequently passed before any nests were well built. This lag is similar to that observed at Great Salt Lake (Mitchell 1977). During the later years of our study, such as 1980, 1981, and especially 1982, nest occupancy increased rapidly in the first half of April, with numbers in our count area increasing by more than five times within one to two weeks (Figure 6.5). It was during these periods that most breeders, especially females, arrived at the colony; many pairs developed quickly and nest building began in earnest.

In all years except the strong ENSO year of 1983, counts of Double-crested Cormorants reached a plateau by late April or early May and remained fairly stable through late June (Figure 6.5). By the latter month, parents began to leave chicks alone at nests. In 1983, like other Farallon breeders, Double-crested Cormorants deserted nests for certain periods during the middle of the season. This shows on Figure 6.5 as steep dips in attendance during late May and early June. Many of these birds reoccupied their nests after deserting, but if they had laid eggs, certainly they lost them. As far as we know they did not relay.

During years of successful breeding, Double-crested Cormorants

TABLE 6.1

Breeding chronology of Double-crested Cormorants, 1971–83

Year	First adults present	First well-built nests	Nests with eggs	Chicks per nest	First visible chicks[a]	First fledglings depart[b]
1971	L Mar.	NA			NA	13 July
1972	M Mar.	L Mar.			NA	15 July
1973	NA	NA			NA	25 July
1974	10 Mar.	M Mar.			L May	25 July
1975	E Mar.	L Mar.	55	1.8	L May	22 July
1976	16 Mar.	L Mar.	90	1.0	8 June	25 July
1977	21 Mar.	M Apr.	90	2.0	16 June	E Aug.
1978	7 Apr.	M Apr.	90	1.5	10 June	31 July
1979	24 Mar.	E Apr.	105	2.4	5 June	27 July
1980	17 Mar.	E Apr.	150	2.0	8 June	28 July
1981	20 Mar.	M Apr.	160	2.0	3 June	L July
1982	12 Mar.	M Apr.	229	2.0	7 June	22 July
1983	15 Mar.	M Apr.	65	0.2	10 June	22 July

NOTE: *E*, early; *L*, late; *M*, mid-; *NA*, data not available.

[a] I.e., date when chicks were large enough to stand and be visible from PRBO house (two to three weeks after hatching).

[b] I.e., date when fledglings were first observed elsewhere than at Maintop colony.

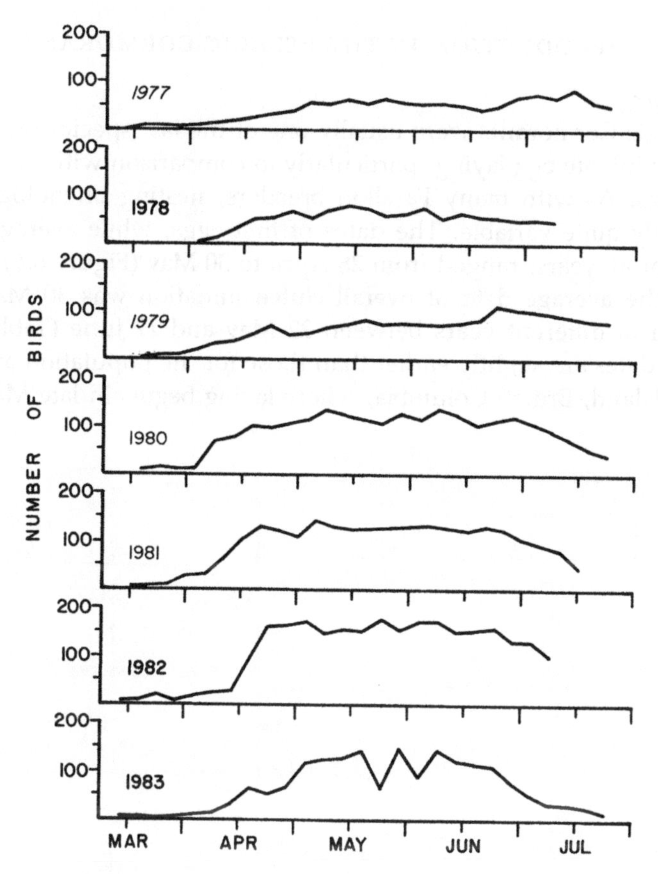

Figure 6.5. Attendance patterns of Double-crested Cormorants at the Maintop colony, 1977–83; censuses made daily from 0700 to 0900 PDT.

continued to feed chicks through August and September, although by the time chicks reached adult size their parents visited the colony only for feedings. As the season progressed, adults apparently roosted away from the Farallones at night, because only small clusters of chicks and an occasional adult were visible in the Maintop colony in the late evenings or early mornings of August and September. Unless adults roosted out of sight on West End, they must have remained on the mainland. The last Double-crested Cormorants finally departed the island by mid- to late September, when the last chicks fledged.

REPRODUCTION IN THE PELAGIC CORMORANT

Egg laying

Pelagic Cormorants were usually one of the last species on the island to initiate egg laying, particularly in comparison with other cormorants. As with many Farallon breeders, nesting phenology was annually quite variable. The dates of first eggs, while averaging 17 May for all years, ranged from 28 April to 30 May (Figure 6.6). Similarly, the average date of overall clutch initiation was 30 May, yet ranged in different years between 22 May and 12 June (Table 6.2). These dates are slightly earlier than those for the population at Mandarte Island, British Columbia, where laying begins in late May and

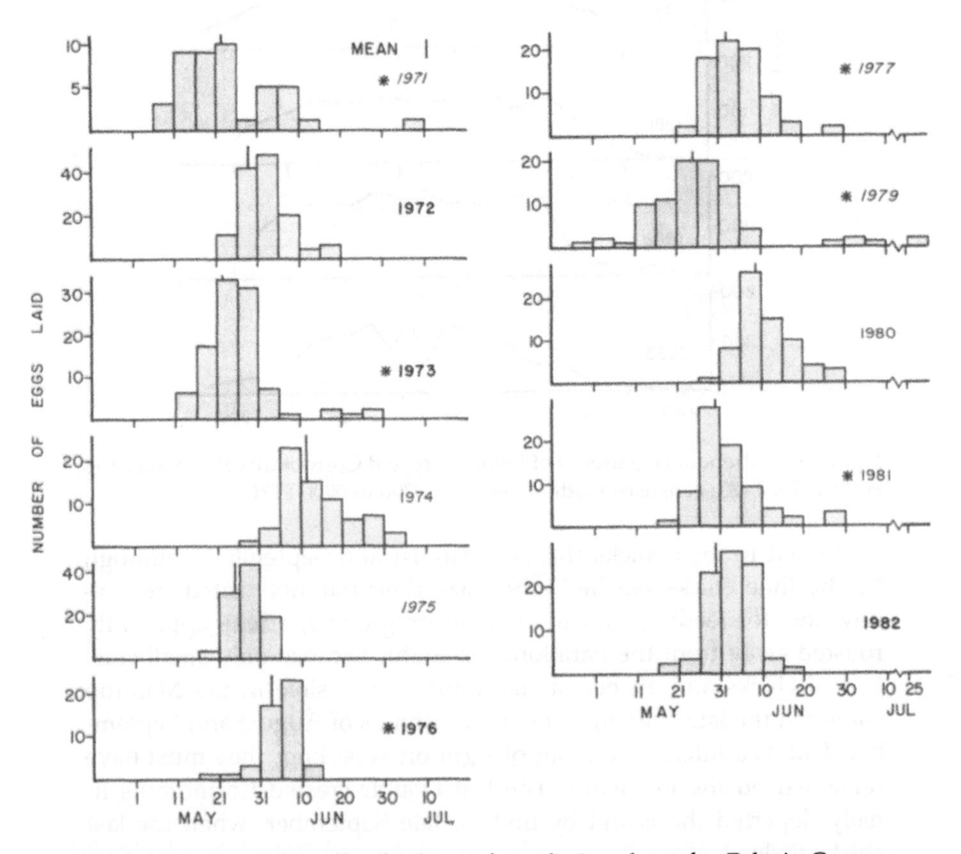

Figure 6.6. Frequency distributions of egg-laying dates for Pelagic Cormorants at the Southeast Farallon study colonies, 1971–82; distributions marked by an asterisk (*) are significantly peaked and skewed (t test, $p < .05$).

TABLE 6.2

Mean clutch-initiation dates, Pelagic Cormorants, 1971–83

1973	*1971*	*1975*	*1979*	**1972**	1981	**1982**	*1977*	**1976**	1980	1974
5/22	5/23	5/26	5/26	5/29	5/31	5/31	6/4	6/4	6/10	6/12

NOTE: Lines connect similar means; SNK test, $p > .05$. No eggs were laid in 1978 or 1983.

peaks by mid-June (Drent et al. 1964). Pelagics on the west side of Vancouver Island, however, lay most eggs between early June and early July (Hatler, Campbell, and Dorst 1978). In southern California, Penhale (1972) observed the first eggs on 18 April. At the Farallones, mean clutch-initiation dates were similar in most years but significantly later in 1974 and 1980 (Table 6.2).

The span of Pelagic Cormorant clutch initiation ranged from as little as two weeks, in 1975, to as great as three months, in 1979 (Figure 6.6). In all years, however, Pelagic Cormorants laid more than 50% of all eggs within less than two weeks. Laying was significantly skewed toward the early season in years with extended laying periods, such as 1973, 1977, 1979, and 1981, primarily because birds continued to build new nests for some time after the laying peak. During 1976, however, egg laying started slowly and most birds laid late. In general, laying dates were positively skewed in years of early egg laying. In years of late egg laying they approximated a normal distribution.

Like Brandt's Cormorants, Pelagics typically laid eggs two days apart, although the interval ranged from one day to the extreme of 24 days, as was witnessed once between third and fourth eggs (Figure 6.7). There was a tendency for intervals to be longer for late-laid eggs.

The Pelagic Cormorant's incubation periods averaged 29.5 days for all eggs (Figure 6.8), but, again as with Brandt's Cormorants, the first eggs in the clutch took longer to hatch than the later ones. The average length of incubation periods for 53 first, 38 second, 17 third, and 2 fourth eggs were, respectively, 30.1 ± 1.0, 29.3 ± 1.2, 28.5 ± 1.6, and 27.5 ± 0.7 days. Differences were significant between first and second eggs and second and third eggs ($t > 2.0$, $p < .05$), but not between third and fourth eggs ($t = 0.85$).

Clutch size

In 1978 and 1983, the two years of extremely warm water and delayed upwelling (Chapter 2), Pelagic Cormorants did not lay eggs. In

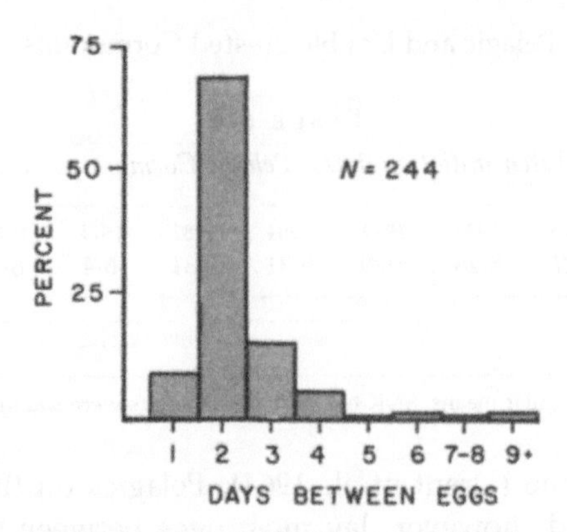

Figure 6.7. Number of days between successive eggs laid by Pelagic Cormorants; all clutch sizes and years combined, 1972–79.

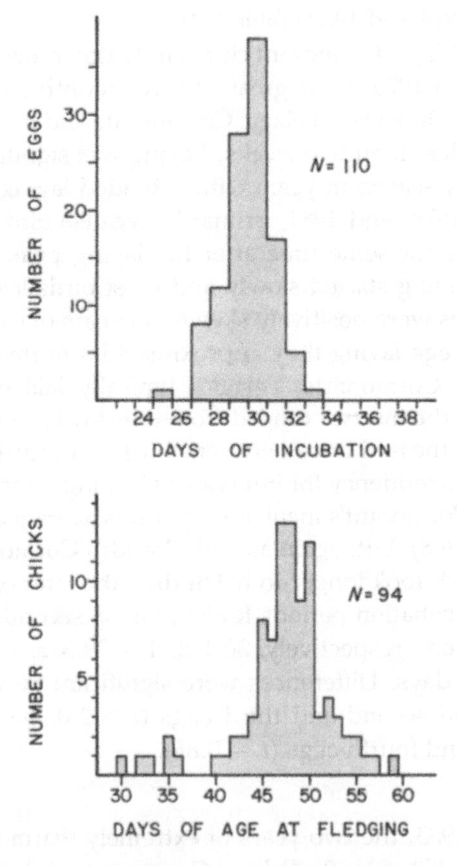

Figure 6.8. Frequency distributions of Pelagic Cormorant incubation and nestling periods; all nests and eggs combined, 1972–79.

1978 this happened despite 83% (n = 29) occupancy of study sites, compared to 24% (n = 34) occupancy of sites in 1983 (Appendixes 6.1 and 6.2). The complete failure to lay eggs during certain years suggests that Pelagics were then energetically incapable of breeding. Presumably their failure to lay was a result of food shortages prior to the laying period that caused their body condition to remain below the point necessary to lay eggs. The regularity of ENSO episodes, as noted in Chapter 2, suggests that most individuals in this population skip breeding several times during their lifetimes, if they survive the lean years.

In a comparison of clutch size in years when eggs were laid, Pelagic Cormorant clutches were fairly consistent (Table 6.3). Only in 1976, another oceanographically aberrant year, did clutches consist of fewer than three eggs. Clutches that year were significantly smaller than those in all other years in which eggs were laid. For all years combined (except nonlaying years), clutch size averaged 3.4 eggs, and over 90% of all clutches contained either three or four eggs. Two-egg clutches constituted a significant percentage of clutches only in 1976 (Appendix 6.1).

Pelagic Cormorants laid progressively smaller clutches as time passed in most breeding seasons. This pattern was especially evident when data from all years were combined (Figure 6.9). Overall, early

TABLE 6.3

Mean clutch size and number of fledglings per laying pair and per successful pair, Pelagic Cormorants, 1971–83

Clutch size

1976	1980	1974	*1977*	*1979*	**1973**	**1982**	1981	*1975*	**1972**	*1971*
2.58	3.05	3.17	3.30	3.37	3.45	3.61	3.65	3.65	3.77	3.82

Fledglings per laying pair

1976	1980	**1982**	1974	**1972**	**1973**	*1971*	*1975*	*1979*	*1977*	1981
0.0	0.0	0.13	0.42	0.60	0.66	1.08	1.81	1.86	2.17	2.43

Fledglings per successful pair

1972	1974	**1973**	**1982**	*1971*	*1975*	*1979*	*1977*	1981
1.11	1.43	1.46	1.50	2.00	2.24	2.48	2.58	2.83

NOTE: Lines connect similar means; SNK test, $p > .05$. No eggs were laid in 1978 or 1983; no nests were successful in 1976, 1978, 1980, or 1983.

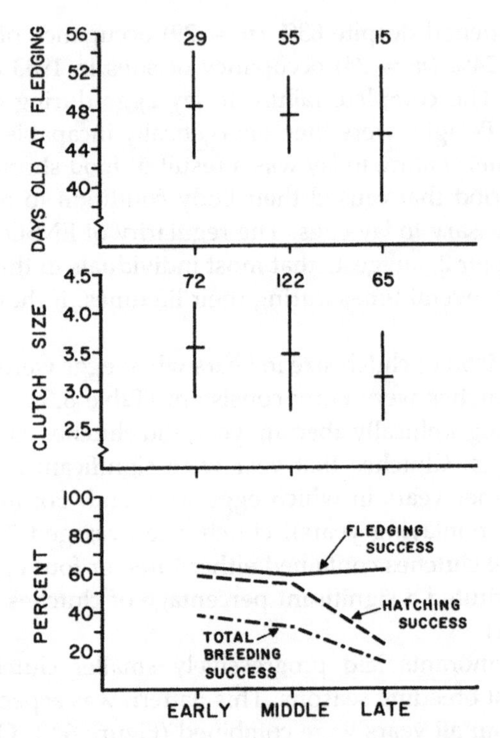

Figure 6.9. Comparison of success, clutch size, and chick age at fledging for early-, mid-, and late-season nesting Pelagic Cormorants, 1971–82. Horizontal lines indicate means; vertical lines, standard deviations; numbers at top, sample sizes.

clutches averaged 3.6 ± 0.6 eggs (n = 72), middle clutches 3.5 ± 0.7 (n = 122), and late clutches 3.2 ± 0.6 (n = 65) eggs.

Breeding success

Of all Farallon species, the Pelagic Cormorant experienced the most extreme annual variation in breeding effort and success (Appendix 6.2). In four of 13 years (1976, 1978, 1980, and 1983), Pelagics either failed to lay eggs or experienced complete nesting failure (this happened again in 1986; Appendix 3.1). In four other years (1972, 1973, 1974, and 1982), fewer than one of five eggs laid resulted in a fledged chick. In contrast, the species did quite well in five years (1971, 1975, 1977, 1979, and 1981), fledging two or more chicks per nest (Table 6.3). During the three years for which we had data on diet, 1975–77 (Chapter 3), the number of chicks produced per pair was highly correlated with the proportion of juvenile rockfish in the Pelagic's diet (r_S = 1.0).

When conditions became unfavorable, Pelagic Cormorants deserted their nests readily, either during incubation or during the nestling period. In fact, desertions were the largest single factor affecting hatching and fledging success. The timing of nest desertions varied among the years of poorer breeding: in 1974 they occurred mostly between 8 and 15 July, in 1976 between 10 and 16 June, in 1980 between 5 and 11 July, and in 1982 during two periods, 25 June to 5 July and 21 to 26 July. During both 1980 and 1982, we noticed that sea-surface temperatures increased 2 to 4° C immediately prior to periods of nest desertion. At the same time as the increase in temperatures during the last week of June 1982, we also noticed a decline in numbers of juvenile rockfish returned to chicks by many island birds. The Pelagic Cormorants seemed particularly sensitive to these changes; adults began spending much longer periods away from nest sites (Chapter 3). During July 1982, all chicks died at 15 of the 17 sites at which chicks had been present.

Similar abandonment and breeding failure were reported by Penhale (1972) for Pelagic Cormorants nesting near Point Buchon, California, about 330 km south of Southeast Farallon. In 1970, 30 pairs in that colony laid 66 eggs, but only two eggs hatched and both chicks died. It appears, therefore, that conditions in at least the southern part of this species' range may frequently inhibit reproduction. This naturally leads to the question of how Pelagic Cormorant populations in California maintain themselves if they regularly experience such unfavorable conditions. Little is known about this species' reproductive biology in the northern part of its range. During a two-year study in Alaska, Baird et al. (1983) noted similar reproductive success during both years.

Through the entire 13 years of study at the Farallones, Pelagic Cormorants hatched only 47.9% (n = 891) of all eggs laid. Eggs lost before term, mostly through nest desertion, amounted to 30.9% of all eggs laid. An additional 12.1% of eggs disappeared on or about their expected hatch date; chicks died either while hatching or a day or two afterward. Nine percent of eggs were incubated well beyond their normal incubation period and did not hatch because of either infertility or embryo death.

In general, Pelagic Cormorants did not lay another clutch if they lost their first one, particularly in years when many abandoned eggs during incubation. Of 80 pairs that lost all their eggs prior to hatching, only five attempted replacements: one in 1971, two in 1979, one in 1981, and one in 1982. In total, these five replacement clutches contained 13 eggs, and from them seven chicks hatched and six fledged.

TABLE 6.4

Spearman correlation coefficients, Pelagic Cormorant nesting, 1971–83

Variable	Row number			
	1	2	3	4
1. Clutch-initiation date	—			
2. Clutch size	.7159[a]			
3. Hatching success	.3545	.3886		
4. Fledging success	.3341	.4000	.7705[a]	
5. Fledglings per laying pair	.4636	.4705	.8636[a]	.9614[a]

[a] $p < .05$.

Within years, the hatching success of late clutches was lower than that of early and mid-season clutches (Appendix 6.3, Figure 6.9). Late birds attained hatching success equal to that of earlier breeders in only three years—1975, 1977, and 1979—all of which were highly successful years for Pelagic Cormorants. Fledging success of late nesters, however, was equal to that of earlier nesters. Mid-season nesters were only slightly less successful in hatching eggs and fledging chicks than early birds. Overall, late-nesting birds produced fewer chicks per pair than other cormorants.

A rank comparison of the major breeding parameters of Pelagic Cormorants from 1971 to 1983 reveals that the number of chicks fledged per pair was largely a function of hatching and fledging success (Table 6.4). Clutch size was a function of clutch-initiation date.

Chick development

Like those of Brandt's (Chapter 5), the chicks of Pelagic Cormorants are altricial, nidicolous, and, when small, dependent on their parents for warmth, protection from gulls, and prevention of harassment by other cormorants. The chicks grow rapidly, reaching adult size and becoming completely feathered in their juvenal plumage in less than two months after hatching. Unfortunately, because of the likelihood of disturbance, we were able to obtain only minimal information on chicks.

The behavior of Pelagic Cormorant chicks is different from that of the other two cormorants because of the species' propensity for nesting on cliffs. Pelagic chicks cannot afford the luxury of wandering from their nest sites prior to attaining flight: they would risk falling off the cliff. Even when approached by humans, at least to the minimal extent to which we disturbed them, they resolutely remained in their nests. Most chicks first departed their nest sites when between 45 and 50 days old (mean 47.4 ± 4.9, range 30–59 days, n = 101;

Figures 6.8 and 6.9). A small number of chicks fell or departed from their nests at a younger age, between 30 and 40 days, but were able to return to their nests and fledge successfully. At Mandarte Island, British Columbia, Drent et al. (1964) recorded very similar ages, ranging from 42 to 58 days, for first departures of Pelagic chicks.

REPRODUCTION IN THE DOUBLE-CRESTED CORMORANT

Egg laying

Although we do not know exact laying dates for Double-crested Cormorants, our observations of the Maintop colony allow some reasonable guesses about chronology. Double-crested Cormorants were the earliest to breed among surface-nesting species at the Farallones. We observed this species sitting on well-built nests as early as mid-March, and in most years first nests were completed and birds appeared to be incubating by early to mid-April (Table 6.1). This is earlier than Double-crested Cormorants at Mandarte Island, where first eggs were laid in late April and most eggs were laid in May (Drent et al. 1964).

This species' unique breeding chronology relative to other Farallon birds may be related to its very different diet, which consists almost entirely of neritic and estuarine fish captured along the mainland coastline and in bays and lagoons (Chapter 3). Ainley, Anderson, and Kelly (1981; Chapter 3) reported that one prey species, Shiner Surfperch, made up 78.6% of the Double-crested Cormorant's diet during the breeding season, whereas juvenile rockfish, the primary item in the diets of most other piscivorous birds at the Farallones, constituted only 1.3% of the diet.

Laying chronologies differing among sympatric species are apparently typical of cormorants. For example, Erskine (1972) and Ross (1977) reported that Double-crested Cormorants nest appreciably later than Great Cormorants in the maritime provinces of Canada. Double-crested Cormorants began laying about one month before Pelagics at Mandarte Island, British Columbia (Drent et al. 1964), and Cody (1973) observed a similar pattern in Washington. The two cormorant species in Great Britain also have different laying schedules (Lack 1945). Despite the measurable differences in laying phenology among Farallon species, however, breeding seasons do overlap appreciably (Chapter 12, Figure 12.1).

Double-crested Cormorants laid earlier in the early 1970's than in the later years of the study. Prior to 1977, the first well-built nests appeared in mid- to late March, whereas from 1977 to 1983 these sites did not develop until early to mid-April (Table 6.1). Also, in both 1971 and 1972 we observed fledgling birds wandering about the island, apparently independent of their parents, in mid-July, at least one week earlier than during any other year (Table 6.1).

Clutch size

We have no information on Double-crested Cormorant clutch sizes. On the basis of the number of chicks we observed—frequently three or four per site—it appears that clutch size at the Farallones is similar to that observed elsewhere. McLeod and Bondar (1953) observed a range from 2.4 to 3.6 eggs per clutch in different years. They also observed some nests with up to nine eggs, possibly laid by more than one female. Other studies recorded three to four eggs (Cline and Dornfeld 1968), 3.2 eggs (Vermeer 1969b), and 3.8 eggs per clutch (Mitchell 1977).

Breeding success

We were equally unable to determine the exact number of chicks fledged each year by Double-crested Cormorants, although we could estimate this quantity on the basis of chicks visible in the colony and the number of chicks banded (Table 6.1). The Double-cresteds appeared less variable in their chick production than the other cormorants, possibly because of their reliance on a different prey source (see Chapter 3). They generally fledged about two chicks per nest, as observed in most other studies (see introductory comments above). Their poorest year was 1983, when they abandoned nearly all nests during two periods in late May and early June. The other years in which they produced low numbers of chicks, 1976 and 1978, also were years of unusual oceanographic conditions. This shows that even the estuarine and nearshore prey of the Double-crested Cormorant may have suffered the deleterious effects of ENSO.

Chick development

We know little about the life of Double-crested Cormorant chicks on the Farallones because of the great distance from which we observed them. As do those of Brandt's Cormorant, chicks of the Double-crested Cormorant wander from their nest sites before fledging, but still use the nest as a spot to meet parents bringing food. Once large enough to leave their nests, Double-crested chicks stand together in large groups within nesting areas on the summit and slopes of Maintop, waiting for their parents to return from feeding trips.

COMPARISON OF BREEDING STRATEGIES

We never observed adult Double-crested Cormorants feeding near the Farallones; even while raising chicks, they daily flew a minimum round trip of 70 km, possibly much more, to estuarine and other coastal habitats to obtain their prey (Chapter 3). The Double-cresteds must make these long flights because a suitable nesting habitat does not exist closer to the mainland coast. Despite their long commute, however, they consistently fledged chicks except during the extreme ENSO year of 1983. The Double-cresteds have a foraging strategy unique among Farallon birds, one that frees them from much of the interannual variation experienced by other populations so dependent on juvenile rockfish for successful reproduction. It is unfortunate that we do not have detailed information on the growth of their chicks, because this character may be strongly influenced by the long periods they stay away from nests on feeding trips. And, like that of the murre (Chapter 8), it is possible that the population of the Double-crested Cormorant now differs in various characteristics, such as the timing of laying and productivity, from the period when its population numbered several thousand and sardines abounded in coastal waters.

In contrast to the Double-cresteds, the Pelagic Cormorants are almost completely dependent for successful breeding on the availability of juvenile rockfish. All years in which they do well are ones when juvenile rockfish dominate diets (Chapter 3). During warm-water years, or years when juvenile rockfish declined during mid-season, Pelagic Cormorants were quick to abandon breeding efforts, presumably because food availability fell below the threshold required for breeding. They were critically handicapped by their inability to use alternate feeding habitats. Despite their reputation as neritic, benthic feeders (Ainley, Anderson, and Kelly 1981, Robertson 1974), Pelagic Cormorants at the Farallones are, most interestingly, strongly dependent on midwater-shoaling juvenile rockfish for successful reproduction.

7

Western Gull

Teresa M. Penniman, Malcolm C. Coulter,
Larry B. Spear, and Robert J. Boekelheide

The Western Gull is a large, white-headed member of the genus *Larus*. It is an ideal subject for study because of its preference for island terraces, which are easily accessible, and its capacity for coexisting with humans. In fact, it is among the six most researched gull species of North America (Southern 1987), with most of this work having been carried out during the last 15 years. The collective knowledge of the reproductive biology of the Western Gull has provided a valuable context in which our results can be viewed.

Western Gulls nest in a narrow zone along the Pacific coast from southwestern British Columbia to central Baja California (American Ornithologists' Union 1983). Within this region, they breed mainly on islands and offshore rocks, although a few mainland colonies also exist (Sowls et al. 1980, Pierotti and Bellrose 1986, Pitman et al. ms, Speich and Wahl 1989). During the nonbreeding season, Western Gulls disperse both north and south of their breeding colonies (reviewed by Coulter 1975; Spear 1988a).

In northern Oregon, the breeding distribution of this species overlaps that of the Glaucous-winged Gull, *L. glaucescens*. From there to southern British Columbia these two species breed in mixed colonies and extensive hybridization occurs (Hoffman, Wiens, and Scott 1978). To the south, in the Gulf of California, the Western Gull is replaced by the Yellow-footed Gull, *L. livens*, which until recently was considered a race of *L. occidentalis* (American Ornithologists' Union 1983). The breeding distributions of the two currently recognized races of the Western Gull, *L. o. occidentalis* and *L. o. wymani*, meet in central California (American Ornithologists' Union 1957). *L. o. occidentalis* nests on the Farallon Islands, and its numbers there constitute about

50% of the California Western Gull population and about 40% of the world population of *L. o. occidentalis* (Table 7.1). This is the single largest colony of any *Larus* gull species on the North American Pacific or Bering Sea coast; only a few Alaskan colonies of the Black-legged Kittiwake, *Rissa tridactyla*, are larger (cf. Sowls, Hatch, and Lensink 1978).

Although gull populations have been much researched, most studies have spanned only a few years. Our research on the Farallones is unique in three regards: (1) we monitored gull breeding for an extended period, 13 continuous years, 1971–83, during which oceanic conditions varied greatly (Chapter 2), (2) we simultaneously collected information on diet during many of those years (Chapter 3), and (3) information collected during the same study period on the diving seabirds breeding at the Farallones provided a comparative measure of seabird response to oceanic variability. These elements of our perspective allowed us to examine the seasonal and annual variability of breeding success of the Western Gull in the context of information on foraging ecology.

Unlike that of many other Farallon seabirds, the reproductive ecology of the Western Gull is fairly well known because of studies by Schreiber (1970), C. A. Harper (1971), Coulter (1973, 1977), Hunt and Hunt (1973, 1975), Pierotti (1976, 1981), Briggs (1977), Winnett (1979), Ewald, Hunt, and Hunt (1980), T. Harvey (1982), Bellrose (1983), and Pierotti and Bellrose (1986). Coulter and Pierotti worked on Western Gulls at the Farallones; Pierotti, Briggs, Harvey, and Bellrose worked on other colonies in central California (Año Nuevo Island, Moss Landing; *L. o. occidentalis*); and Schreiber, Harper, Hunt et al., and Winnett on colonies in southern California (San Nicolas Island, Bird Rock, and Santa Barbara Island; *L. o. wymani*).

TABLE 7.1

Estimated numbers of Western Gulls nesting in North America, 1965–85

Area	Breeding pairs	Sites	Source
Washington	4,000	11+	Speich & Wahl 1989, Spear et al. 1987
Oregon	900	9	Varoujean 1979
California	25,470[a]	159	Sowls et al. 1980
Mexico	> 11,150	> 5	DeLong & Crossin ms
Total	41,520	184+	

NOTE: Most recent data.

[a] The Farallon population (12,750 pairs) constitutes 50% of this figure.

METHODS

Censuses of the entire Farallon breeding population were conducted in nine years: 1970, 1972, and 1980–86. Gulls were counted individually, either through binoculars or with the naked eye. The 1970 and 1972 counts did not include birds on West End, which is isolated from the main island by a narrow surge channel (Chapter 1; Figure 7.1), and were conducted over a three-day period. More recent counts were conducted on just one day. Total population estimates were adjusted for birds not present during the count period. We made these adjustments using information from study plots where we knew the number of active nests. We counted the number of birds present in each plot and determined the percentage of breeding birds absent during the count period.

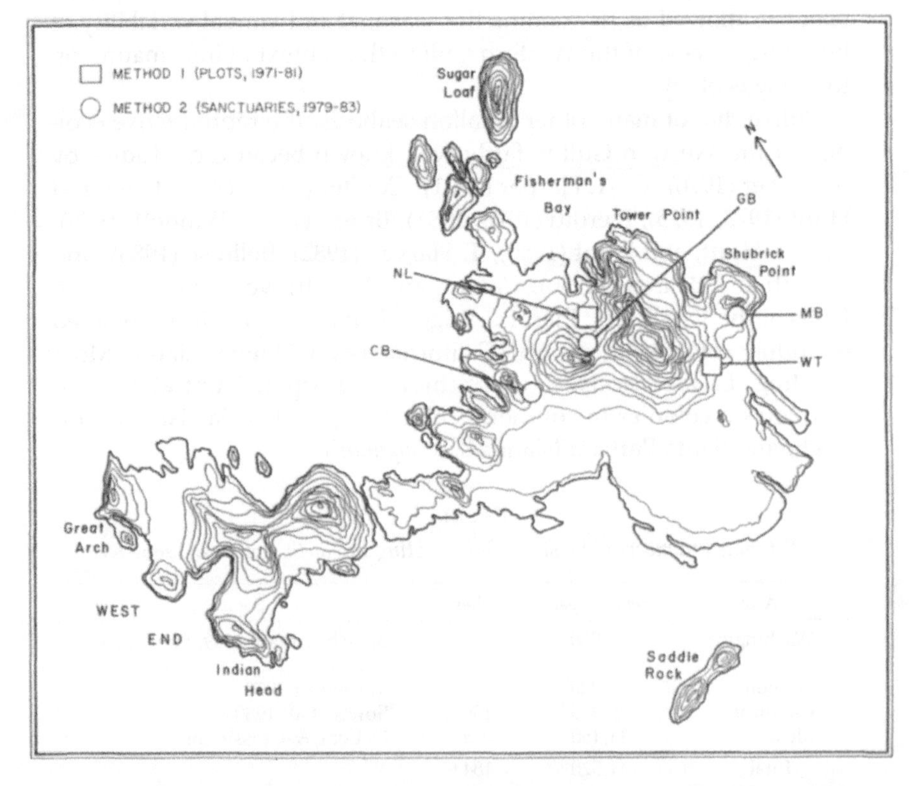

Figure 7.1. Locations of plots NL and WT and sanctuaries CB, GB, and MB used in the study of Western Gulls, 1971–83.

Diurnal attendance patterns were monitored daily from 18 April 1971 to 24 April 1972. Several times each day we counted the number of adults in three plots established for this purpose. Results were added for the plots and then averaged for five three-hour periods over seven-day intervals: 0600 to 0900, 0901 to 1200, 1201 to 1500, 1501 to 1800, and 1801 to 2100. During November, December, and January, it was too dark after 1800 to make counts. Evening attendance patterns were also monitored in one plot from May 1979 to December 1983 by counts of adults at dusk.

During 1972, 1974, 1978, 1979, and 1983, we measured length and breadth of eggs and estimated volume by using the equation developed by F. W. Preston (1974):

$$\text{volume} = \pi/6 \times \text{length} \times \text{breadth}^2.$$

In these calculations, we included only three-egg clutches and only those for which the laying sequence was known. The sample size for 1972, 15 clutches, was insufficient for within-season comparisons but sufficed for other analyses. We made comparisons of length, breadth, and volume, but because patterns were similar, we limit discussion to egg volume.

For monitoring of breeding phenology and success, early, intermediate, and late nesters were categorized by the method used for Brandt's Cormorants (Chapter 5). We pursued two methods of data collection, method 1 (plots) and method 2 (sanctuaries). We began using the first method in 1971 by establishing two study plots (plots WT and NL; Figure 7.1). Boundaries were marked with wooden posts. Beginning 20 April, we checked plots at three-day intervals and mapped all nests containing eggs. Nest were marked with small numbered wooden stakes. Laying order within a clutch, when known, we indicated by marking eggs with felt-tip pens. Chicks were web-punched at hatching with individual punch codes. When chicks were sufficiently large, we placed a metal USFWS band on one leg and a plastic colored band on the other, using a different color or leg combination each year. We ended visits to a plot when all chicks had been banded. At the end of the breeding season, plots and neighboring areas were thoroughly searched for dead banded chicks. Those not found were assumed to have fledged. We excluded from analysis the few nests in which eggs or chicks were stolen by adults as a result of our disturbance in the colony. Both plots were followed from 1971 to 1980, and plot WT was monitored for one additional year, 1981.

In 1979, we turned to method 2 by delimiting two "sanctuary" areas (plots CB and MB; Figure 7.1) that were observed only from blinds by using binoculars and patience to determine the contents of nests. In 1981, we added a third area (plot GB). Boundaries and nest locations were defined on maps. Observations were made at three-day intervals in 1979 and 1980 but daily from 1981 to 1983. Observations continued into mid-August or until all surviving chicks in a plot had fledged.

Concern for the possibility that observer interference lowered breeding success was our main motivation for changing methods (see Hunt 1972, Robert and Ralph 1975, but also Pierotti 1982). The terrain and location of sanctuaries MB and GB were similar to those of plots WT and NL, respectively. We overlapped use of MB and WT during three years, 1979–81, and then compared mean dates of clutch initiation, clutch size, and hatching and fledging success. No significant differences existed between results from the two techniques, indicating that results were comparable for between-year analyses.

All four plots had cover for chicks, particularly in the form of rocks and rocky outcrops; the southeast plots contained more rocky substrate, whereas the northern plots had scattered rocks on loose soil. Most of the vegetation (primarily Farallon Weed) had dried up by the onset of egg laying; during the pre-egg period vegetation was more dense in the northern plots than in the southeastern plots, the MB area having very little vegetative cover. The southeastern plots were more sheltered from the prevailing northwest winds; the terrains of these plots were more heterogeneous than those of the northern plots.

The terrain and location of the CB sanctuary (Figure 7.1) differed considerably from those of any of the other four areas; it was flat and had very little rock cover and little protection from the wind. Therefore, data from CB were not included in interannual comparisons, although they were used to examine breeding chronology relative to nesting density. Thus, all our information on nesting biology is based on data from the WT and NL plots, 1971–80, and from the MB and GB sanctuaries, 1981–83.

Chicks were weighed during five years, 1972–75 and 1977. In all years except 1972, we examined the weights of chicks at hatching, at 20 days, and at 40 days to investigate annual variation. In 1972, chicks were followed only to about day 25; therefore, data from that year were not included in some of the analyses. Weight at 40 days is 88% of final weight (Coulter 1977), and we have used this as a measure of final weight. In these analyses we consider only chicks that survived to fledging. We fitted the weight data for each individual chick by using a logistic equation. To investigate differences between years, we combined data from individual chicks for each year and examined two parameters, mean growth rate constant (*K*) and mean asymptotic weight.

For the plots, dead chicks were grouped into two categories on the basis of age and whether they were banded: dead before banding (age <16 days) and after banding (>15 days). The sanctuary data included only nests where dates of laying and dates of death were known. Because we did not have individual chicks marked, we used the average hatching date of eggs in a clutch when calculating a chick's age at death.

FARALLON POPULATION

The Western Gull population on the Farallones has been greatly affected by human activities (Ainley and Lewis 1974). The earliest estimate, 20,000 gulls, was extrapolated from Heermann's (1859) account. Earlier, numbers were likely smaller as a result of large pinniped populations using what subsequently became gull nesting habitat: these pinnipeds were exterminated in the early 1800's (Doughty 1974; Chapters 1 and 12). After humans and domestic animals arrived in the mid-1800's, the gull population dropped to a low of 6,000, as a result of concerted efforts to reduce gull numbers (gulls "competed" with humans for murre eggs). Emerson (1904) and Dawson (1911) noted that gulls bred primarily in small colonies on West End (Figure 7.1) and on the steep ridges of South Farallon. Gull numbers were not documented between the early 1900's and the late 1950's, but during that time the population rebounded.

Counts from 1959 to the present have exhibited a remarkable constancy, ranging from 22,000 to 25,500 breeding birds. This stability is apparently due to a balance of recruitment and mortality and a low incidence of emigration (Spear et al. 1987). Recruitment is a function of territory availability. Relatively stable populations of Western Gulls have also been reported at the southern California colonies (Hunt et al. 1981), but the factors affecting population structure there are not necessarily the same as those operating at the Farallones. In the Farallon colony, increased mortality did result from the ENSO of 1982–83; although population size did not change, turnover in the population of banded birds we observed between 1979 and 1986 increased (Spear et al. 1987; PRBO, unpubl. data). The higher mortality with little apparent change in breeding numbers indicates that a pool of reproductively mature individuals was available, and from that pool gulls took over vacated breeding territories. The report by Sowls et al. (1980) of an increase to 32,000 is an overestimation.

The gulls' nesting distribution on the island has shifted and expanded dramatically since it was first mapped in 1959 (Bowman 1961; Figure 7.2). Increasing populations of pinnipeds have forced breeding gulls out of peripheral areas (Chapter 1, Figure 1.4), and expanding populations of cormorants and murres have also caused shifts (Chapter 12). Today approximately 35% of the nesting territories are on the southwest marine terrace (Figure 7.2), a flat area with little rock cover. The rocky slopes on the main island provide nesting habitat for 45% of the pairs. The remaining territories are distributed on the island's perimeter (13%), on small islets (3%), and on West End (4%).

Nesting densities (no. sites / 100 m²) differed among plots but, as

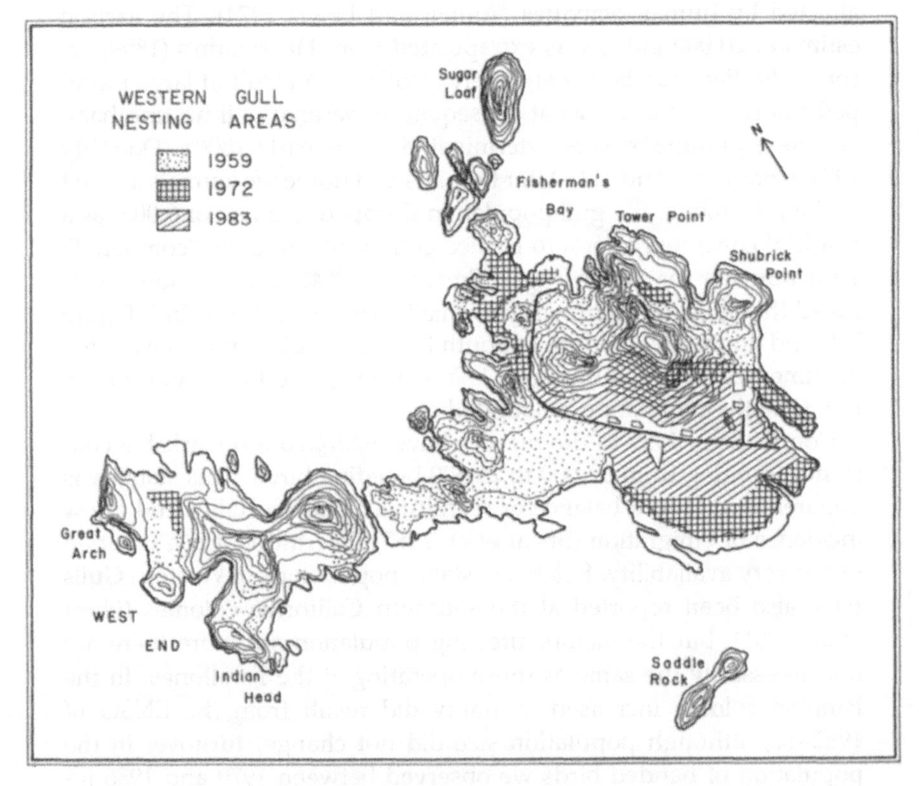

Figure 7.2. Expansion of gull nesting area over a 25-year period, 1959–83; data for 1959 from Bowman (1961).

expected in a saturated habitat, varied little annually. The southeast plots, WT and MB, had consistently higher densities (7.1 and 8.2, respectively) than the north plots, NL and GB (5.7 and 7.0, respectively); densities were even lower in plot CB (5.0). Densities were significantly different between WT and NL ($t = 3.97$, DF = 16, $p < .001$) and between MB and GB ($t = 3.83$, DF = 4, $p < .02$). Differences in nesting density concurrent with a pool of nonbreeding adults (Spear et al. 1987) is further evidence that recruitment is limited by the availability of suitable nesting sites. Porter and Coulson (1987) noted that nonbreeding kittiwakes are recruited into the breeding population when "attractive" central ledges are created, but avoid using available peripheral sites. The rocky substrate and leeward location of the southeast plots (see Methods) either may be more attractive (see Pierotti 1982, 1987, for similar results in Herring Gulls, *L. argentatus*) or may provide more visual boundaries, thus allowing for nesting at greater densities (Burger 1977).

Western Gulls at other colonies nest in much lower densities (Pierotti 1976). Densities on the Farallones are greater than those at Santa Barbara Island by a factor of 10 to 60 (Hunt et al. 1981), depending upon area and year. Apparently, the rich marine prey base of the Gulf of the Farallones is so favorable for breeding that it outweighs any potential cost of overcrowding, i.e., greater need for territorial defense.

OCCURRENCE PATTERNS

Nonbreeding distribution

We banded about 2,000 gull chicks each year beginning in 1970. Many banded birds have been recovered, but because the dispersal patterns of this population have been documented elsewhere, we will not repeat all the details here (see Coulter 1975, Spear 1988a). The following patterns are evident: (1) after the breeding season, the majority of fledglings move north; (2) they then move south for their first winter, and north in the following spring; (3) this pattern is repeated as individuals age, but the distance moved away from the island becomes less and less; and (4) during movements in its prebreeding years, an individual finds a spot to which it returns year after year during the species' postbreeding period—Spear (1988a) termed this locality the "vacation spot." Western Gulls in southern California do not appear to disperse as far as Farallon gulls (Hunt et al. 1981) but follow a similar pattern of shorter movements with age.

Colony attendance

The Western Gull is practically a year-round resident on the Farallones (Figures 7.3 and 12.1). In late July, at the end of the breeding season, numbers of adults begin to decline, and by mid-September all adults and most young have departed. During early fall, adults visit their vacation spots along the mainland coast (Spear 1988a), but by late October or early November they begin to reoccupy territories. Individuals whose vacation spots are relatively close may alternate between there and the island. Numbers increase gradually through the next several months, and by mid-March the entire breeding population is consistently present on a daily basis. First arrival dates varied among years by one month (Figure 7.3), and visitation by large numbers of gulls varied even more: ten-day means of the gull population exceeded 100 by 15 November in 1980 but not until 5 February in 1983 (a period of extremely warm water; Chapter 2). In other years, peak attendance was reached in December and January. The duration of the pre-egg occupancy period is longer at the Farallones than at other colonies, a pattern consistent with the apparent crowding and important requirement for securing a territory. Site occupation on Santa Barbara Island begins in January (Hunt et al. 1981), and at Moss Landing gulls return just prior to egg laying (Bellrose 1983, but see T. Harvey 1982 for different results).

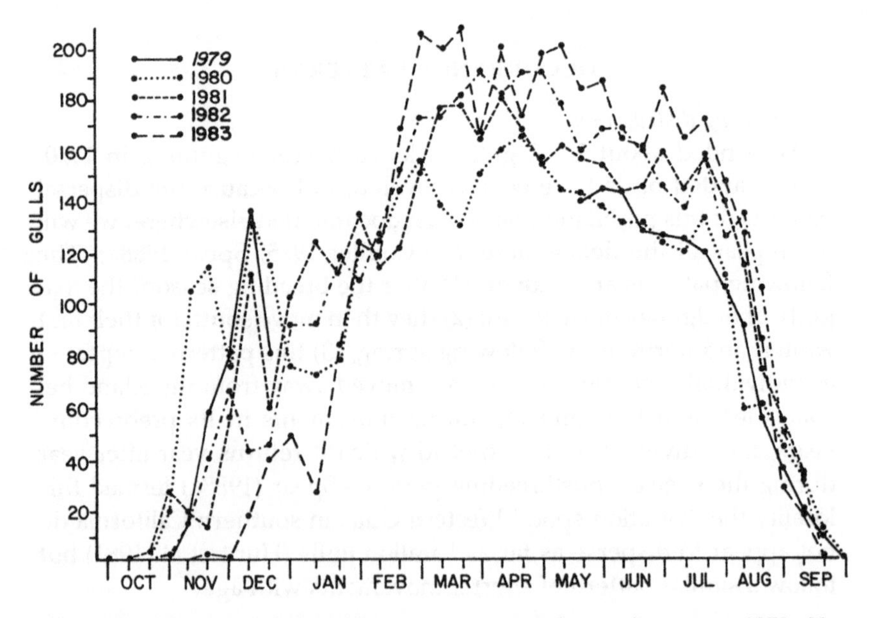

Figure 7.3. Numbers of gulls present on territory during five years, 1979–83.

Diurnal attendance changed with the stage of breeding (Figure 7.4). From November to January, gulls were on territory in the late evening, night, and early morning, but were generally absent during mid-day. It was not until late January of most years that gulls occupied territories in the late afternoon or early evening. Although this began the trend of daytime occupation, often only one member of a pair was present. During April, just before eggs were laid, females and males were on territory 94 and 55% of daylight hours, respectively (Pierotti 1981). Thus, the lower mid-day numbers probably reflected an absence of males. Female gulls on the Farallones spent considerably more time on territory during this period than do those on Santa Barbara Island (Pierotti 1981, Hunt et al. 1984). This difference was likely due to the increased intruder pressure observed on the Farallones, where it is more cost-effective to defend a site than to oust an established intruder (Pierotti 1981).

When egg laying began, the partial mid-day exodus ended. One gull of every pair remained on territory to incubate eggs and, later, to protect young chicks. By July, adults spent less time on territory and more time foraging, returning to the colony only at night. This phase of activity accounted for the reduction in numbers at all hours, except the evening, beginning the week of 27 June to 3 July.

Reestablishment of territories during the winter and mid-day occupation during the prelaying period probably helped gulls to hold a breeding site for the coming season. Diurnal occupation varied annually depending on food availability (Pierotti 1981), and the observed variation in winter occupation patterns may have been similarly related.

EGG LAYING

If all plots among the 13 years are considered, laying dates of first eggs ranged from 20 to 27 April (Figure 7.5), and mean dates of clutch initiation (first clutches) ranged from 3 to 14 May (Table 7.2). Such a small range in laying dates was in sharp contrast to all other Farallon seabirds. In several species, phenology was a function of variation in sea-surface temperatures during the prelaying period (see other chapters). Mean temperatures in January, February, and March were only weakly associated with mean laying dates in the Western Gull between 1971 and 1983 (Spearman rank correlations, $p < .2$).

The mean length of the laying period for all years was 40.0 ± 5.5 days (range 25 to 46 days). Mean clutch-initiation dates and length of the laying period corresponded closely with those found at other

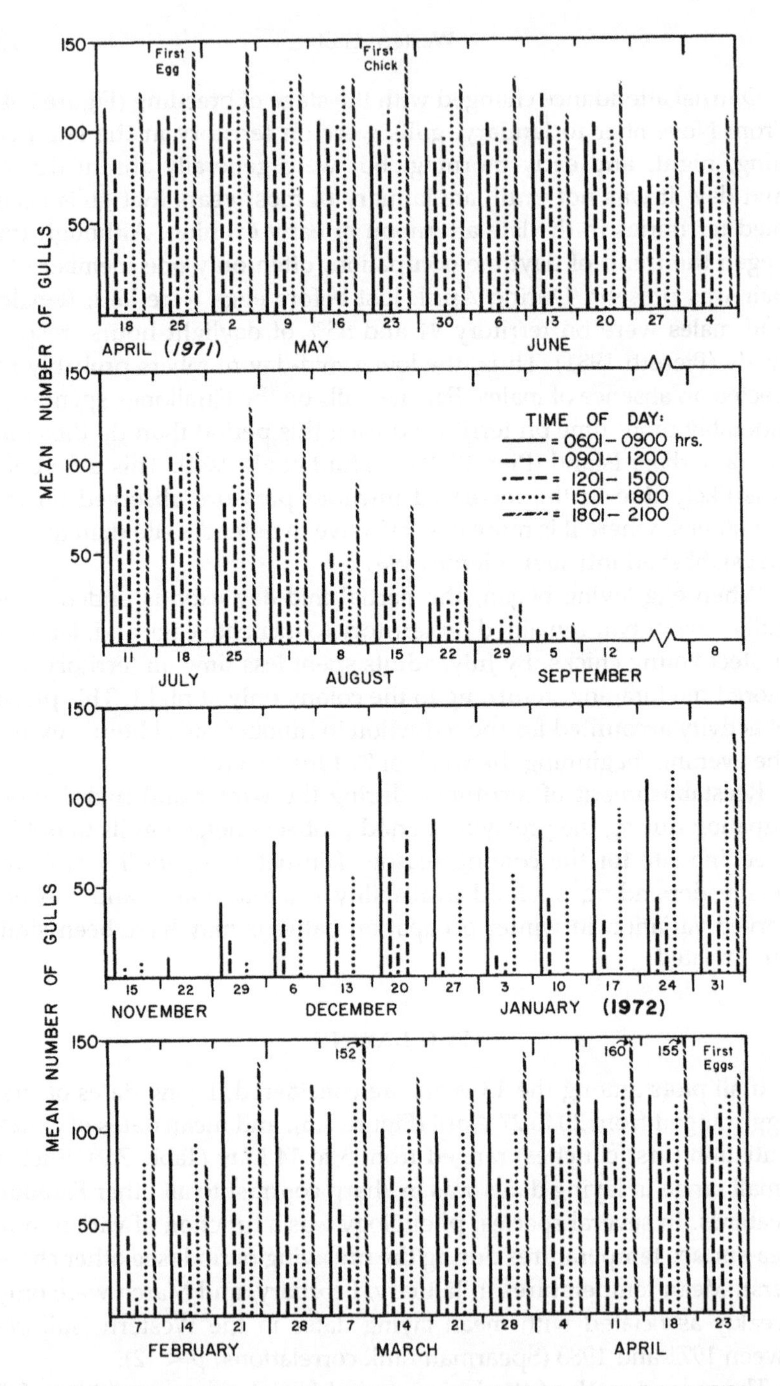

Figure 7.4. Diurnal patterns of territory occupation, by time of year, Western Gulls; weekly means (excepting the one leap-day), April 1971–April 1972. No gulls on territory between 5–11 September and 15–21 November 1971.

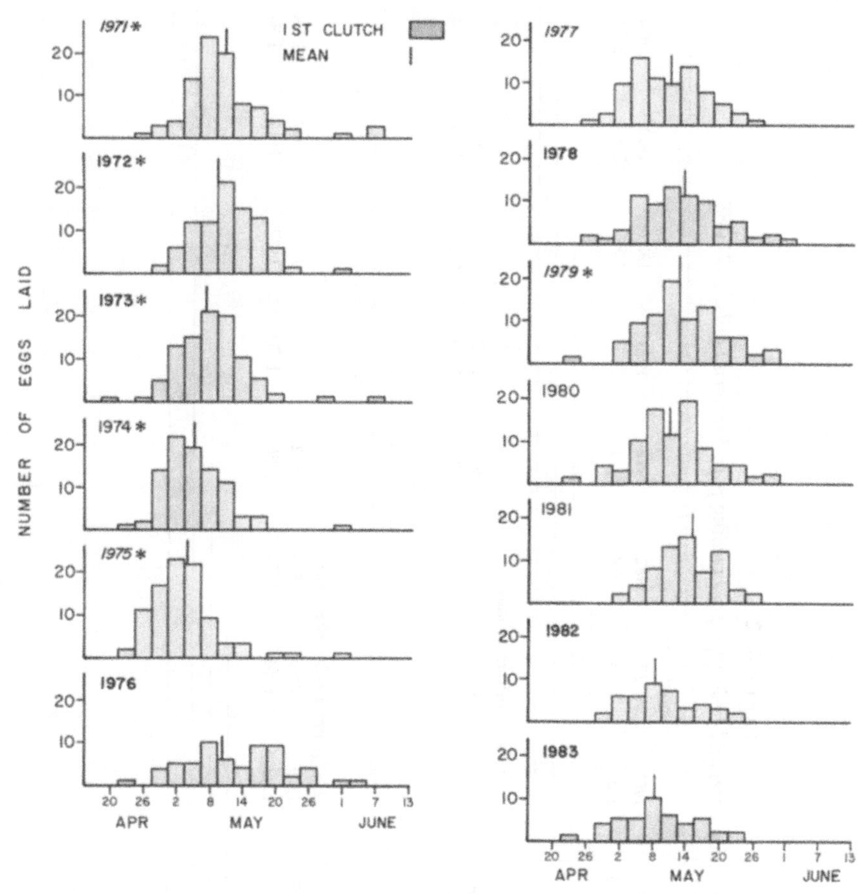

Figure 7.5. Frequency distributions of gull egg-laying dates, all plots and sanctuaries combined, 1971–83; distributions marked by an asterisk (*) are significantly peaked and skewed (t test, $p < .05$).

Western Gull colonies (Schreiber 1970, C. A. Harper 1971, Hunt et al. 1981, Bellrose 1983, T. Harvey 1982). The low variability in laying phenology within and among California colonies suggests that the onset of egg laying may be governed more by changes in day length than by local environmental factors.

Laying phenology was related to nesting location. During the period 1971–80, dates of clutch initiation consistently averaged earlier in plot WT than in plot NL (Mann-Whitney $Z = -6.53$, $n = 885$, $p < .001$). Plots GB and MB represented the nesting habitat of NL and WT, respectively (see Methods), and the same pattern emerged for MB with respect to GB during the period 1981–83 ($Z = -2.43$,

TABLE 7.2

Mean clutch size, clutch-initiation date, and fledglings per breeding pair, Western Gulls, 1971–83

Clutch size

1978	1981	**1976**	*1979*	1980	**1983**	*1977*	**1982**	**1972**	**1973**	*1975*	*1971*	1974
2.51	2.53	2.57	2.58	2.58	2.61	2.66	2.69	2.71	2.77	2.81	2.82	2.84

Clutch-initiation date

1975	1974	**1973**	**1983**	**1982**	**1972**	**1976**	*1977*	*1971*	1980	**1978**	1981	*1979*
5/3	5/5	5/8	5/8	5/8	5/9	5/10	5/10	5/11	5/12	5/13	5/13	5/14

Fledglings per breeding pair

1983	**1978**	1981	**1976**	*1977*	*1979*	1980	**1973**	**1972**	*1971*	*1975*	**1982**	1974
0.65	1.04	1.08	1.09	1.38	1.42	1.44	1.54	1.56	1.70	1.72	1.81	1.91

NOTE: Lines connect similar means; SNK test, $p > .05$.

$n = 129$, $p < .02$). For each plot, the mean clutch-initiation date (over all years) was negatively associated with the gulls' average density in that plot ($r_S = -.900$, $n = 5$, $p < .10$); gulls nesting at higher densities initiated egg laying earlier. The increased social stimulation that occurs in more crowded nesting areas may induce birds to lay earlier (Darling 1938, but see review by Gochfeld 1980). Briggs (1977) noted earlier laying in the larger and denser subcolonies of Western Gulls at Año Nuevo Island. Differences in laying phenology in a population of Herring Gulls was related to interactions among habitat, density, and quality of birds (Pierotti 1982).

Egg size

Among the 654 eggs measured, length averaged 70.1 ± 2.7 mm, breadth averaged 48.8 ± 1.5 mm, and calculated volume averaged 87.4 ± 7.3 cm^3. The first egg within the clutch was usually the largest, and the third was almost always the smallest (Appendix 7.1). Average volume was greatest in 1972 and 1974 and was least in 1978 and 1983, two years of extremely warm water (Appendix 7.2). The ranking of eggs by volume and laying sequence also varied among years (Appendix 7.3).

During all years, the third egg ranked smallest in 87% of clutches ($\chi^2 = 249.77$, DF = 1, $p < .001$). The first egg was larger than the second in over 70% of clutches in 1978 and 1983 ($\chi^2 = 51.84$, DF = 1, $p < .001$), but there was less difference between eggs in 1972, 1974, and 1979 ($\chi^2 = 4.91$, DF = 1, $p < .03$). In 1972, 1974, and 1979, the third egg averaged 92 to 95% of the size of the first, whereas during 1978 and 1983, the third egg was 87 and 88%, respectively, of the size of the first. Third eggs of heterosexual gulls on Santa Barbara Island averaged 93% of first eggs (Hunt 1980), similar to the Farallon average of 91% for all years combined (Appendix 7.1). Within-clutch variation at the Farallon and Santa Barbara colonies contrasts with the lack of variation at the Moss Landing colony (Pierotti and Bellrose 1986).

The differences in egg volume among years were least among first eggs and greatest among third eggs. First eggs in 1978 and 1983 averaged 95% of the volume of first eggs in 1974, the year during which mean egg volume was largest. In 1978 and 1983, third eggs averaged 89 and 90%, respectively, of the volume of 1974 third eggs.

During 1974 and 1979, volumes were larger early in the season (Appendix 7.3), but in 1978 and 1983, when eggs were smaller, no within-year variation in egg volume occurred. This pattern was also evident when eggs were analyzed according to laying sequence and time of season (Appendix 7.3). The sample size in 1972 was too small to allow seasonal comparisons. The lack of within-season variation during 1978 and 1983 was proximately a function of the relatively small first eggs that year. Ultimately, this pattern likely resulted from the low availability of food during those years (see Chapter 3).

Within Western Gull clutches, the first egg is usually the largest and the last egg is usually the smallest (Briggs 1977, Coulter 1977, Hunt et al. 1981, but see Pierotti and Bellrose 1986; for other gull species see Slagsvold et al. 1984). The chick that hatches from the third and smallest egg hatches last, grows more slowly than either the first or second chick, and suffers higher mortality (Parsons 1970, 1975, Coulter 1977). This phenomenon among gulls has been interpreted as a brood-reduction strategy (Coulter 1973, 1977, O'Connor 1978, Hahn 1981; see also Chapter 5). On the Farallones, any disadvantage related to size differences among eggs appears to be accentuated during years of reduced food availability. These observations, as well as those by Pierotti and Bellrose (1986) and W. V. Reid (1987), indicate that breeding adaptations such as the clutch/egg-size strategy of gulls is probably not as simple or as inflexible as interpreted in the past.

Clutch size

Clutch size for Western Gulls averaged 2.7 ± 0.6 eggs for the 13-year period 1971–83 (Table 7.3); three eggs were laid in 73% of all nests. Years when clutches were larger tended to be years of plentiful food (1972, 1974, and 1975); two of the smallest average clutch sizes occurred during 1976 and 1978, two warm-water years of scarce food (see Chapters 2, 3). All clutches combined, the average number of eggs was 2.8 from 1971 to 1975, but clutches then became smaller, averaging 2.6 eggs in the next eight years, 1976–83 (Table 7.3). Coulson and Thomas (1985) also detected longer-term shifts in clutch sizes of Black-legged Kittiwakes, possibly associated with a decline in fish stocks. There is no direct indication of a decline in food availability for Farallon gulls (see other chapters), nor is there evidence of a change in body condition of laying females (as indicated by egg volumes). On the other hand, the frequency of years of anomalously warm ocean temperatures (and corresponding changes in the food web) was much greater in the second, eight-year period (Chapter 2).

For all clutches combined, late-laying birds produced smaller clutches than did either middle or early nesters (Table 7.3). This pattern was also observed at Santa Barbara Island (Hunt et al. 1981) and has been noted for many species of gulls. Whether the reduced clutch

TABLE 7.3
Mean Western Gull clutch size in relation to laying date, 1971–83

Year	Clutch size Early	Middle	Late	Mean total (± SD)
1971	2.9	2.9	2.6	2.8 ± 0.4
1972	2.7	2.7	2.6	2.7 ± 0.5
1973	2.8	2.9	2.4	2.8 ± 0.5
1974	2.9	2.9	2.7	2.8 ± 0.4
1975	2.8	2.8	2.8	2.8 ± 0.4
1976	2.9	2.7	2.0	2.6 ± 0.6
1977	2.9	2.8	2.0	2.7 ± 0.6
1978	2.7	2.7	2.1	2.5 ± 0.7
1979	2.8	2.5	2.4	2.6 ± 0.6
1980	2.8	2.6	2.0	2.6 ± 0.6
1981	2.7	2.7	2.0	2.5 ± 0.7
1982	2.9	2.7	2.7	2.7 ± 0.6
1983	2.8	2.7	2.4	2.6 ± 0.6
Mean total	2.8	2.7	2.4	2.7 ± 0.6

NOTE: *Early, Middle,* and *Late* denote relative position during laying period. Lines connect time periods having similar values; SNK test, $p > .05$. Annual variation significant: $F = 3.57$, DF = 12,1001, $p < .001$.

size of late-laying birds is related to the physiological condition of laying females, age of females, or seasonal fluctuations in food supply appears to vary depending upon the species and the study (see reviews by Winkler and Walters 1983, W. V. Reid 1987). During 1972, 1975, 1979, and 1982, little within-year variation in clutch size was evident. In other years, and especially those of warm water, however, clutches initiated during the last third of the laying period were smaller than those initiated earlier. There was no apparent correlation between existence of seasonal variation and annual clutch size or initiation date (see Table 7.2).

Mean annual values for clutch size and date of clutch initiation were inversely correlated. When clutch initiation was early, average clutch size was larger, and when initiation occurred later, clutch size was smaller ($r_S = -.704$, $p < .01$). Average clutch size was also positively associated with average egg volume ($r_S = .901$, $p < .05$, $n = 5$).

For nearly all gull species, the most common clutch size is three eggs (W. V. Reid 1987). Although this is true in other Western Gull populations (Schreiber 1970, C. A. Harper 1971, Bellrose 1983, T. Harvey 1982), clutches larger than three eggs are not uncommon in the southern California colonies (Hunt and Hunt 1973, 1977, Hunt et al. 1981). Supernormal clutches were not observed on the Farallones, nor have any female–female pairings (Pierotti 1981) or a skewed sex ratio in the breeding population (Spear et al. 1987) been observed. The intense competition for breeding territories may prohibit territory acquisition by female–female pairs (Spear et al. 1987).

Incubation period

The modal interval between successive eggs was two days for two- and three-egg clutches combined, but the interval between laying of successive eggs was longer in two- than in three-egg clutches (Table 7.4). The time required to complete a three-egg clutch averaged 4.5 days. During the period 1981 to 1983, there were no significant between-year differences in clutch completion times, which averaged 4.4, 4.6, and 4.4 days, respectively.

A laying interval of about two days is similar to that for most large larids (Paludan 1951, Vermeer 1963, Hunt and Hunt 1977). Differences in intervals between two- and three-egg clutches likely relate to the condition of the female. A female that produces a two-egg clutch may be less physiologically fit than one that lays a three-egg clutch.

TABLE 7.4

Mean egg-laying intervals in Western Gull clutches, 1981–83

Eggs in clutch	Egg interval[a]	Laying interval[b] (days)	n
2	1–2	2.8 ± 0.9	46
2 or 3	1–2	2.4 ± 0.8	126
3	1–2	2.2 ± 0.6	80
3	2–3	2.3 ± 0.7	80
3	1–2, 2–3	2.2 ± 0.6	160
3	1–3	4.5 ± 0.9	80

[a] I.e., interval being measured: e.g., 1–2 indicates interval between first and second eggs.

[b] Significant difference in mean laying intervals, two-egg vs. three-egg clutches; t test, DF = 124, $p < .001$. Means are shown ± SD.

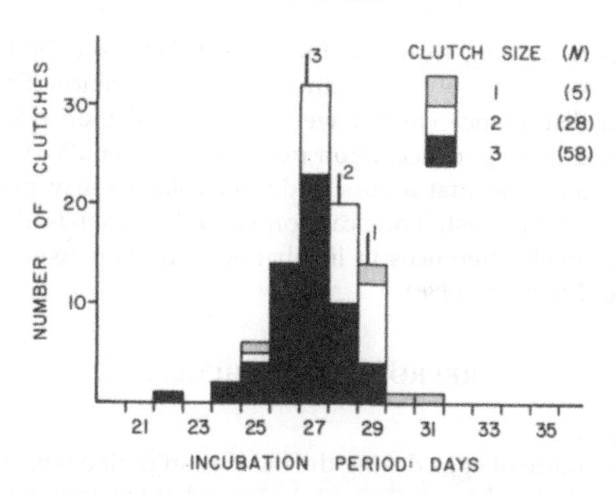

Figure 7.6. Frequency distributions of incubation periods for gull eggs in clutches of different sizes, 1981–83; vertical lines show means for indicated clutch sizes.

A less fit female may also require more time between the formation of successive yolks. In studies of known-age seabirds, young individuals are usually responsible for most small clutches (reviewed by Ryder 1980, Pugesek 1981, Ainley, LeResche, and Sladen 1983). Pierotti and Bellrose (1986) observed Western Gulls laying at shorter intervals at Moss Landing than at any other colony; an abundant food supply coupled with low energy expenditures for territory defense may shorten the time required for yolk formation (but see below, *Replacement clutches*).

Incubation periods were measured during three years, 1981–83, when nests were checked daily; those nests not observed on any one day were excluded from analyses. The average incubation period for eggs in three-egg clutches was 26.7 ± 1.3 days (Figure 7.6). Within-season differences were not evident: the incubation period in early clutches averaged 26.5 ± 1.2 days, compared to 27.0 ± 1.4 days in late clutches. These results are closely similar to those observed at San Nicolas and Año Nuevo islands (26.2 and 26.6 days, respectively, Schreiber 1970, Briggs 1977) but contrast with a longer period on Santa Barbara Island (28.8 days, Hunt et al. 1981) and a shorter period at Moss Landing (25.7 days, Pierotti and Bellrose 1986; 25 days, T. Harvey 1982).

The 1981–83 Farallon data also contrast with the results of work on the Farallones in 1970 (28.2 ± 1.4 days, Coulter 1973). The difference may reflect the techniques employed. The early data were collected by observers walking through study areas daily; the 1981–83 data were also collected daily but without any interference. The greater amount of disturbance could have affected incubation effectiveness and resulted in longer incubation periods (see Beer 1963). Also, the increased confusion that occurs in denser colonies may prolong the time spent off the nest. Thus caution should be exercised when interpreting small differences in incubation periods between colonies (Pierotti and Bellrose 1986).

REPRODUCTIVE SUCCESS

Hatching success

Disappearance of eggs during the incubation period was the largest contributor to hatching failure and also exhibited the most annual variability. We examined annual variation in the timing of egg loss by means of three categories: (1) disappearance of eggs before they were scheduled to hatch (< 30 days), (2) disappearance at hatching, and (3) failure to hatch (= addled). On average, 10% of all eggs laid disappeared before they were scheduled to hatch, with values ranging from 2% in 1982 to 17% in 1979. Eggs that disappeared at hatching or never hatched constituted only 5 and 6%, respectively, of total hatching failures.

The percentage of chicks hatched per egg laid ranged from 69% in 1979 to 88% in 1982 (Appendix 7.4). Success during nine years was within 4% of the average, 76%, for the entire period. During three years, 1973, 1974, and 1982, hatching success was high, whereas in 1979, when many eggs disappeared prematurely, it was considerably reduced. Hatching rates and causes of egg mortality were similar to those found at San Nicolas Island (Schreiber 1970), Bird Rock (C. A. Harper 1971), and Año Nuevo Island (Briggs 1977) but were different from those at Santa Barbara Island (Hunt et al. 1981) and Moss Landing (Bellrose 1983, T. Harvey 1982), where hatching success was lower. At Moss Landing, mammalian predation significantly reduced hatching success; at Santa Barbara Island, a larger proportion of eggs failed to hatch, possibly because of disturbance (Hunt et al. 1981).

Hatching success was affected by several factors, one of which was clutch size. Hatching success of eggs in three-egg clutches was 72%, compared to 66 and 36% for two- and one-egg clutches, respectively. Other studies have documented the positive correlation between clutch size and hatching success in Western Gulls (Schreiber 1970, C. A. Harper 1971, Hunt et al. 1981). The same is true for other seabirds that lay multiple-egg clutches, e.g., guillemots (see Chapter 9).

Another factor affecting hatching success was seasonal. Hatching success declined as the season progressed (Appendix 7.5). During five years, 1973–76 and 1979, little seasonal pattern was evident. Laying dates, clutch size, and hatching success appeared to be interrelated. With laying date held constant, hatching success was more strongly correlated with clutch size ($r = .485$, DF = 1011, $p < .001$) than was laying date with clutch size held constant ($r = .221$, DF = 1011, $p < .001$).

Hatching success was also affected by nesting location on the island. Overall, success was consistently lower in the north-side plot (NL) than in the southeast plot (WT) for the ten-year period 1971–80 ($\chi^2 = 13.88$, DF = 1, $p < .05$), but only during 1972 and 1977 did significant within-year differences occur. Differences in hatching success between the two sanctuary plots, MB and GB, were not significant. Timing of egg loss varied little among the four plots and, therefore, the observed differences did not appear to be related to predation or quality of the adults; they may, however, have been an artifact of the gulls' laying earlier in plot WT than in plot NL (see above).

Breeding success

The percentage of chicks fledged per egg hatched varied annually and ranged from 34 to 82% (Appendix 7.4). The number of chicks fledged per pair averaged 1.4 ± 1.1 and ranged from 0.6 to 1.9. Differences were significant between years (Table 7.2). Fledging success was lowest in 1976, 1978, 1981, and 1983, but was similar in other years. The number of chicks fledged per pair was also low during the same four years, three of which were ones in which ocean temperatures were anomalously warm (Chapter 2).

We compared annual variation in fledging success with variation in the proportion of fish and garbage in the diet. To determine diet we analyzed regurgitations we collected while handling chicks (Chapter 3). Fledging success was positively correlated with percentage of fish ($r_S = .613$, $n = 9$, $p < .05$) and negatively correlated with percentage of garbage in the chicks' diet ($r_S = -.742$, $n = 9$, $p < .03$). Changes in diet composition have been linked in other gull studies to changes in various reproductive parameters. On Santa Barbara Island, a decline in anchovy abundance was paralleled by a decline in the number of breeding Western Gulls, and growth rate, a predictor of chick survival (Hunt and Hunt 1976a), was affected by the amount of fish in the diet (Hunt and Butler 1980). In a population of the Herring Gull, overall reproductive success was lowest among individuals that specialized on garbage (Pierotti and Annett 1986).

At the Farallones, a high proportion of garbage in the diet was also related to the timing of chick mortality. On average, 74% of all mortality occurred among chicks under 16 days old. Years when garbage figured more prominently in the diet were those when a greater proportion of chicks died at this age (Mann-Whitney $U = 20$, DF = 4,5, $p < .01$).

Fledging success was related to several other factors. Success was significantly greater among gulls initiating clutches early than among those laying later (all years combined). Early nesters fledged more offspring than either mid-season or late nesters (Appendix 7.5). Because the number of young fledged is related in part to initial clutch and brood size, we held these factors constant; date of clutch initiation was still an important factor ($r = .641$, DF = 1009, $p < .001$).

As noted previously, overall hatching success was related to clutch size. A similar relationship existed between fledging success and initial brood size. Fledging success was 74, 67, and 57% for three-, two-, and one-chick broods, respectively.

Fledging success was also related to nesting location. Gulls in the southeast plots, WT and MB, were significantly more successful than those nesting in the north plots, NL and GB (WT vs. NL, $\chi^2 = 11.31$, DF = 1, $p < .001$; MB vs. GB, $\chi^2 = 9.53$, DF = 1, $p < .005$). Not only did chick survival rates differ between plots, but the timing of chick mortality and age at which chicks died also appeared to be related to nesting location. Among chicks in plot MB (1981 to 1983), 62% of deaths occurred during June and 38% occurred during July. In plot GB, however, 93% of all deaths occurred during June and only 7% occurred during July. Differences in timing of chick mortality by plot and month were significant ($\chi^2 = 16.5$, DF = 1, $p < .001$). A result related to this difference in timing is that chicks in the plot MB were appreciably older at death than were those in plot GB (20.4 vs. 8.4 days, respectively).

Spear et al. (1987) noted that at least 50% of chick deaths in June were due to starvation, whereas in July 54% of chick mortality was due to attacks by adults. This pattern indicates that the later mortality observed in the southeast plots could be a function of the greater nesting densities in these areas. Where nests are more dense, mobile chicks have less room to err by crossing boundaries. If the rate of chick mortality in June was constant between plots, then the additional July mortality in the southeast plots would reduce reproductive success further. Yet, gulls nesting in plot MB had greater fledging success, suggesting perhaps that they were better able to provide for the young during the "starvation" period.

Replacement clutches

Loss of an entire clutch occurred in 9% of nests, ranging by year from 2% in 1974 and 1982 to 14% in 1978. Replacement clutches, however, were observed in only seven of 13 years. Birds that lost entire clutches tended to be late breeders who initiated laying an average of seven days later than the mean date for original clutches (16 vs. 9 May, respectively). Among the 88 pairs that lost entire clutches, only 12% relaid. The average initiation date of replacement clutches was 2 June, which approached the late end (10 June) of the laying period for original clutches (see Figure 7.5).

The size of replacement clutches averaged 2.1 eggs. This compared closely with the average size of original clutches (2.0 eggs) of these birds. Both values, however, were considerably smaller than the average for all original clutches, 2.7 eggs. For the nine nests in which dates of original egg loss and dates of clutch replacement were known, the interval between the loss and the laying of first replacement eggs ranged from 12 to 18 days, with an average of 14.4 days. This was a little longer than the ten to 11 days required for rapid yolk development (Roudybush et al. 1979) but similar to observed intervals at the Moss Landing colony (Bellrose 1983). One of the sanctuary pairs that had an original three-egg clutch took a full 14 days to lay a second three-egg clutch, considerably longer than the average 4.5 days for original clutches (see above).

Replacement clutches did not fare well. Hatching success, 48%, fledging success, 45%, and the number of chicks fledged per nest, 0.4, averaged much lower than respective values for all eggs (see Appendix 7.4). Compared to first clutches initiated at the same time as replacement clutches, the latter were slightly more successful. Among first clutches initiated during the same period as replacements, hatching success was 30%, fledging success was 44%, and the number of chicks fledged per nest was 0.2. Most failed breeders maintained their territory throughout the duration of the breeding season.

Chick development

Western Gull chicks fledge at about seven weeks and disperse at about ten weeks of age (Spear, Ainley, and Henderson 1986). At hatching they weighed an average 64.8 ± 7.2 gm, and they fledged

46 to 60 days later at 650 to 1,200 gm. Hatching weights differed between years (Table 7.5). During most years, chicks grew at similar rates, but in 1977 they reached asymptotic weights faster (Figure 7.7). The differences between years were significant for weights on days 20 ($F = 6.37$, DF $= 4,59$, $p < .001$) and 40 ($F = 4.52$, DF $= 3,32$, $p < .01$). When 1977 data were excluded from analysis, differences were not significant ($p > .2$).

During June and July 1974 and 1977, gulls fed nearer the island than during the same period in 1975 and, especially, 1973 (see Chapter 3, Figures 3.21 and 3.22). This pattern exactly matched the yearly rankings of chick weights at day 40 as well as of growth-rate constants (Table 7.5). Furthermore, the number of gulls feeding near the island during the chick period was much higher in 1977, the year of highest growth rate, than in the other three years.

When we compared fitted growth curves for 1973, 1974, 1975, and 1977 (Table 7.5), no differences in asymptotic weights were apparent, but the growth constant (K) varied significantly and correlated with weights at day 40. Weights at day 40 are usually about 88% of asymptotic weights (Coulter 1977). Within broods, differences in growth constants between the heaviest and lightest chicks (at hatching) suggest that initially lighter chicks develop more slowly than their siblings. On Santa Barbara Island, Western Gull chicks hatching from third eggs also continue to grow after their two siblings reach the asymptote (Sayce and Hunt 1987).

On the basis of a between-year comparison, asymptotic weights and growth rates did not vary among single-chick broods (F test, $p > .05$), perhaps because the sample was small. Within multiple-chick broods, the chicks heaviest at hatching showed little between-

TABLE 7.5

Mean growth parameters for Western Gull chicks, 1972–77

(gm)

Year	Hatching weight[a]	Asymptotic weight[b]	Growth constant (K)[c]	Weight, day 40[d]
1972	61.0 ± 6.3 (37)			
1973	65.6 ± 6.7 (21)	937 ± 173 (9)	0.117 ± 0.025 (9)	812 ± 149 (11)
1974	70.1 ± 6.6 (22)	888 ± 135 (11)	0.150 ± 0.017 (11)	859 ± 128 (13)
1975	65.1 ± 6.4 (14)	895 ± 57 (5)	0.135 ± 0.019 (5)	819 ± 20 (5)
1977	64.4 ± 5.9 (11)	1,037 ± 120 (9)	0.154 ± 0.026 (9)	1,029 ± 139 (7)

NOTE: All chicks combined. Means are shown ± SD; n in parentheses.
[a] Differences among years significant: $F = 7.001$, DF $= 4,100$, $p < .001$.
[b] Differences among years not significant: $F = 2.27$, DF $= 3,30$, $p > .05$.
[c] Differences among years significant: $F = 5.47$, DF $= 3,30$, $p < .01$.
[d] Differences among years significant: $F = 4.52$, DF $= 3,32$, $p < .01$; with 1977 excluded, $p > .2$.

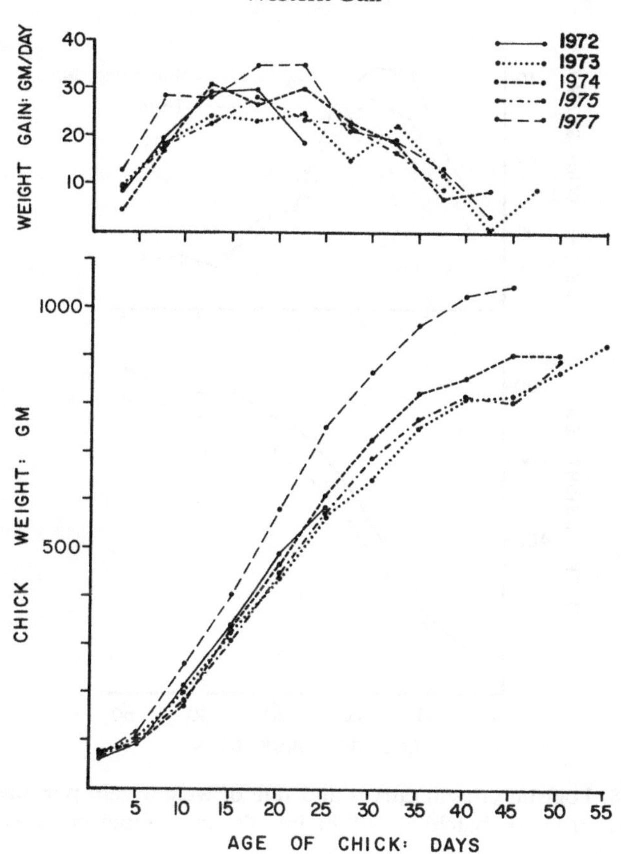

Figure 7.7. Logistic growth curves and rate of weight gain per day for Western Gull chicks (ANOVA; see Table 7.5). Dots are mean values; all chicks combined for five separate years during 1972–77.

year variation in either asymptotic weight or growth rate among four years (1973, 1974, 1975, and 1977; F test, $p > .05$). For chicks that were lightest at hatching, asymptotic weights did not differ in 1973, 1974, and 1975 ($F = 1.92$, DF = 3,8, $p > .10$), although growth rates did ($F = 5.98$, DF = 3,8, $p < .02$; see also Figure 7.8).

During the nestling period, growth rates provide a measure of the food available to breeding adults, whereas hatching weight is closely related to egg size (Coulter 1973). Western Gulls responded to increased availability of local marine prey with faster growth rates. Growth rates also vary annually at the Santa Barbara Island colony of Western Gulls (Hunt et al. 1981), where reduced rates correlated with lower chick survival.

Western Gull chicks progress at different rates through develop-

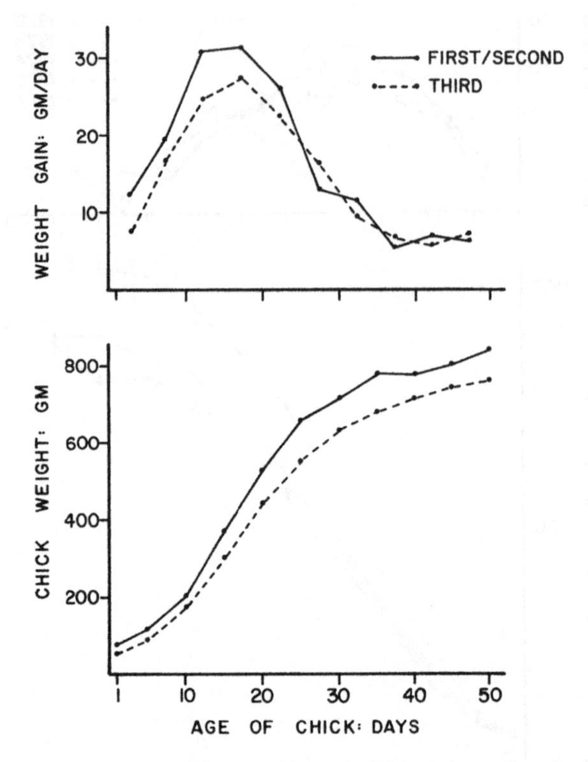

Figure 7.8. Logistic growth curves and rate of weight gain per day for gull chicks, by order of hatching within broods, all sample years combined, 1972–77.

mental stages such as those when tarsal growth is maximal and when feather growth is commencing (Coulter 1977). Annual variation in the development rate may explain similar asymptotic weights among chicks whose weights differed at day 40. The lack of variation in asymptotic weight suggests that fledging size among gulls may not vary much. Chicks that were heaviest at hatching showed little variation in growth rate, but chicks that were lightest at hatching were more affected by variation in food supply. For surviving chicks, an effect of poor food availability might be to cause an extended nestling period as observed in some other Farallon species (see Chapters 9 and 10).

THE WESTERN GULL BREEDING STRATEGY

Western Gulls on the Farallones exemplify the adaptability of species in the genus *Larus*. No other surface-nesting Farallon species is as catholic in its choice of nesting habitat or as willing to nest near people. In part, however, this apparent tolerance may be a function of the necessity of securing a nesting territory, the availability of which is very limited.

Another element of the plasticity exhibited by gulls is their ability to fall back on a predictable food supply of offal and garbage when marine food is not readily available (Spear 1988a). As a consequence of this ability, except for the storm-petrels, the Western Gull was the only Farallon species that did not experience total or near-total breeding failure during the study period. Offal and garbage were not the perfect substitutes, however, because fledging success decreased as the percentage of garbage in the chicks' diet increased. A high percentage of garbage implied a decrease in quality of diet (Hunt 1972, Pierotti and Annett 1986) as well as a decrease in food availability because of the greater foraging distances required (70 km or more round trip to mainland garbage dumps). Both these factors likely contributed to the reduced breeding success when garbage predominated in diets.

Another component of gulls' plasticity or opportunism was their ability to exploit the abundant live prey. The number of young fledged per nest was a function of food availability over the entire breeding season, expressed through effects on egg volume, clutch size, fledging success, and growth, and values of all of these variables increased during years of plentiful food. Among Farallon species, only the cormorants could also exploit abundant food by increasing reproductive output to a significant degree.

Early nesters had greater egg volumes, clutch sizes, hatching success, and fledging success. They also fledged more young per pair than mid-season or late nesters. By each measure except fledging success, late nesters did the poorest. Within-season variation of other breeding parameters offers clues as to why early nesters experienced greater success. During years of abundant food, clutch size tended to be greater, and, although early nesters produced slightly larger eggs, there was little within-clutch variability in egg volume. When food was less abundant, average clutch size was reduced, within-season variability of clutch size was greater, eggs were smaller, and within-clutch differences in egg volumes were accentuated. Thus, early nesters demonstrated a greater ability to take advantage of favorable feeding conditions by laying larger eggs in good years and larger clutches in poor years. In these ways they maintained a reproductive advantage over mid-season and/or late nesters during the extremes of feeding conditions. It is likely that early nesters comprised largely older, more experienced, and generally higher-quality breeders (PRBO, unpubl. data, Coulson 1966, 1968, Coulson, Duncan, and Thomas 1982, Coulson and Porter 1985, Pugesek and Diem 1983).

Other studies have explored the effects of habitat and of differences in density and timing on reproductive success in gulls. Usually, the differences are much greater than what we have found on the Farallones. Either marginal habitat and the lower-quality breeders that would be expected to occupy it do not exist on the Farallones, or only high-quality birds can secure a territory at all. Of significance is that even small annual and seasonal variations in reproductive effort represent a sensitivity and adjustment to fluctuations in the marine environment. Gulls nesting in areas that superficially seem similar in habitat and nest density differ in reproductive output. These results point to the value of habitat analyses and independent measurements of quality of individuals within different habitats; these are current and future directions of research on this population of Western Gulls.

8

Common Murre

Robert J. Boekelheide, David G. Ainley, Stephen H. Morrell, Harriet R. Huber, and T. James Lewis

The Thick-billed Murre (*Uria lomvia*) and Common Murre are quintessential Northern Hemisphere seabirds, distributed widely throughout the North Pacific and North Atlantic oceans. Though the two species are sympatric in a significant portion of their respective ranges, Thick-billed Murres are concentrated in colder arctic waters and Common Murres in subarctic waters (Hunt, Eppley, and Drury 1981, R. G. B. Brown 1985). They also differ in diet and nest-site selection, the Thick-billed eating more invertebrates and nesting on narrower ledges (Sergeant 1951, Belopol'skii 1961, Tuck 1961, Spring 1971, Swartz 1966, A. J. Williams 1974, Bradstreet and Brown 1985, Squibb and Hunt 1983). On the basis of presently available information it appears that the two species split during an early Pleistocene glaciation, when *U. lomvia* differentiated from the parent stock in an arctic refugium (Storer 1952, Udvardy 1963, Bédard 1985). The Thick-billed Murre does not breed on the Farallones.

Murres frequently breed in spectacular colonies numbering tens or hundreds of thousands of birds (Tuck 1961, Sowls et al. 1980). At the peak of the annual breeding period, colonies resemble beehives as thousands of murres swarm back and forth between nests and the sea. Like many seabirds, murres spend much of their lives on the ocean, where they are one of the most numerous and visible components of coastal and pelagic communities. For example, Sanger (1972) estimated that large alcids, mostly Common Murres, account for 74% of the total number and 88% of the biomass of seabirds in California coastal waters during winter; the prominence of murres was confirmed by the studies of Briggs et al. (1987).

Much of what we know about the breeding ecology of murres comes from a handful of detailed studies. Two pioneering Soviet studies (Uspenski 1958, Belopol'skii 1961) described seabird communities in the Barents Sea, where murres are not only the primary breeding species, but are also an important source for the commercial harvest of eggs. Other information is available from Tuck (1961), who reviewed the genus *Uria*, in addition to presenting his own studies on Thick-billed Murres from the Canadian Arctic and Newfoundland. Gaston and Nettleship (1981) and Gaston et al. (1985), also working in the Canadian Arctic, and Hunt, Eppley, and Schneider (1986), working in the Bering Sea, present richly detailed studies on the same species. Much less work has appeared on the Common Murre. At Skomer Island, in Wales, Birkhead (1977a,b, 1978a,b) and Birkhead and Hudson (1977) studied many aspects of Common Murre population dynamics and breeding ecology. M. P. Harris and Birkhead (1985) summarized what is known about Common Murre breeding biology in Atlantic populations. More recent information on Common Murres in eastern Canada is available from Piatt and McLagan (1987) and McLagan and Piatt (ms). One of the few published breeding studies from the Pacific Basin is that by Swartz (1966), who reviewed all species in the Cape Thompson, Alaska, seabird community, where again murres were the major component.

Like many seabirds, murres balance low fecundity (one egg per clutch) with relatively high breeding success and extended longevity. Their breeding system is noteworthy for two reasons. First, they breed in extremely dense colonies, with adjacent sites so close that neighboring birds usually touch one another. Average density in flat areas is about 20 pairs per square meter (M. P. Harris and Birkhead 1985). By nesting so densely they achieve some protection from aerial predators, particularly gulls and corvids, but this high density also requires alteration of territorial behaviors to accommodate close spacing, especially through well-developed appeasement displays (R. A. Johnson 1941, Pennycuick 1956, Birkhead 1977b, 1978b, Mahoney and Threlfall 1982). Second, murre chicks are semi-nidifugous. That is, they leave the nest site at 20 to 25% of adult weight (S. R. Johnson and West 1975, Birkhead 1977a, Hunt, Eppley, and Schneider 1986) and, unable to fly, they swim with one parent to feeding areas at some distance from the colony. This saves breeding adults time and energy, freeing them from flights to and from the colony to feed large chicks, and provides flexibility in exploiting different feeding areas (Gaston and Nettleship 1981; Chapter 3). The parent that goes to sea with the chick is nearly always the male (Birkhead 1976, Hunt, Eppley, and Schneider 1986, PRBO, unpubl. data), a phenomenon that

TABLE 8.1

Estimated numbers of Common Murres nesting in western North America, 1977–79

Area	Birds	Sites	Source
Alaska	5,000,000	142	Sowls, Hatch & Lensink 1978
British Columbia	3,000	2	Manuwal & Campbell 1979
Washington	11,950	11	Manuwal & Campbell 1979
Oregon	168,500	21	Varoujean 1979
California	406,150[a]	20	Sowls et al. 1980
Total	5,589,600	196	

NOTE: Most recent data.

[a] The Farallon population (46,000 birds) constitutes 11% of this total.

raises questions about parental effort within pairs and the food-finding and chick-feeding abilities of the different sexes (see Wanless and Harris 1986).

Eight subspecies of Common Murres are currently recognized, with three in North America (Storer 1952), but the validity of some of these taxa is suspect (Bédard 1985). The Farallon Islands are the type locality for the subspecies *U. a. californica*, whose breeding range extends from northern Washington to the southernmost breeding colony of Common Murres in the world, Hurricane Point, Monterey County, California. Formerly, murres also bred at Prince Island in southern California, but this colony has not been occupied since 1912 (Hunt et al. 1981).

Outside Alaska, the largest community of breeding Common Murres in western North America is at the Farallon Islands (Table 8.1). During the 1979–80 count period used by Sowls et al. (1980), two murre colonies within California exceeded 100,000 birds, six ranged between 10,000 and 100,000, ten ranged between 1,000 and 10,000, and one contained less than 1,000 birds, making a total California breeding population of over 400,000 murres. South of Alaska, only the Arch Rock colony in Oregon and the Castle Rock colony in Del Norte County, California, rival the Farallon population in size. As noted below and in Chapter 1, the Farallon murre colony alone numbered 400,000 individuals in the 1850's.

METHODS

We made detailed observations of breeding murres at Shubrick Point, the easternmost corner of Southeast Farallon Island (Figure

1.3). During winter 1971–72, a small wooden blind was built near the summit of Shubrick Point. It overlooked a colony of what was then fewer than 500 pairs of murres; by 1982 this subcolony contained over 1,500 pairs. From the blind, we looked directly down on breeding murres and Brandt's Cormorants, plus scattered nesting sites of Pelagic Cormorants and Tufted Puffins. A murre study plot was established on the slope about 15 m below the blind. The plot covered an area of approximately 25 m^2 from 1972 to 1976, but we then reduced it to 15.2 m^2 as densities of breeding murres increased. No other colonies were accessible to us without our causing great disturbance to murres and other nesting birds.

Each year, from 1972 to 1981, we followed all pairs within the plot to determine nesting density, dates of laying, hatching, and fledging, and reproductive success. We made daily observations from mid-April until the last chick fledged during July or August. Observations each day lasted as long as was necessary to determine the presence or absence of eggs or chicks at each egg-laying site. Sometimes this effort required up to six hours per day. A black-and-white photograph of the study plot covered by an acetate overlay was used to map sites. Sites were numbered consecutively each year as eggs were laid. By this method, 116 to 173 sites were followed annually. To investigate within-season variability in reproductive parameters, we compared sites grouped by whether eggs were laid early, midway, or late in the nesting period, defining temporal groups by the criteria given for Brandt's Cormorants (Chapter 5).

In the period 1982–85, we followed only a selection of sites that were part of a separate study on parental investment. Chosen prior to egg laying, these were occupied by adults recognizable by distinctive physical characteristics. Nearly all such sites lay within the boundaries of the former study plot.

On a few occasions, we witnessed murres losing eggs immediately after they were laid. These females laid their eggs on steep ground, where they rolled downhill soon after exiting the female's cloaca. Undoubtedly such events also occurred when we were not present, and, therefore, our numbers overestimate hatching success to some degree. Gaston, Noble, and Purdy (1983) missed an estimated 3.9% of total eggs laid because of egg loss prior to detection. They observed their study plots for fixed amounts of time every day (one or two hours), whereas we observed our plot as long as was necessary to determine the status of every site. They calculated that three hours per day were adequate to predict within 5% the total number of eggs laid. Therefore, we feel our consistent observation effort throughout the study supports our annual comparisons.

We gathered information on attendance patterns in two ways. First, from December 1971 to August 1972, using a 20× telescope, we regularly counted the number of murres present in a plot on Saddle Rock, which lies about 200 m off Southeast Farallon (Figure 1.3). We counted several times each day, then combined counts into three-hour periods by week. These data illustrate how murre attendance changed diurnally and seasonally. Second, each day during two successive winters, 1980–81 and 1982–83, we tallied the presence or absence of murres on the entire island, logging whether murres visited nest sites and if they remained all day.

Besides using nesting density within our Shubrick Point plot to assess changes in population size, we also made one-day counts of the number of murres on the entire island each year between 1979 and 1982. The 1979 count was conducted on 6 May from a boat drifting around the island, and because this count was conducted prior to egg laying, when murre attendance is variable and less indicative of the total breeding population (Lloyd 1975), it was not used to estimate the breeding population that year. In contrast, the 1980–82 counts occurred in early June, when almost all birds had completed laying. The 1980–82 counts were made from the lighthouse, from blinds, and from boats, whichever offered the best vantage point for each colony. During these counts, we also determined the number of paired and single birds at known egg-laying sites in our study plot, plus the number of loafing birds away from sites. These figures were used to calculate a correction factor (as discussed by Birkhead and Nettleship 1980), which was subsequently used to estimate the total breeding population from the overall counts. Because of time and weather constraints, these counts were not replicated within years, as suggested by Lloyd (1975) and Wanless et al. (1982). We feel, however, that they do show relative differences in the breeding population between years.

Since murre chicks leave the breeding colony prior to achieving flight or independence from their parents, fledging is defined in this chapter as the date of departure of the chick from the ledge (see Glossary). Nearly all murre chicks departed the colony in the evening just prior to or after dark, so we usually did not note their absence until the day following departure.

FARALLON POPULATION

Changes in the Farallon murre population during the last 150 years demonstrate how resilient a seabird population can sometimes be, given a chance for recovery. Ainley and Lewis (1974) estimated that 400,000 murres bred at the Farallones in the middle of the nineteenth century. Their estimate was based on the number of eggs removed by commercial eggers in one year. Despite continual pressure from eggers during the latter half of that century (during which an estimated 14 million murre eggs went to market in San Francisco; see Doughty 1971), various authors from 1856 to 1895 described the murre population as "countless," far surpassing all other bird populations combined. By 1910, only 20,000 murres remained. During the twentieth century the murre population continued in a depleted state owing to perennial oil spills and intense human disturbance. During this time, the shells of murre eggs also became thinner, most likely because of the effects of organochlorine compounds (Gress, Risebrough, and Sibley 1971). Shell thinning, however, had no known impact on the murres' breeding performance. By the late 1950's, the population reached a low of 6,000 birds.

The murres increased through the next decade, reaching a peak of 88,000 on Southeast Farallon in 1982. The number of murre sites within our study plot increased during the same period (Figure 8.1). From 1972 to 1982, the number of sites with eggs increased every year except in 1978, a year of warm water and poor reproductive success for most Farallon birds (see Chapters 2 and 12). Years with the highest percentage of increase from the previous year, 1975, 1977, and 1979, were years when juvenile rockfish predominated in murre diets (see Table 3.1). From 1979 to 1982, our study plot approached saturation, thereby becoming less of an indicator of actual population growth. Counts of murres on the entire island during those four years show that the population changed much more than the plot densities indicate (Table 8.2). Following a "crash" in the murre population at Skomer Island, Wales, Birkhead and Hudson (1977) also observed a rapid recovery over two years.

In 1982 and 1983, murres experienced the combined effects of one of the strongest ENSO's on record (Chapter 2) and a large increase in fishermen's use of gill nets in central California. Even though 90% of murre study sites were occupied during the early spring of 1983, at only half of these were eggs laid (Figure 8.1). This illustrates that large segments of the population may not breed if conditions are not favor-

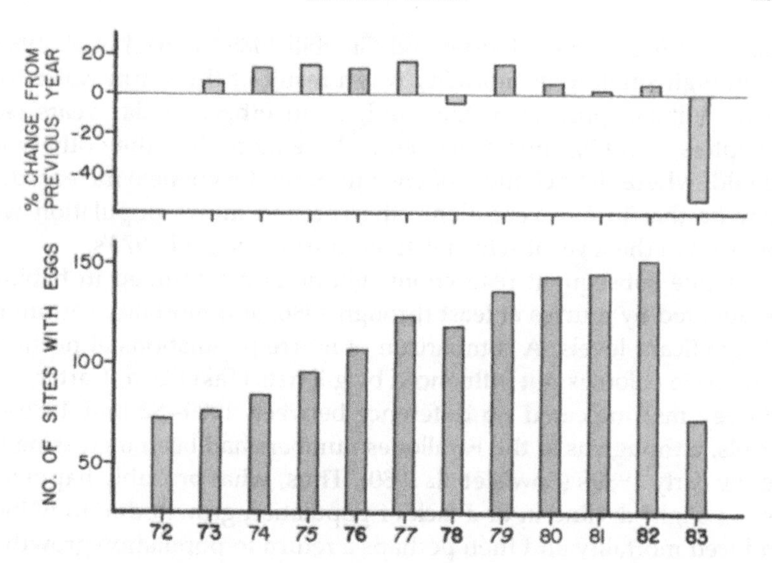

Figure 8.1. Number of Common Murre breeding sites with eggs in the Shubrick Point study plot, 1972–83.

able, as discussed by Lloyd and Perrins (1977). Additionally, the California Department of Fish and Game (P. Wild pers. comm.) estimated that 22,000 seabirds, mostly murres, drowned in gill nets in central California during the first nine months of 1983; the total for the entire year may have reached 30,000 (PRBO estimate). Murre mortality was also high in areas along the northern California and Oregon coast not

TABLE 8.2

Common Murre breeding populations on the South Farallon Islands, 1979–85

Date	West End	Southeast Farallon	Total counted	Correction factor[a]	Estimated population
6 May *1979*	20,400	11,350	31,750		
7 June 1980	14,035	16,000	30,035	1.64	49,257
8 June 1981	20,190	24,910	45,100	1.66	74,866
9 June **1982**	23,390	27,310	50,700	1.57	79,599
May **1983**					42,000[b]
1 July 1984	13,070	12,945	26,015	1.14	29,657
18 June *1985*	8,160	15,150	23,310	1.61	37,529

[a] I.e., correction for missing breeders. Correction factors for 1980–85 determined by the number of occupied laying sites within the study plot during the count; the 1979 count took place before egg laying and, therefore, is not considered an accurate assessment of the number of birds breeding that year.

[b] Rough estimate based on proportion of birds that attempted breeding in the study plot, and the number of sites occupied as a fraction of that occupied the previous year; most sites abandoned by mid-May.

fished with gill nets (Hodder and Graybill 1985, Stenzel et al. 1988). Such high single-year mortality was a result of the warm water that year. Suitable prey were few, and, as in other, similar years (see Chapters 2 and 3), murres moved inshore more than they otherwise would, where their chances of encountering the gill nets increased. It may be that in one year alone, the Farallon murre population was reduced to the level at which it stood during the mid-1970's.

Despite subsequent restrictions, gill netting continued in habitats frequented by murres at least through 1986, and mortality continued at significant levels. A comparison of murre populations at northern California colonies not influenced by gill nets (Takekawa, Carter, and Harvey ms) indicated no difference between 1980–82 and 1985–86 levels, although as at the Farallones numbers had been increasing up to the early 1980's (Sowls et al. 1980). Thus, what probably happened was a slight decline in or a lack of population growth due to ENSO-induced mortality and then perhaps a return to population growth in 1985 and 1986. Applied to the Farallones, this scenario, in the hypothetical absence of gill netting, might have produced a population in 1985 approximately equivalent to that of 1982 (Table 8.2).

Compared with Southeast Farallon, the history of the murre population on the North Farallones is essentially unknown. The north islands include four small sea stacks, all of which rise precipitously from the ocean. Despite their inaccessibility, some egging occurred on the north islands, but not to the same extent as on Southeast Farallon (P. White pers. comm.). Although undoubtedly affected by oil spills, North Farallon murres were spared the intense human disturbance experienced by murres at the south islands during the nineteenth and twentieth centuries. We censused the North Farallon murres by boat on 20 June 1982. Our count for all four islets was 25,940, and our estimate for the total breeding population was 40,000. On 4 June one year later, in the midst of ENSO, we counted only 1,400 murres. As of 1982, the fourth largest murre population in California resided on the North Farallones.

OCCURRENCE PATTERNS

Nonbreeding distribution

We have not banded large numbers of murres at the Farallones, and, therefore, our understanding of their movements and behavior outside the breeding period has been derived largely by piecing together information from other sources, namely, the data presented in

Chapter 3 as well as those of Briggs et al. (1987). After chicks fledge in July and August, adults remain entirely at sea for at least three months, most likely dispersing along the California coast (see Figure 12.1). During these months chicks complete their growth and adults molt their flight feathers, becoming flightless for several weeks (45 to 50 days for east Atlantic murres; Birkhead and Taylor 1977). Father–chick pairs likely remain in central California; they are commonly seen through August and September in Monterey Bay and the Gulf of the Farallones. Females, and males not accompanying chicks, are more mobile before beginning wing molt, but most probably remain in California waters also. The diet of California murres in the late summer and early fall is much more diverse than during the spring and early summer (PRBO, unpubl. data).

It is likely that most of the breeding population remains within a one- or two-day flight from the Farallones the year round. Indeed, Briggs et al. (1987) encountered high densities of murres within 100 km of the Farallones during winter. This pattern is consistent with that observed by Mead (1974) and Birkhead (1974), who reported highest proportions of band recoveries of wintering adult murres in Britain within 200 km of breeding colonies. Since Common Murres do not breed until they are four or five years of age, and probably do not visit the breeding colony until age two or three (Birkhead and Hudson 1977), first-year birds may disperse significant distances from the islands, as do first-year Brandt's Cormorants and Western Gulls (see Chapters 5 and 7). In Britain, Mead (1974) and Birkhead (1974) found that first-year murres disperse much farther than older ones.

Island distribution

At the South Farallones, Common Murres nest in many discrete colonies ranging in size from fewer than ten to over 6,000 pairs (Figures 1.3 and 8.2). The largest colonies are on terraces and gradual slopes facing north and northeast, such as on Lighthouse Hill and within Pelican Bowl. Other important colonies are on shoreline promontories, such as those on the far western edge of West End and at Tower Point. Significant numbers of murres also nest on narrow ledges of cliffs and steep slopes on offshore islets, for example, the north and northwest sides of Sugarloaf and Aulon Islet. Smaller colonies exist in some unusual sites, such as those within Great Murre Cave and on top of Great Arch. Murres do not nest in areas of chronic human disturbance, such as near major trails and human habitation. The intense human disturbance experienced by the murres during the past 140

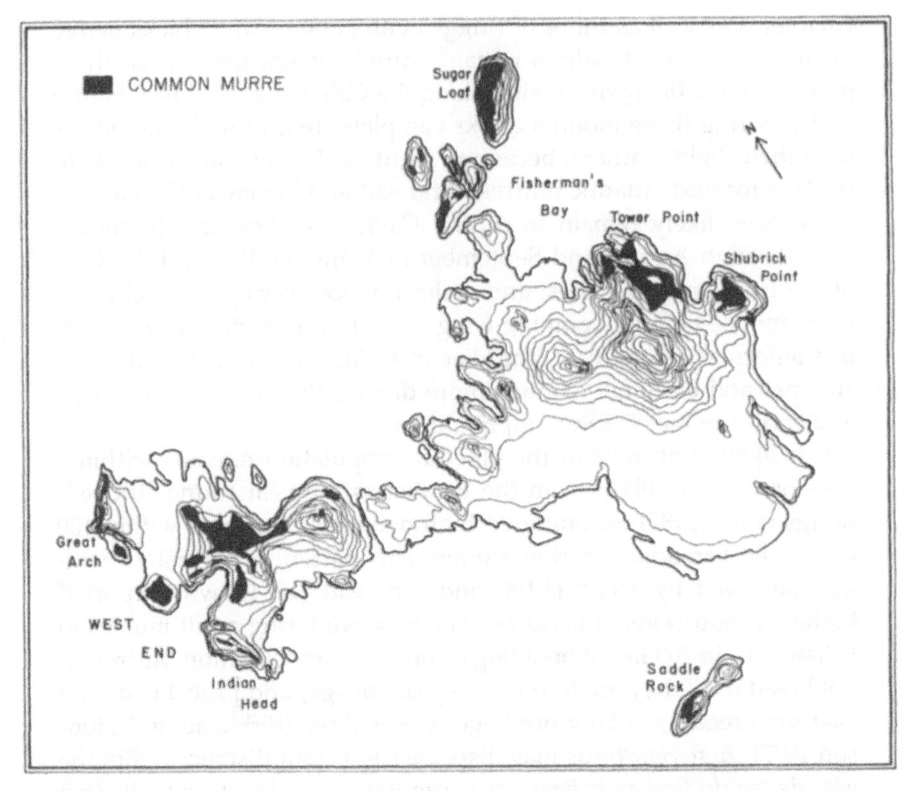

Figure 8.2. Common Murre nesting areas on the South Farallon Islands; shaded areas were used during at least one year between 1977 and 1982.

years is certainly a major factor influencing their present distribution on the island. In recent years, such sources of disturbance have been confined largely to the leeward portions of Southeast Farallon, so disturbance is at a minimum.

The Common Murre, Western Gull, and Brandt's Cormorant are the primary surface-nesting species on the Farallones. These three species nest together within the largest murre colonies on Lighthouse Hill, on Shubrick Point, and at West End. Through their early occupation of nest sites during fall and winter, their high nesting densities, and the commotion of their breeding activities, murres have been able to supplant both cormorants and gulls from nesting areas, despite their smaller body size. The nesting success of Brandt's Cormorants is in fact reduced in areas where murres also nest (Chapter 5). At present, nesting space does not limit this murre population, although it may have been an important factor when the murres numbered several hundred thousand (see Chapter 12).

Through the 1970's, as the population tripled in size, murres expanded their nesting areas mostly by filling in established colonies and by occupying adjacent suitable habitat. Only a few new colonies were founded, particularly during the last five years of the study. Murres tended to establish these new sites within Brandt's Cormorant colonies, perhaps because of the additional protection from gull predation. For example, in 1979 we noticed four or five murres wandering and occasionally standing at sites within our Brandt's Cormorant study colony near Sea Lion Cove (Colony I, Chapter 5). This colony lies several hundred meters from any other murre colony and had not contained nesting murres for at least the previous decade. In 1980, three eggs were laid by murres in this colony and two chicks were fledged; by 1982 at least seven pairs nested there. In 1981 and 1982 we observed murres inspecting several other colonies where only Brandt's Cormorants had nested previously.

During the 1982 and 1983 breeding seasons, the number of California Sea Lions (*Zalophus californianus*) using the Farallones increased dramatically (Huber et al. 1985). The sea lions' increase was likely related to ENSO effects on prey species in southern California and to the general increase of their population throughout their range (see Heath and Francis 1983). At the Farallones, as sea lions became more numerous, they hauled out higher and higher above the intertidal zone, invading several murre colonies and causing significant egg loss. This disruption of breeding was especially important in the large colony at Fertilizer Flat, on the northeast side of Lighthouse Hill (Figure 1.4). Several hundred sea lions sprawled over the lower half of the colony for several weeks in May and June of both years, displacing nearly all murres in the area. As most of these murres were still on eggs, it is possible that they moved to other areas and relaid. On several occasions individual sea lions wandered through other colonies of incubating murres. Prior to human exploitation of pinnipeds in the early nineteenth century, these animals certainly could have had a major influence on the distribution of nesting seabirds at the Farallones (Warheit, Lindberg, and Boekelheide 1984; see Chapter 12).

Colony attendance

Murres become increasingly visible near the island in mid-October. Their first landfalls occurred between 22 and 29 October during every year except two: 1976, when they arrived on 6 November, and 1983, when they arrived on 16 December (see also Figure 12.1). Their first arrivals are exciting events: the murres concentrate at dawn in large flocks, rapidly flying around the islands, and then approach the cliffs warily, circling many times before landing. Once a few finally land, others pour from the sky until the colonies are again filled with calling murres.

Arrival dates of murres at the Farallones are much earlier than those in northerly breeding areas affected by sea ice (Tuck 1961) but are similar to those of murres in the British Isles (Greenwood 1972, Mead 1974, Birkhead 1974). Earlier attendance at more southerly colonies possibly occurs because these murres can find food throughout the fall and winter within a day's flight of their colonies. The abnormally late landfall in 1983 provides possible support for this hypothesis. Extreme ENSO conditions during fall 1983 (Chapter 2), probably through effects on prey populations, presumably caused murres to forgo visits to the breeding ledges. On the other hand, extended prelaying occupation of ledges could also be a response to competition for nesting space, although at present population levels space is certainly available (see Chapter 12).

Through most winters murres visited nest sites in an irregular pattern and were generally absent on days with high winds or heavy rains. Birkhead (1978a) attributed a similar pattern in Britain to the possibility that murres find it difficult to forage in heavy seas. In addition, we noticed that they rarely came ashore on days with large swells, regardless of rain or wind. This may also support the findings of Gaston and Nettleship (1981), who observed that attendance prior to egg laying was affected by changes in barometric pressure. Particularly in winter, large swells in the Gulf of the Farallones are associated with the passage of storm systems (see Chapter 2), when air pressure fluctuates considerably.

In winter 1971–72, murres sporadically visited the Saddle Rock colony from December to February, but attended more consistently in March and April just before the egg-laying period (Figure 8.3; see also Lloyd 1975). In Table 8.3, we compare murre landfalls during a relatively "normal" winter (1980–81) with those during an ENSO winter (1982–83). Murres came ashore on 72% of days in 1980–81 but on only 7% of days in 1982–83. Such a difference in attendance patterns was likely a function of both lessened prey availability and increased storm frequency in 1982–83.

Attendance was consistently higher in the morning than after noon throughout winter and spring, as reported by Lloyd (1975) for Com-

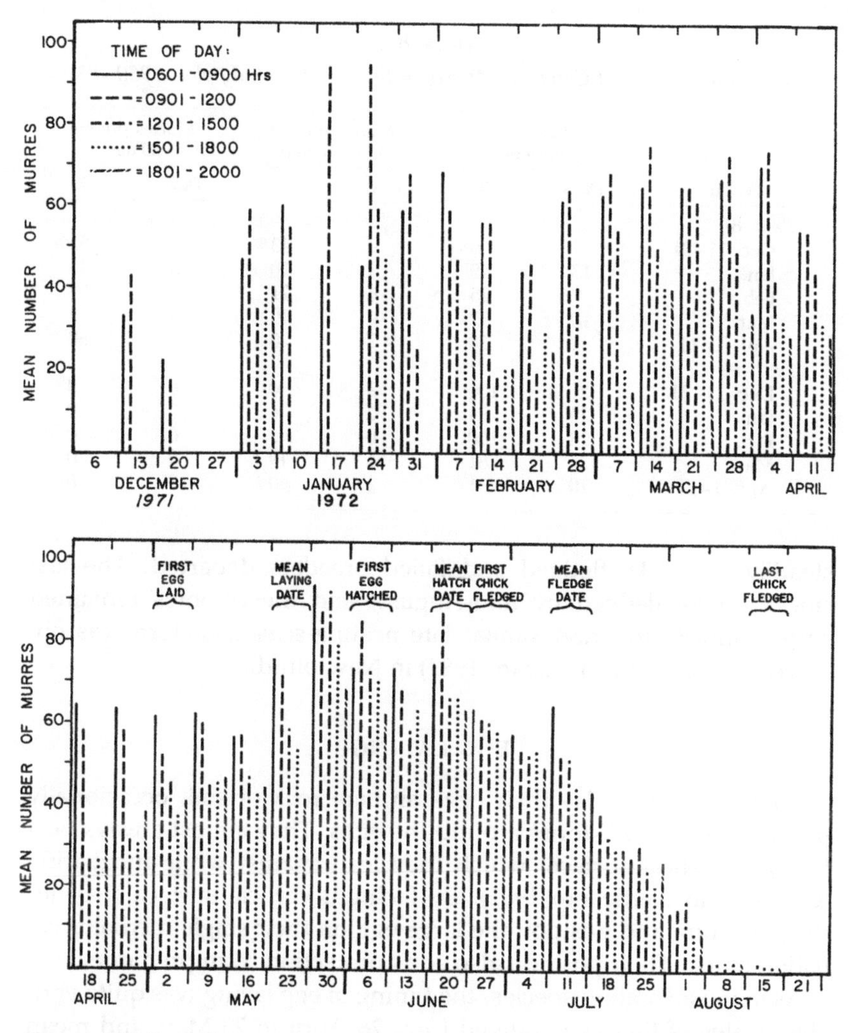

Figure 8.3. Attendance patterns of Common Murres at the Saddle Rock study plot; seven-day periods (excepting the one leap-day), December 1971–August 1972. No murres present mid-August to late autumn.

mon Murres at Skokholm (Figure 8.3). Murres frequently arrived at sites in the early morning but departed by mid-day (see also Table 8.3). They occupied some colonies on certain days from dawn to dark but on others for only a few hours.

During counts in 1972, peak numbers occurred from late May to mid-June, when most birds were incubating eggs (Figure 8.3). Following the median hatching date in June, colony attendance dropped

TABLE 8.3

Winter attendance of Common Murres in the Farallon study plot, 1980–83

	Murres not present		Murres present morning only		Murres present all day	
Period	Days	%	Days	%	Days	%
1980–81						
Dec. 11–31	1	5%	9	43%	11	52%
Jan.	13	45	3	10	13	45
Feb.	7	25	6	21	15	54
Mar.	10	32	6	19	15	48
Apr. 1–11	2	20	2	20	6	60
1982–83						
Dec. 11–31	21	100	0	0	0	0
Jan.	29	94	2	6	0	0
Feb.	28	100	0	0	0	0
Mar.	25	81	6	19	0	0
Apr. 1–11	10	91	1	9	0	0

steadily as chicks fledged and failed breeders departed. The last chicks finally fledged by mid-August, and the colonies remained empty until October. A similar late nesting-season pattern was observed by Piatt and McLagan (1987) in Newfoundland.

EGG LAYING

Common Murres lay their eggs on bare rock or soil, occasionally placing a few small stones around the eggs. They almost always nest facing a vertical surface, which they lean against while incubating (Gaston and Nettleship 1981). By facing a vertical surface they lift their sternum and position the large egg on their single medial incubation patch. Their position also adds protection from predators.

As in most Farallon species, the timing of egg laying was quite variable. Dates of first eggs ranged from 26 April to 23 May, and mean laying dates ranged from 9 May to 9 June (Figure 8.4). The mean laying dates for most years were statistically similar. The two years with the latest mean dates, 1978 and 1983, were significantly different from each other and from all other years; egg laying was also late in another set of ENSO/warm-water years, 1972–73 (Table 8.4). The two earliest years, 1979 and 1981, were the only years in which egg laying commenced in April and in which the modal date of laying occurred during the first ten days of May.

Mean laying dates occurred earlier, in general, as the study progressed, especially if one excludes the unusually late years 1978

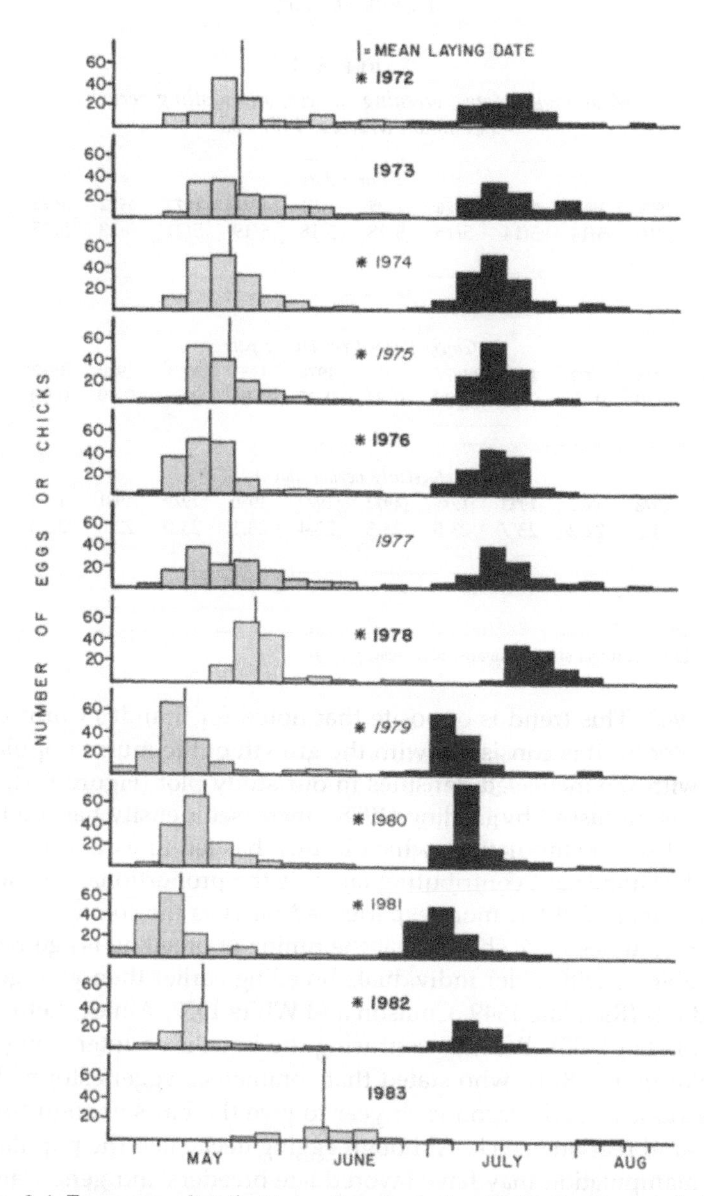

Figure 8.4. Frequency distributions of murre laying (light shading) and chick-exodus (dark shading) dates at the Shubrick Point study plot, 1972–83; during years indicated by an asterisk (*), laying was significantly peaked and skewed (t test, $p < .05$).

TABLE 8.4

Mean laying dates, breeding success, and nestling periods, Common Murres, 1972–83

Laying date

1981	*1979*	1980	**1982**	**1976**	*1975*	1974	*1977*	**1973**	**1972**	**1978**	**1983**
5/9	5/10	5/14	5/14	5/15	5/18	5/18	5/19	5/21	5/23	5/25	6/9
———	———										
		———	———	———							
					———	———	———				
								———	———		

Chicks fledged per laying pair

1983	**1978**	**1972**	**1973**	*1977*	*1979*	**1976**	1981	1974	1980	**1982**	*1975*
0.05	0.70	0.78	0.79	0.80	0.84	0.85	0.88	0.88	0.89	0.90	0.91
	———	———	———								
		———	———	———	———	———	———	———	———	———	———

Nestling period (days)

1976	**1978**	**1982**	**1973**	*1975*	**1972**	*1977*	1980	*1979*	1981	1974
25.2	24.5	24.2	23.7	23.5	23.5	23.4	23.1	23.0	22.6	22.3
	———	———	———							
		———	———	———	———	———				
				———	———	———	———	———	———	
							———	———	———	———

NOTE: Lines connect similar means; SNK test, $p > .05$.

and 1983. This trend is opposite that noted for Brandt's Cormorants (Chapter 5). It is consistent with the growth of the murre population and with the increased densities in our study plot (Figure 8.1). Possibly, as discussed by Darling (1938), increased density has led to increased social stimulation, which in turn has led to earlier breeding. Another important contributing factor is the proportional increase in the number of older, more experienced birds as the colony matured. Several studies have shown that the timing of breeding is age-related in seabirds, with older individuals breeding earlier than younger individuals (Richdale 1949, Coulson and White 1957, Ainley, LeResche, and Sladen 1983). Possibly confusing the issue is an interesting note by Nordhoff (1874), who stated that commercial eggers stopped egg collections in mid-season each year to give the birds enough time to relay and rear one chick. Although egging decimated the population, this manipulation may have favored late breeders and genetically altered the population's laying phenology.

Despite annual variation, egg laying within years was closely synchronous. In all years except 1977, more than 50% of original clutches were laid within a ten-day period, even though laying usually extended four to six weeks (Figure 8.4). Laying dates were significantly peaked except during the late years 1973, 1977, and 1983, and were significantly skewed toward the early season in all years except 1983 (two-tailed t test, $p < .05$). This differs from the patterns observed in Thick-billed Murres by Gaston and Nettleship (1981), who found that mean laying dates were similar but laying synchrony was variable within the three years of their study; in three subsequent years, 1978, 1981, and 1984, however, laying dates differed somewhat, though not to the degree exhibited by Farallon murres (A. Gaston pers. comm.). It is possible that more northerly populations of both Thick-billed and Common murres must lay eggs within a narrower period each year because of the short season available for reproduction (see Chapter 12). McLagan and Piatt (ms) noted little variation in breeding phenology among several colonies in Newfoundland studied for several years.

Neither dates of first eggs nor mean laying dates correlated well with mean monthly sea-surface temperatures for January through April (Spearman rank correlations, $p > .05$). In other Farallon species, such a correlation existed (see Chapters 7, 9, and 10). Two years, 1972 and 1974, stand out as having relatively cool sea-surface temperatures during winter, yet murre laying dates in those years were later than average. On the other hand, sea-surface temperatures in 1981 were very warm, yet mean laying dates then ranked earliest among years. Ocean conditions apparently did affect the synchrony of egg laying (Figure 8.5). The distribution of egg dates was more skewed toward the start of the laying period during years when the spring transition (Chapter 2) was especially sharp. The correlation is significant, however, only if two outlying years, 1977 and 1983, are excluded from the analysis ($r_S = .6477$, $p < .05$). Actually, three of the five years in which values fell farthest from the calculated regression line were the major warm-water years, 1973, 1978, and 1983 (Chapter 3). The observed relationships suggest that differences in timing of the onset of intense upwelling may have an important role in adjusting the murres' breeding phenology but that other factors are involved as well. Additional data would likely validate the regression.

Egg color

Murre eggs come in an astonishing variety of colors and patterns (Tuck 1961, Gaston and Nettleship 1981), which the birds apparently use to identify their own eggs (R. A. Johnson 1941). At the Farallones in the period 1973 to 1983, we recorded background colors on eggs in our plot, mostly to help identify those that rolled away from sites ($n = 1,288$). A compilation of colors showed that dark green or blue

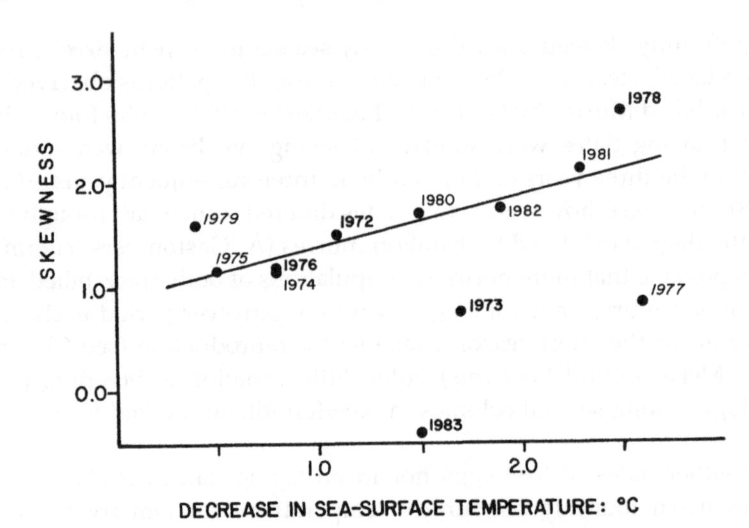

Figure 8.5. Relationship between degree of skewness in the frequency distribution of Common Murre egg-laying dates and the drop in sea-surface temperature during the spring upwelling period, 1972–83; excluding 1977 and 1983, $y = 0.5x + 0.89$, $r = .648$ (see text).

eggs accounted for slightly greater than half the eggs in eight years: 53 ± 3%. During three years, 1978, 1982, and 1983, the ratio of pale eggs (white or light green-blue) to dark eggs differed significantly from the usual, approximately 50:50 ratio observed also by Gaston and Nettleship (1981: their plate 14, table 53; their color classes 1–3 appear similar to our white, light green-blue, and tan categories, and their classes 4 and 5 appear similar to our dark category). During those three years the ratios were 60:40 (n = 121) and 63:37 (n = 224), respectively. The only other samples in which the proportion of light eggs exceeded 50% were those for 1976 (52%, n = 150) and replacement eggs in all years (56%, n = 32). These patterns indicate (1) that the strong pigmentation of eggs could be a function of the murres' eating more strongly pigmented prey early in the breeding season (i.e., euphausiids in the usual prelaying period of most years, March–April, as observed by direct samples in 1985 and 1986; see Chapter 3) or (2) that during some years (i.e., the warm-water years such as 1976, 1978, and 1982–83) the euphausiid component of the murre diet (or the diet of murre prey) is reduced.

Incubation period

Incubation periods were statistically similar in all years but one. From 1973 to 1983, incubation periods ranged on average from 32.2 ± 1.6

days to 32.7 ± 1.3 days, but in 1972 they averaged 31.4 ± 1.9 (SNK test, $p < .05$). In total, incubation periods averaged 32.4 ± 1.4 with a range of 26 to 39 days (n = 1,202; Figure 8.6). The mean value is virtually identical to that calculated by M. P. Harris and Birkhead (1985). The small number of eggs with particularly long or short incubation periods is consistent with Gaston and Nettleship's (1981) observations of Thick-billed Murres. The authors attributed outlying periods to observer error, but the occurrence of outliers in both studies plus our daily close observations suggest that murre incubation periods may sometimes exceed the average by a wide margin. This is surprising but also curious, because murre eggs are incubated almost continuously. Similarly, we have no explanation for the slightly shorter incubation periods in 1972.

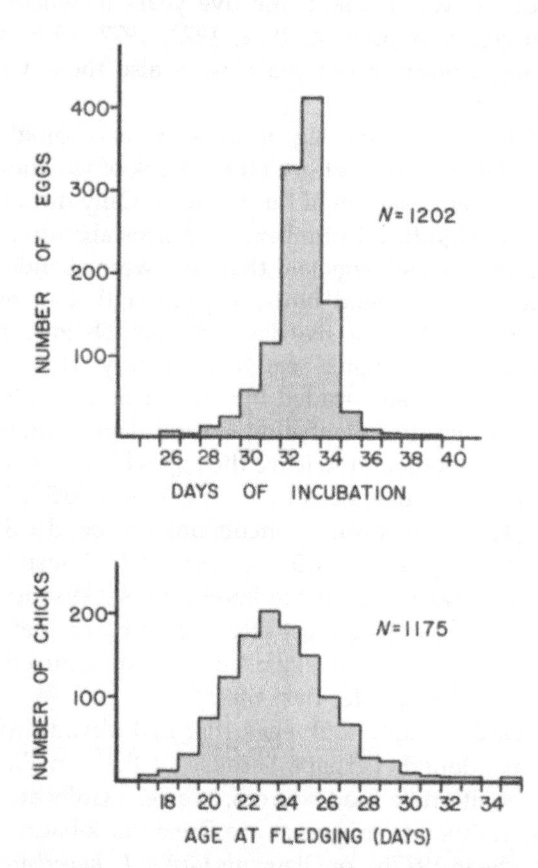

Figure 8.6. Frequency distributions of Common Murre incubation and nestling periods; all nests combined, 1972–83, including first as well as replacement eggs.

REPRODUCTIVE SUCCESS

Hatching success

Hatching success was consistent from year to year (Appendix 8.1). Excluding the year of strong ENSO, 1983, success for first eggs ranged from 79.5 to 92.7% and averaged 84.7%, rates similar to those observed elsewhere (M. P. Harris and Birkhead 1985, McLagan and Piatt ms). No statistical differences existed among years other than 1983, when hatching success at 31.7% was lower than in all others (t tests following arcsin transformation, $p < .05$). The murres experienced high rates of hatching in some years when other species, especially cormorants, commonly abandoned their nests (1976, 1978, 1980, and 1982; Chapters 5 and 6). A relationship between time of laying and hatching success was evident: the five years in which the murres' hatching success was poorest, 1972, 1973, 1977, 1978, and 1983 (all but one being a warm-water year), were also those with the latest mean laying dates.

Despite laying only one egg, murres are exceedingly resolute in maintaining their breeding effort, and the lack of variation in hatching success is one manifestation of this strategy. Only during the poorest year, 1983, did significant numbers of murres abandon their eggs (a minimum of 12% of all eggs laid that year were abandoned). By following uniquely plumaged birds, we noted that many incubation shifts in 1983 lasted three to five days, presumably in response to long periods during which mates searched for prey. This contrasts with other years, when mates traded incubation duties at least once per day. In 1983, even after incubating several days without food, most murres appeared reluctant to leave the egg when their mates arrived.

Overall, 12.2% of first eggs (n = 1,506) and 30% of replacement eggs (n = 54) were lost during incubation (Appendix 8.1). An additional 3.2% of all eggs (n = 1,560) never hatched despite being incubated full term. Most eggs lost to known causes disappeared by rolling away (Table 8.5). This most often occurred accidentally as birds shifted positions or turned eggs, but it also occurred occasionally when they were fighting for nest sites.

Western Gulls usually took eggs that had already rolled away or had been abandoned. Perhaps because of their smaller size or the availability of alternate food sources, Western Gulls are not as much of a predation threat to murres as are Great Black-backed Gulls, *Larus marinus* (Birkhead 1977b), or Glaucous Gulls, *L. hyperboreus* (Daan and Tinbergen 1979), at other murre colonies. During the strong ENSO of

TABLE 8.5

Causes of Common Murre egg loss, 1972–83

Cause of loss[a]	n	Percentage of eggs: Not hatching	Percentage of eggs: Lost[b]
Rolled away from nest	21	8%	39%
Wedged in rocks	2	1	4
Broken	7	3	13
Dislodged by fighting murres	2	1	4
Dislodged by Brandt's Cormorant interference	2	1	4
Gull predation	6	2	11
Abandoned	14	6	26
Subtotal	54	22%	101%
Disappeared, cause unknown	145	58%	
Never hatched (incubated past day 39)	50	20	
Total	249	100%	

[a] First and replacement clutches combined.
[b] Excluding eggs lost due to unknown causes, or after normal incubation period; total exceeds 100% because of rounding.

1983, however, Western Gulls, hungry in their own right, became much more brazen and often grabbed murres by their tails or wings and dragged them off their eggs. Egg predation was facilitated by the wide spaces between nesting murres that summer: fewer than 50% of the egg-laying sites used in 1982 were used in 1983, and distances between neighbors widened rapidly as many failed breeders departed. This, along with the absence of cormorants, permitted gulls easy access to the murres remaining on eggs.

The percentage of eggs not hatching within the expected incubation period, presumably because of infertility or embryo death, was less than 5% in all years except 1979 (Appendix 8.1, Table 8.5). In the latter year, 9% (n = 135) never hatched, for reasons unknown, although we believe that an unusually high proportion of younger birds attempted to breed that year, and, in accord with results from studies of known-age seabirds, these inexperienced birds would be expected to be ineffective breeders. These figures are generally within the range of 3.6 to 7.5% observed by Gaston and Nettleship (1981) for Thick-billed Murres at Prince Leopold Island.

On several occasions murre eggs took circuitous journeys. Once, a murre egg accidently rolled into one of our Brandt's Cormorant study nests, where the cormorants had lost their own eggs the day before. The egg hatched after being incubated by the cormorants for 26 days. The foster parents tried to feed the murre chick by gaping and regurgitating food, but, not surprisingly, the chick did not respond and died after two days. Lewis (1929) also observed a murre egg in a Double-crested Cormorant nest in Quebec but did not follow its fate. On another occasion, we observed an egg that rolled away from its mother immediately after being laid. It reached a spot downhill where it sat unattended for about 30 minutes. A passing gull swallowed the egg whole, stood in the colony for several minutes with a hugely distended esophagus, then regurgitated it. The egg again rolled down the slope, this time colliding with a standing murre. The murre promptly climbed on the egg and began incubating it. Unfortunately, we do not know its eventual fate.

Overall, hatching successes for eggs laid early and midway in the laying period did not differ ($\chi^2 = 1.73$, DF = 1, $p > .1$) but were statistically greater than for late eggs ($\chi^2 \geq 25.03$, DF = 1, $p < .01$; Figure 8.7). Early eggs had the highest hatching success in seven of 12 years, middle eggs in four years, and late eggs in only one year, 1983. Interestingly, early eggs had poorer hatching success than middle eggs in four of five years when egg laying was earliest—1981, 1979, 1982, and 1976. This suggests the possibility of increased risk for eggs laid before the majority are laid. One factor that might in-

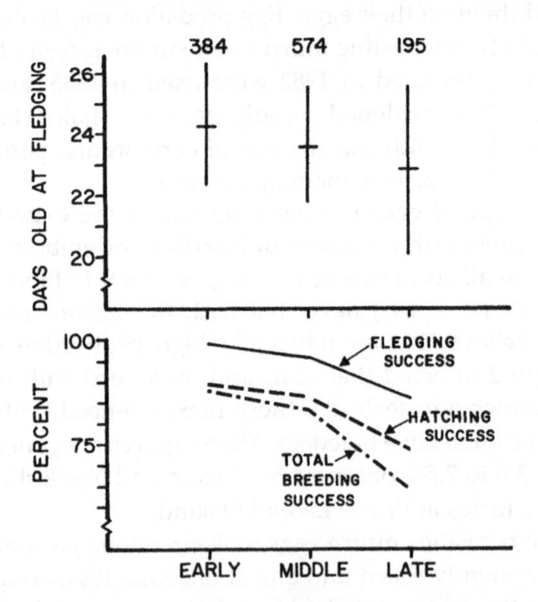

Figure 8.7. Comparison of success and chick age at fledging for early-, mid-, and late-season Common Murres, 1972–83. Horizontal lines indicate means; vertical lines, standard deviations; numbers at top, sample sizes.

crease risk early in the nesting period is the greater frequency of territorial disputes that occurs then.

Replacement eggs

A second egg was laid at 31.8% of all sites (n = 170) where the first egg disappeared before hatching, and more late eggs than early eggs were lost—7.3 and 8.9% of early and middle first eggs were lost, compared to 21.6% of late first eggs. The overall replacement rate agrees closely with the 30% rate observed in Thick-billed Murres (Gaston and Nettleship 1981) but is lower than the 52% observed in Common Murres elsewhere (Birkhead and Hudson 1977). Like these authors, we also observed that replacement eggs were much more likely at sites where the first egg was laid early in the nesting period: 82% of lost first eggs were replaced, compared to 33% of middle eggs and 8% of late ones. Since we did not have marked birds, we do not know if the second egg that appeared at a given site was laid by the same individuals that laid the first one. We assume, however, that in most cases second eggs were replacements for lost eggs. Neither we nor Gaston and Nettleship (1981) ever observed more than one replacement egg laid at any site. Tuck (1961) reported that 11% of his birds laid two replacement eggs, and Uspenski (1958) also observed a few second replacements. In the mid-1800's, egg harvesters believed that murres would readily lay several clutches of eggs, and that is one reason why the murre population on the Farallones declined!

The period between loss of the first egg and appearance of a second averaged 15.1 ± 3.4 days (n = 50, range 13 to 23 days). This generally falls within the range of 12 to 18 days required for rapid yolk development in the Common Murre (Roudybush et al. 1979) and agrees with values found by Hedgren and Linnman (1979) and Birkhead and Hudson (1977) for Common Murres and by Uspenski (1958) and Gaston and Nettleship (1981) for Thick-billed Murres.

Over half of all replacement eggs (n = 54) were laid in only three years: 1973, 1977, and 1981. In the latter two years this seemed due to a combination of relatively high loss of early and middle eggs plus favorable conditions for egg laying continuing into the latter part of the laying period.

Breeding success

Murres were extremely successful in raising their chicks, especially in 1979, when all 112 of the chicks hatching from original clutches fledged (Appendix 8.1). Only in the years of late laying, 1978 and 1983, did fewer than 93% of chicks survive to leave the colony. Like hatching success, fledging success was statistically similar in most years (t tests following arcsin transformation). Again, 1983 (15%) differed from all other years; additionally, 1978 (85%) was similar to only 1973 (93%), 1976 (94%), and 1981 (94%). Except for the few outlying years, fledging success at the Farallones, and at a site in the Baltic Sea (Hedgren 1980), appears to be higher than rates observed in Britain, Newfoundland, or the Pribilof Islands (Birkhead and Hudson 1977, McLagan and Piatt ms, Hunt, Eppley, and Schneider 1986).

For the entire study period, chicks fledged more successfully from eggs laid early than from those laid mid-season ($\chi^2 = 6.36$, DF = 1, $p < .02$), and middle-season chicks fared better than late chicks ($\chi^2 = 24.28$, DF = 1, $p < .01$; Figure 8.7). Actually, in eight of the 12 years all chicks fledged from eggs laid early, and half of all the chicks that failed to fledge from early eggs did so in only one year, 1983. Middle chicks had the highest fledging success in only two years, 1974, when nearly all chicks fledged, and 1983, when the only chicks that fledged were middle chicks. In no year did chicks from late eggs have the highest fledging success.

Hedgren and Linnman (1979) and Birkhead and Nettleship (1981) found that chicks from late eggs weighed significantly less than early and middle chicks. Such a weight disadvantage could lead subsequently to increased mortality of late chicks even if they do successfully depart the breeding ledges. Lloyd (1979) explained higher chick mortality in late-breeding Razorbills, *Alca torda*, by the fact that late breeders are young, inexperienced birds; the same is true in other seabirds, for example, Adélie Penguins, *Pygoscelis adeliae* (Ainley, LeResche, and Sladen 1983). In contrast, we observed in 1983 that early eggs were at greater risk from predation, a trend also noted by Gaston et al. (1985). In accord with this pattern, during 1983 egg laying was less synchronized than during any other year. Thus, the protection from predation afforded by synchronous, colonial nesting was weakened (A. Gaston pers. comm).

In accord with the high fledging success, noted above, the mean number of chicks fledged per site was statistically similar in all but a few years: the rate for 1978 was similar only to those for 1972, 1973, and 1977, and the rate for the extreme ENSO episode of 1983 was lower than those for all other years (Table 8.4). At the Farallones, early breeders fledged the greatest number of chicks per pair (0.9 ± 0.3), although the number was not statistically different from that for middle breeders (0.8 ± 0.5; $t = 1.427$, DF = 1162, $p > .1$). The number of chicks fledged per pair by middle nesters was statistically greater than that of chicks fledged by late breeders (0.6 ± 0.5; t = 5.871, DF = 1055, $p < .001$).

Hatching and fledging success was lower for replacement eggs than for first-clutch eggs (Appendix 8.1). Most replacements were laid well after the median laying date and had lower success, similar to that of late first eggs. In total, chicks fledged from 48% of replacement eggs (n = 54). This compares with 88.0% of early eggs (n = 449), 82.9% of middle eggs (n = 715), and 63.7% of late first eggs (n = 342).

The murres' productivity being so consistent is particularly interesting in the context of the environmental variability they experienced and the much greater variability in productivity exhibited by other Farallon species. To be sure, murres are not quite comparable because we measured "fledging success" when chicks were less than half grown rather than when fully grown, as for other species. On the other hand, the rapid growth of the murre population during our 12 years of study indicates that "true" fledging success, and subsequent survival of fledglings, were both high. Unfortunately we were not able to measure chick growth rates, for although equivalent numbers of chicks fledged per site in most years, it is possible they did not fledge at similar weights or body condition, as noted for some of the other Farallon species (see Chapters 7, 9, and 10; also Hunt, Eppley, and Schneider 1986).

The high rate of fledging success exhibited by Farallon Common Murres was similar to that at a site in the Baltic Sea between 1974 and 1977 (Hedgren 1980). Murres at other localities appear to have somewhat lower but nevertheless equally consistent breeding success. Birkhead and Hudson (1977) reported that 72% of Common Murre pairs in Wales fledged chicks, ranging from 62 to 74% in three years, and figures for a site in Newfoundland during five years were similar to these (McLagan and Piatt ms). Gaston and Nettleship (1981) found that between 68 and 79% of Thick-billed Murres fledged young, also during three years.

Catastrophic nesting failure, such as we observed during the 1983 ENSO, apparently occurs infrequently in murres. Vermeer, Cullen, and Porter (1979) and J. Rodway (*fide* A. Gaston pers. comm.) noted low chick production by Common Murres at Triangle Island, British Columbia, in 1977 and 1984, respectively, presumably due to the disappearance of their usual prey. Similarly, M. P. Harris (1984) hypothesized that a decline in the nesting success of murres in the Lofoten Islands, Norway, was due to decreased food availability. Many authors mention bad weather, especially hard rain and wind, plus atypi-

TABLE 8.6

Spearman correlation coefficients, Common Murre nesting, 1972–83

	Row number		
Nesting variable	1	2	3
1. Clutch-initiation date			
2. Hatching success	−.7063[a]		
3. Fledging success	−.5734	.4528	
4. Fledglings per laying pair	−.7378[a]	.9038[a]	.6345[a]

[a] Significant at $p < .05$.

cal sea-ice conditions, as major influences on the phenology and success of murres breeding in the Arctic (Uspenski 1958, Belopol'skii 1961, Birkhead and Nettleship 1981).

Rank correlations among yearly mean laying dates, hatching success, fledging success, and the number of chicks fledged per pair showed that the number fledged was significantly correlated with all other parameters (Table 8.6). Hatching success showed the highest correlation, suggesting that events during the incubation period may be most important in determining the success of murres prior to the chicks' departure from the colony. Timing of laying was also significantly correlated with hatching success (as noted above, the five years of latest breeding also had the poorest hatching success). Fledging success significantly correlated only with the number of chicks fledged per pair, but this seems consistent with the small degree of variability in fledging success between years.

Chick development

As with cormorant chicks (Chapters 5 and 6), we were unable to gather information on growth of murre chicks. This was due primarily to our efforts to avoid disturbing not only the murres but also other species. We were able to document some aspects of murre chicks' lives, however, such as mortality prior to departure from the colony and the amount of time they stayed at nest sites.

Within our study plot, 75% (n = 79) of chicks that did not successfully depart disappeared with no known cause of death. We observed another 11% dead at their sites, 13% taken by Western Gulls, and one chick forced away from its site and killed by an adult murre. Eight of the ten instances in which we witnessed chicks taken by gulls occurred in 1983, when, as discussed above, widespread nesting failure and large next-neighbor distances permitted easy access for hungry gulls. Particularly in 1983, but also in 1978, we observed chicks alone at sites while both parents foraged at sea. These chicks were undoubtedly more vulnerable to gull predation. Western Gulls, however, were mostly scavengers or kleptoparasites during the murres' chick period, picking up dropped food or stealing it whenever possible (L. Spear pers. comm.).

Overall, the age at which chicks died or disappeared was not skewed toward either young or old chicks. Among those for which we knew dates of death, 26% ($n = 64$) were one to five days old, 25% were six to ten days old, 26% were 11 to 15 days old, and 22% were older than 15 days. Age of death was clustered toward older birds in only one year, 1978, when 80% ($n = 18$) of deaths occurred among chicks at least 11 days of age. The reverse was true during another warm-water year, 1983, when 82% ($n = 11$) of chicks died when younger than ten days. Most other studies (Uspenski 1958, Tuck 1961, Gaston and Nettleship 1981) observed highest mortality among young chicks, usually those less than five days of age.

The first chicks usually departed in late June or early July, with average first departures occurring on 30 June if the extremely late year 1983 is excluded (Figure 8.4). Mean departure dates usually fell in the first half of July. As did laying dates, mean departure dates advanced as the study progressed, the late years 1978 and 1983 again discounted. Like that of laying and hatching, the distribution of dates of departure during most years was skewed toward the beginning of the season. Departure dates were slightly less clumped than laying or hatching dates.

The number of days that murre chicks remained at nest sites (i.e., the nestling period) averaged 23.5 ± 2.4 days for the entire study ($n = 1{,}175$; Figure 8.6). This is somewhat longer than the 18 to 21 days in the literature summarized by Hedgren and Linnman (1979) but is consistent with the more recent literature summarized by Gaston (1985). Uspenski (1958) felt that chicks at more northerly colonies departed at a younger age because of the shortened season available for breeding. At the Farallones, nestling periods varied somewhat between years, ranging from 22.3 ± 3.4 days in 1974 to 25.2 ± 2.7 days in 1976, a value different from all others (Table 8.4). Noticeable within-season variation in ages at fledging also occurred, with earlier-hatching chicks remaining longer than late-hatching ones (Figure 8.7). Nestling periods ranked by year did not correlate well with either fledging success ($r_S = 0.216$, $p > .5$) or the number of chicks produced per pair ($r_S = .445$, $p > .1$). Both Tuck (1961) and Greenwood (1964) suggested that, for Thick-billed Murres in the Arctic, longer periods at nest sites indicate slower chick growth.

Another explanation for differences in the length of nestling periods is the influence that strong winds apparently have on the timing of chick departures (Gaston and Nettleship 1981; Figure 8.8). During June and July, seasonal northwesterlies blow consistently (Chapter 2); gusts exceed 50 km/h and occasionally reach 80 km/h. Chicks ready to fledge begin to accumulate during these windy periods, so that as soon as wind speed drops many chicks depart. This was most dramatic in 1975, when a gale persisted from 10 to 12 July and quickly abated on the 13th. No chicks fledged on 10 or 11 July, but from 12 to 14 July, 71% ($n = 99$) of all the chicks that fledged that year went to sea.

As has been the case in all other studies, we observed a strong diurnal pattern in the timing of chick departures. All Farallon chicks

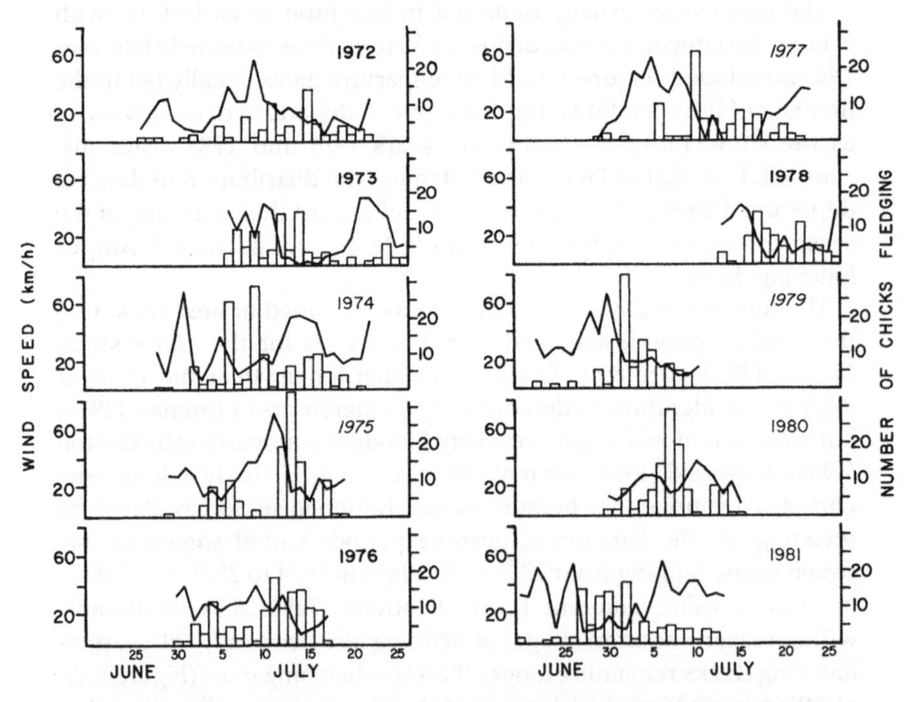

Figure 8.8. Comparison of wind speed and Common Murre fledging dates, 1972–81. Columns denote number of chicks fledging during evening from the Shubrick Point study plot; solid line denotes the wind speed at 1900 hrs.

left in the evening or after nightfall (2000 in late June) and rarely left their nests before 1800. Greenwood (1964) and Daan and Tinbergen (1979) suggested that synchronous fledging "swamps" predators. According to L. Spear (pers. comm.), Western Gulls at the Farallones rarely attempt to capture jumping murre chicks. Following departure from ledges, father–chick pairs immediately swam away from the island. From boat transects in the Gulf of the Farallones (Chapter 3) we observed the largest concentrations of parents with chicks inside the 15-fathom (27-m) depth contour during the late summer. Similar concentrations occur throughout central California coastal waters, especially in Monterey Bay (K. Briggs pers. comm.). The overall annual cycle of the Farallon murres is depicted in Figure 12.1, with an estimate of continued dependency of chicks on parents after they have departed the islands.

THE MURRE BREEDING STRATEGY

Common Murres have proved to be exceptionally successful breeders at the Farallones. They consistently take to sea, as three-week-old chicks, 70 to 90% of the eggs they lay, despite annual variation in prey sources. Their success is affected in only the leanest years. Had the 1982–83 ENSO, one of the strongest on record, not occurred during this study, we may have concluded that murres were impervious to many of the annual changes that so strongly affected other Farallon species. Even during 1983, however, the murres were remarkably determined to maintain their breeding commitment once they had laid eggs, sometimes spending three to five days on incubation shifts before abandonment. Such a commitment, in spite of a relatively low reproductive rate, may have contributed to the rapid population increase observed during the 1970's. Unlike cormorants, murres are not able to capitalize on years of exceptionally abundant food by increasing the number of chicks they raise (see Chapter 12). It may be that murres are physiologically better equipped than cormorants to cope with the occasional necessity of long turns at incubation; indeed, two to three days, a period almost unknown at the Farallones, is the normal incubation shift for members of mated pairs at Digges Island (A. Gaston pers. comm.).

Like most Farallon diving birds, murres rely heavily on juvenile rockfish as feed for their chicks (Chapter 3). The rockfish appear to be an easily obtainable prey, available close to the island in large numbers. In years of moderate to high rockfish availability, murres are virtually assured successful breeding. The number of chicks fledged per pair was related to the proportion of juvenile rockfish in their diet between 1973 and 1983 ($r_S = .582$, $p = .06$). Yet the murres' consistent breeding success seems also largely due to their ability to switch successfully to other prey, such as anchovies, smelt, and squid, when rockfish are not available. Exercising this plasticity of behavior, of course, is dependent on the high though more distant availability of these prey species. The murres' plasticity in foraging range appears to be the key to their consistent success; in contrast to species such as the Pelagic Cormorant or Pigeon Guillemot, murres have a propensity to forage in more habitats and over a much wider area (Chapter 3). At some colonies, murres may fly great distances to obtain prey, largely because of a scarcity of suitable colony sites. For example, at Prince Leopold Island, Thick-billed Murres regularly fly 60 to 100 km, and sometimes as much as 150 km, in search of food during the incubation and nestling periods (Gaston and Nettleship 1981, R. G. B. Brown 1985). Similarly, breeding Common and Thick-billed murres feed up to at least 60 km from colonies at Cape Thompson, Alaska (Swartz 1966). On the other hand, murres at Skomer and in Newfoundland forage within 10 km of breeding cliffs (Birkhead 1976, R. G. B. Brown 1985). Thus, Farallon murres exhibit almost the entire range of foraging distances observed elsewhere. Greater distances, however, increased overall foraging effort (Chapter 3), and, not surprisingly, if breeding success was low it was so during years when the birds flew long distances for food.

Murres displayed annual variation in only one very important parameter: the timing of laying. The fact that their laying dates were significantly delayed in the two years of extremely warm water, 1978 and 1983, and advanced in certain years of cold water shows the murres' sensitivity to oceanographic conditions and resulting vagaries of prey availability. Variation in laying phenology may also affect the murres' consistent breeding success, for it results in breeding efforts' being timed to food availability. On the other hand, most other Farallon species (as detailed in various chapters) demonstrated equivalent or even greater variability in the timing of egg laying but were less consistent in reproductive success (storm-petrels excepted; Chapters 4 and 12). It is particularly interesting that regardless of when the murres laid eggs, their efforts were synchronous once laying had begun. This synchrony is perhaps due to their extremely social breeding system, in which all activities come under the closest scrutiny of neighboring birds.

Despite being one of the most southerly colony sites of murres in the world, the Farallon Islands and surrounding waters amply provide the nesting conditions and food sources required by Common Murres for successful breeding. Historical records testify to the species' numerical dominance on the Farallones and in the California Current avifauna (Briggs et al. 1987). It is unfortunate that we were not able to view this colony closely when it contained several hundred thousand birds, for it is quite possible that some of the breeding characteristics we observed during this study were different then. Formerly, the murres may have placed greater pressure on food sources within the Gulf of the Farallones, a subject discussed in Chapter 12. In essence, we viewed a depleted population in a recovery phase of growth. We hope it will be given a chance to approach its former size, to reveal other of its characteristics.

9

Pigeon Guillemot

David G. Ainley, Robert J. Boekelheide,
Stephen H. Morrell, and Craig S. Strong

The Pigeon Guillemot, *Cepphus columba*, and the Black Guillemot, *C. grylle*, two species that are closely related in taxonomy, behavior, and ecology, are among the best-known members of the family Alcidae and among the better-known seabirds. Without much effort, we can think of a host of research papers and ten dissertations written about these two species. Thus, in terms of information available, guillemots are in a class different from most other Farallon seabird species (the Black Guillemot does not breed at the Farallones). Contributing to this well-studied position are the species' tendencies to nest in small, loose colonies, to lay eggs in shallow cavities that are often located in easily accessible, gradually sloped talus, and to feed in nearshore waters often within view of the colony. This all goes to indicate that guillemot breeding ecology, in contrast to that of most other seabirds, is easily observed by biologists.

Guillemots are dramatically colored. In breeding plumage, they are entirely jet black, except for the bold white patches on the upper surface of the wing and the intense red of their legs, feet, and mouth lining. In nonbreeding dress, their black plumage is profusely flecked with white. Like other auks they are compactly feathered and proportioned. Adult Black Guillemots weigh about 400 gm and Pigeon Guillemots weigh about 450 gm. Storer (1952) noted that California and Oregon guillemots are slightly larger than Pigeon Guillemots elsewhere, perhaps as a response to the rough waters of this upwelling region. The larger size might also be related to interspecific competition for nesting cavities (Chapter 12). Such a pattern in size, however, is also exhibited by Leach's Storm-Petrel, which is largely sympatric with the guillemot, but in the storm-petrel the larger size of California and Oregon birds is consistent with the colder sea temperatures of the region (Ainley 1980).

The Pigeon Guillemot nests from southern California north and west around the Aleutian arc to northern Japan and north to the Bering Strait. On the Asian coast adjacent to Manchuria and south to Korea, it is replaced by the little-known Spectacled Guillemot, *C. carbo*; north of the Bering Strait, thence east and west to the Atlantic, it is replaced by the Black Guillemot (Udvardy 1963, Bédard 1985).

METHODS

We began our observations of guillemots in 1971. That year we found nests opportunistically as eggs were laid or chicks hatched. In all years, we concentrated our efforts on the southern slopes of Lighthouse Hill and the surge channels on the southern border of the marine terrace (Figures 1.3 and 9.1). Numbers were painted on rocks adjacent to burrow entrances. After 1971, we narrowed the search area to include all nests confined to the southwestern slope of Lighthouse Hill. Depending on year, the number of nests in this area varied between 80 and 100.

To determine laying date, we visited all nest sites every other day beginning 1 May. Throughout the breeding season, we made our visits between 1400 and 1600, because this is when the fewest adults frequent nesting areas (see below) and because this standardized the time of day when chicks were weighed. After the second egg of a clutch was laid, we did not revisit that site specifically until 27 days later, just before hatching should have begun. When the first egg of a nest hatched, we returned daily to weigh chicks. After 1979, we checked for clutch initiation and hatching at weekly intervals, and we weighed chicks only when their expected fledging date approached. This change in procedure was designed to reduce our efforts in gathering information on this species, but without compromising certain between-year comparisons.

During 1971 and 1972, we marked eggs by using a felt-tip pen with indelible ink in order to distinguish between first- and second-laid eggs of clutches. We discontinued this practice (1) to reduce significantly the amount of time we spent in the nesting area, so reducing disturbance, and (2) because our findings were the same as Drent's (1965), namely, that virtually all first-laid eggs hatched before second-laid eggs (99%; $n = 162$). Thereafter, we assumed that the first-

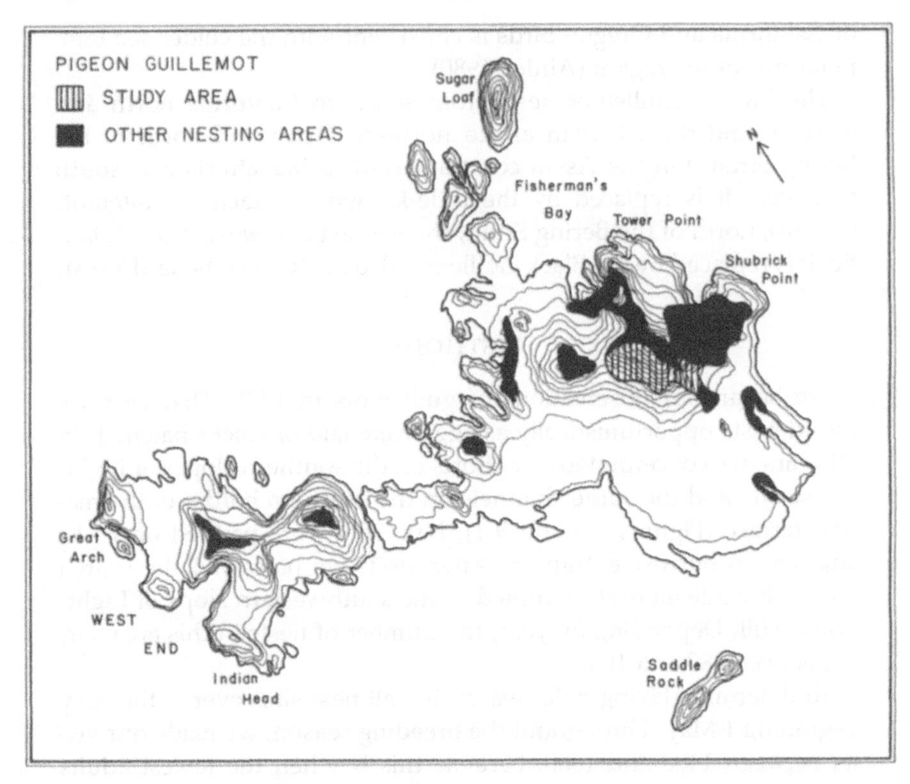

Figure 9.1. Major nesting areas of Pigeon Guillemots on the South Farallones, and location of the study area, 1971–83.

hatched chick came from the first-laid egg. At about seven days of age, chicks were banded with USFWS bands, and we assumed that the larger chick was always the older, an assumption consistent with Koelink's (1972) findings that growth rate is similar among young chicks regardless of brood size or hatching order. In other words, one would not expect the younger chick to surpass the older one in size during the first week. After 1971, we made no attempt to band adults found in their nest cavities, because doing so increased desertion and increased chances of eggs being broken.

Besides the USFWS bands, chicks were given a plastic spiral band, colored specifically to year of hatching. Through 1977, we banded as many guillemot chicks as possible, including those from a large number of nests not in the area of intensive study.

We were concerned that our visits to the study plot should not disturb the birds, significantly altering their behavior and productivity.

The procedures outlined above were designed with reduction of disturbance in mind. Of course, we walked near many nests, without looking in, on our way to check others. In many cases this could have constituted disturbance under Cairns' (1980) criterion of walking within 15 m of a nest. To determine whether our interference was affecting results, we compared two plots. The first was our intensive study area; the second was a sample of about 50 nests discovered in 1971. We visited the latter only twice in 1972, 1973, and 1974, at the peak of hatching and late in the nestling period just before chicks began to fledge. We found brood size and overall productivity in both plots to be essentially the same (Appendix 9.1). In the comparison of nest contents, the plots differed only during 1974, when the study area had disproportionately few one-chick broods. Why the nests differed in proportion of single-chick broods during that year is unclear. We conclude that our level of interference in the study plot did not disturb guillemots enough to reduce reproductive success. Besides, if the rate of turnover among the adults nesting in the plot was low, those birds may have become accustomed to us over the years.

To determine the seasonal patterns in which guillemots occupy the island, we counted individuals in that portion of the study area visible from our quarters several times daily from dawn to dusk throughout the 1973 breeding season. To determine between-year differences in the population of the entire island, we made counts of birds on land and rafting in the water on several evenings during mid-April. Much of the island and its adjacent waters were visible from atop Lighthouse Hill, from which these surveys were made. Such counts were similar to those recommended by Cairns (1979). We took the maximum count as an estimate of the size of the nesting population, under the assumption that younger, nonbreeding guillemots arrived later (D. A. Nelson 1982; Table 9.1).

FARALLON POPULATION

Through 1982, the Farallon breeding population of Pigeon Guillemots remained at a plateau of about 1,000–1,100 pairs, except for the lower numbers present during the year of extremely warm water, 1978 (Table 9.1). In 1983, the population decreased again, and only partially rebounded from 1984 to 1986. Earlier in the present century, the population may have been smaller because of oil pollution and disturbance from occupation of the island by several human families and their associated livestock (see Chapter 1; Ainley and Lewis 1974). At

TABLE 9.1
Spring arrival and estimated numbers of Pigeon Guillemots, Southeast Farallon Island, 1971–86

Year	Date first seen[a]		Breeding pairs	Percent change
	Offshore	Onshore		
1971	17	22		
1972	2	18	1,000	
1973	12	16		
1974	9	16		
1975	10	17	1,000	
1976	3	?	1,000	0
1977	1	18	1,000	0
1978	1	13	500	−50%
1979	3	20	1,000	+100
1980	7	14	1,000	0
1981	5	19	1,100	+10
1982	3	11	1,100	0
1983	2	28	< 50	−95
1984			750	+1,400
1985			650	−13
1986			625	−4
Mean date ± SD (days)	5.8 ± 4.9	17.8 ± 4.5		

[a] All dates in March.

1,000 pairs in 1982, it was one of the largest known colonies of the species and was near in size to or greater than, respectively, all Washington or all Oregon colonies combined (Table 9.2). Only four other colonies, numbering about 2,000 pairs each and all in Alaska, are larger than that at the Farallones (Sowls, Hatch, and Lensink 1978). The Farallon population before 1983 constituted 16% of the total number nesting in California.

On the Farallones, as at most sites where guillemots breed, nests are scattered along the lower elevations, where talus rocks are large. Nests also occur in crevices in solid rock along the crests of ridges and in rock walls built to widen certain pathways. Not only is the population currently larger than it was earlier in the century, but it may be larger than before the mid-1850's, when intense human use of the area began and Rhinoceros Auklet and Tufted Puffin populations suffered as a consequence. The latter two species are increasing in number on the Farallones, and are displacing guillemots (see Chapters 11 and 12). Therefore, we may find a smaller population of guillemots breeding in the future.

Guillemots nest in cavities such that incubating birds, if they choose, are just about able to see out. As Storer (1952) put it, "their principal requirement seems to be a roof over their head." At Cooper Island, Alaska, and the Norde Ronner Islands, Denmark, where Black Guillemots nest under boards and other human debris (Divoky, Watson, and Bartonek 1974, Asbirk 1979), this statement can be taken almost literally. Actually, most cavities are slightly deeper than the length of an adult to allow chicks, once they hatch, to be out of reach of gulls looking for prey. Nevertheless, at the Farallones, even in cavities suitably deep for puffins and Rhinoceros Auklets, adult guillemots lay and incubate their eggs near the entrance.

The availability of suitable cavities likely limits the size of the guillemot population breeding at the Farallones, as may be the case elsewhere (W. Preston 1968). At the Farallones, a relatively constant number of suitable cavities remains unused or used sporadically (see below). Most of these are within a few meters of sites being used, and the same pairs may alternate irregularly among them. Evidence for the limiting availability of nest sites is twofold (D. A. Nelson 1982, 1987, PRBO, unpubl. data). First, guillemots occupied a number of artificial nest cavities (about ten) not long after their creation. The same phenomenon has been observed, and similarly interpreted, for guillemot populations at Kent Island, Bay of Fundy, by W. Preston (1968) and at Cooper Island, Alaska, by Divoky, Watson, and Bartonek (1974; G. Divoky pers. comm.). Second, at the Farallones, appreciable numbers of nonbreeding individuals occur in nesting areas throughout the breeding season. These birds were color-banded as chicks, and they are seen year after year. D. A. Nelson (1982) considered guillemots old enough to be physiologically mature (i.e., older than two years) to be part of a "floating" reserve of birds capable of breeding but unable to do so for lack of a nest cavity. Observations supporting this type of evidence are available for other guillemot pop-

TABLE 9.2

Estimated numbers of Pigeon Guillemots nesting in North America, 1960–82

Area	Breeding pairs	Sites	Source
Alaska	100,000	363	Sowls, Hatch & Lensink 1978
British Columbia	3,245	41	Drent & Guiguet 1961, Vermeer et al. 1983
Washington	1,990	33	Manuwal, Wahl & Speich 1979
Oregon	316	12	Varoujean 1979
California	6,962[a]	172	Sowls et al. 1980
Total	112,513	621	

NOTE: Most recent data.
[a] The 1982 Farallon population (1,100 pairs) constitutes 16% of this total.

ulations also. W. Preston observed a large number of older nonbreeders frequenting the breeding areas on Kent Island. Asbirk (1979), working with a rapidly expanding population, noted a paucity of nonbreeding birds. In contrast, Kuletz (1983) noted a large and annually fluctuating proportion of nonbreeders in the population at Naked Island, Alaska, where there appeared to be more than enough nesting cavities. Kuletz did not know the ages of these nonbreeders, but she ascribed the annual fluctuations in their numbers to food availability (pers. comm.).

OCCURRENCE PATTERNS

Seasonal attendance

Guillemots first appear in waters adjacent to the island during March (Table 9.1). Often, large numbers appear on the first day of their return. Even if not, the population rises sharply after the first bird arrives and quickly reaches a plateau, usually in about two or three weeks. The plateau is maintained through July, except for a brief peak during June (Figure 9.2; see also D. A. Nelson 1982, figure 4). This second peak may represent an influx of immature birds; Nelson noted that one- and two-year-olds arrive during June and early July. After July, the size of the guillemot population declines rapidly. Few remain to September and none are present in October, except for the odd fledgling remaining in nearby waters (Chapter 12, Figure 12.1). A notable exception to this pattern occurred during 1983, when hundreds of birds arrived in mid-July, although none bred. The gen-

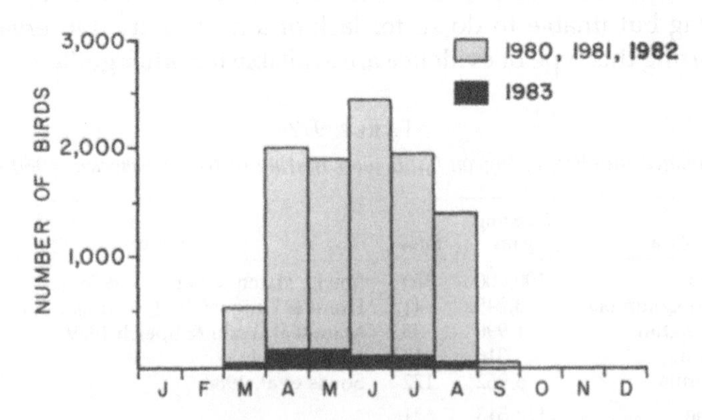

Figure 9.2. Numbers of guillemots present, by month, and means of monthly estimates for two sets of years, 1980–83.

TABLE 9.3
Recoveries of Farallon-banded Pigeon Guillemots, by age, month, and latitude, 1970–82

Latitude (°N)	Period						Total
	Jan.–Feb.	Mar.–Apr.	May–Jun.	Jul.–Aug.	Sept.–Oct.	Nov.–Dec.	
First-year birds[a]							
35–37°				34	4	2	40
38–42°				8	1	2	11
43–49°					1	2	3
Total				42	6	6	54
Older birds[b]							
35–37°		5	2	2	1		10
38–42°	1	2	2	2			7
43–49°		1		1			2
Total	1	8	4	5	1		19

NOTE: All birds were banded as chicks at Southeast Farallon.
[a] I.e., birds 12 months old or less. Monthly frequencies not similar among zones ($\chi^2 = 14.229$, DF = 4, $p < .01$).
[b] Monthly frequencies similar among zones ($\chi^2 = 4.302$, DF = 8, $p > .8$).

eral pattern of seasonal occupation by the Pigeon Guillemot at the Farallones differs from that of the Black Guillemot at Kent Island. The latter population, which is apparently also limited by available nesting space, reaches its annual peak early in egg laying and then gradually declines throughout the nesting period, presumably because of attrition (W. Preston 1968).

After fledging, young Farallon Pigeon Guillemots move north at least as far as British Columbia; a few move south, but only as far as Monterey Bay (Table 9.3). The movement can be very rapid; banded fledglings have been recovered in Oregon and Washington within weeks of their departure from the Farallones. During July and August, 61% of recovered fledglings were found within 150 km of the Farallones, and the remainder were found to the north. After August, however, about 75% of recoveries occurred from extreme northern California north, and after October one-third of recoveries occurred from Washington northward. Since Audubon Society Christmas Bird Counts rarely record guillemots in California or Oregon, but often record larger numbers in Washington and British Columbia (see also Manuwal, Wahl, and Speich 1979), much of the adult population may also move north considerable distances for the nonbreeding period. An adult, banded as a chick on Southeast Farallon, has been found breeding in British Columbia.

Rates of recovery of fledglings have varied little from year to year, except during 1972, 1976, and 1981, when larger numbers apparently died within a few months of leaving the island (Table 9.4). The first of these years began a two-year warm-water episode; 1976 was a warm-water year, and 1981 preceded the intense 1982–83 warm-water event (Chapter 2). There was, therefore, a likely connection between the survival of young birds after their departure and the oceanographic conditions and prey abundance they encountered. Too few banded adults were available early in the study period for interannual variability in mortality to be assessed, but in later years large numbers of adults apparently died in 1980 and 1983. Ten of 16 of the banded adults found dead were picked up during the warm-water years of 1973, 1976, 1978, and 1983.

Daily attendance

The largest numbers of adults are present on the nesting slopes during the morning and evening (Figure 9.3). At the beginning and end of the season, adults can be found on land only during those times. The same pattern of occupation occurs at Mandarte Island (Drent 1965) and has been observed in the Black Guillemot as well (Bergman 1971, Cairns 1984). Morning is apparently the time of day

TABLE 9.4

Recoveries of Farallon-banded Pigeon Guillemots, 1970–83

	Fledglings		
Year	Banded[a]	Recovered (%)[b]	Adults recovered
1970	181	7 (3.9%)	
1971	284	11 (3.9)	0
1972	96	12 (12.5)	0
1973	167	6 (3.6)	1
1974	144	5 (3.5)	0
1975	107	2 (1.9)	0
1976	97	4 (4.1)	1
1977	107	1 (0.9)	0
1978	19	0 (0.0)	1
1979	137	0 (0.0)	1
1980	59	2 (3.4)	3
1981	80	4 (5.0)	1
1982	64	1 (1.6)	1
1983	0		7
Total	1,542	55 (3.6)	16

NOTE: These are birds found dead, mostly on beaches. All birds were banded as chicks on Southeast Farallon.

[a] Beginning in 1979, chicks were banded in a smaller sample of nests.

[b] These are recoveries of fledglings during the same calendar year in which they were banded.

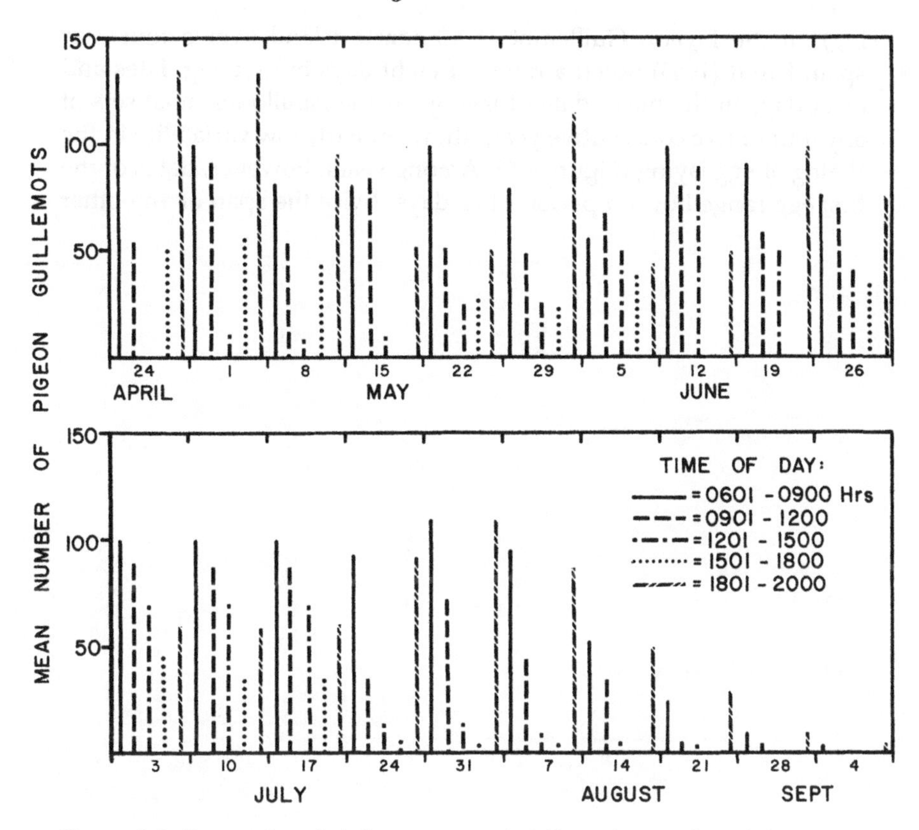

Figure 9.3. Seasonal and daily patterns of cliff occupation by adult Pigeon Guillemots; seven-day periods, 1973. No guillemots present after dates shown.

when guillemot chicks are fed most frequently (Koelink 1972), although feeding rate peaks in late afternoon as well (Cairns 1981). Kuletz (1983) noted a small influence of the tidal cycle on attendance during the breeding season at Naked Island: fewer birds fed during high tides.

EGG LAYING

In each guillemot study that has extended beyond one year, results have indicated little variability in the timing of egg laying. For example, in the Black Guillemot at Kent Island over five years, W. Preston (1968) noted a range of 12 days in first egg dates, whereas in Hudson Bay over three years, Cairns (1984) noted a range of five days in first hatching dates and a range of seven days in the median date of hatching (he used hatching dates to describe the timing of breeding); in the Pigeon Guillemot at Mandarte Island over a four-year span, Drent (1965) noted a range of eight days in first egg dates and of 11 days in the mean date of laying. At the Farallones most sets of any four or five consecutive years show similarly low variability in the timing of egg laying (Figure 9.4). Among years, however, dates of the first egg ranged over a period of 22 days, twice the span of any other

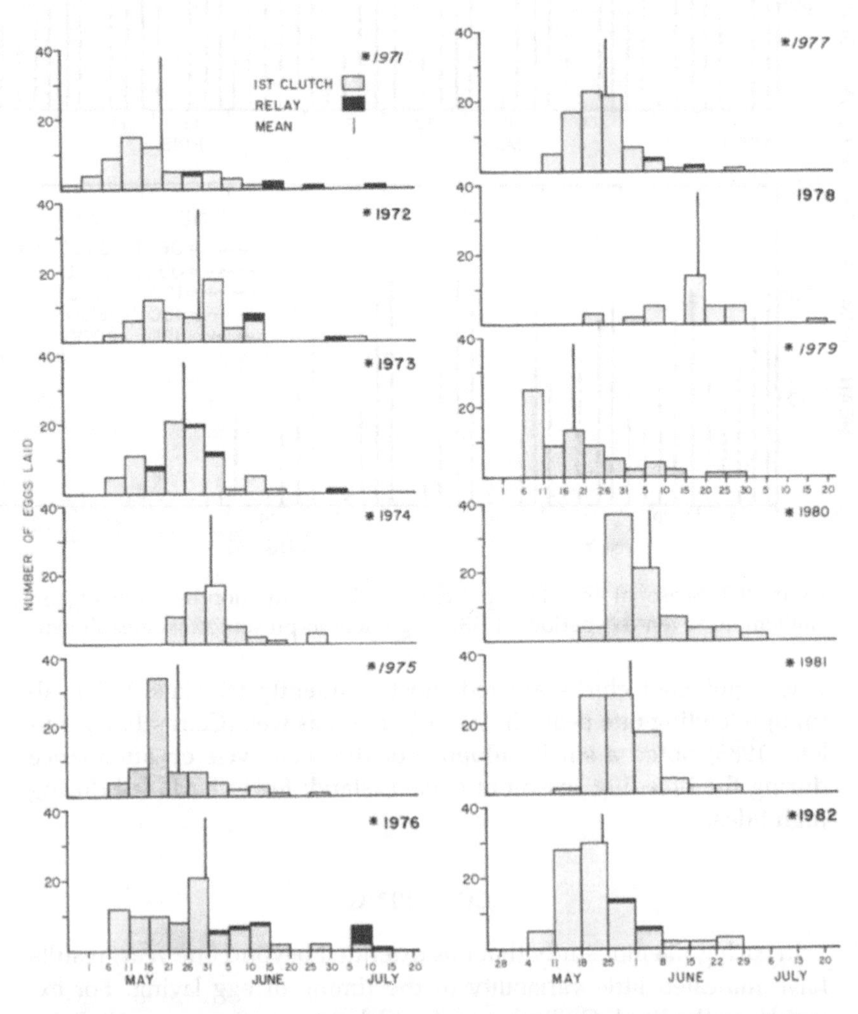

Figure 9.4. Timing of egg laying in the Pigeon Guillemot, by year, 1971–82; in years denoted by an asterisk (*), laying was significantly skewed and peaked (t test, $p < .05$). After 1979, nest checks changed to once per week; no eggs laid in 1983.

TABLE 9.5

Mean dates of Pigeon Guillemot first-clutch initiation, 1971–82

1979	*1971*	*1975*	**1982**	**1973**	*1977*	**1972**	**1976**	1981	1974	1980	**1978**
5/20	5/21	5/23	5/24	5/26	5/27	5/29	5/29	5/31	6/1	6/4	6/17
—	—	—	—								
	—	—	—	—	—						
				—	—	—	—	—			
						—	—	—	—	—	

NOTE: Lines connect similar means; SNK test, $p > .05$.

study, and the mean date of laying varied over a span of almost a month, i.e., 28 days. As in other populations, laying was usually skewed toward the beginning of the season, and replacement clutches produced a small peak at the end. Only the frequency distribution of egg dates during 1978 was not significantly skewed (or peaked; *t* test, $p > .05$, Figure 9.4).

All years combined, the average date of clutch initiation was 28 May, with a range in means between 20 May and 17 June. Egg laying was noticeably late during four years: 1974, 1978, 1980, and 1981 (Table 9.5). However, no relationship was apparent between the timing of egg laying and any of the environmental variables discussed in Chapter 2. Because the Farallon population winters far to the north, it could be that conditions there may affect the timing of reproduction to a significant degree. Certainly, local conditions could play a role as well; for instance, D. A. Nelson (1982, 1987, pers. comm.) noted that guillemots arrived at the Farallones at about the same date in 1979 and 1980, but laying was delayed in the latter year. Furthermore, there appeared to be an inverse relationship between synchrony (skewness) of laying and the intensity of the spring transition, the change from the Davidson Current to the upwelling period (Chapter 2). The smaller the decrease in sea-surface temperature during the transition, the more skewed and synchronous laying became (Figure 9.5). An exception to this pattern was evident during 1977, when the spring transition was abrupt and intense, but laying was highly skewed and peaked. As almost all years of clumped egg laying, including 1977, were ones in which juvenile rockfish were prevalent in guillemot diets (see Chapter 3), the relationship between egg laying and spring transition may be mediated by prey availability. At Naked Island, Kuletz (1983, pers. comm.) also noted a relationship between early laying and prevalence of *Ammodytes* in the diet.

Combining dates for all years of study (Figure 9.6) indicates that, in general, initiation of laying at the Farallones was the same as at Man-

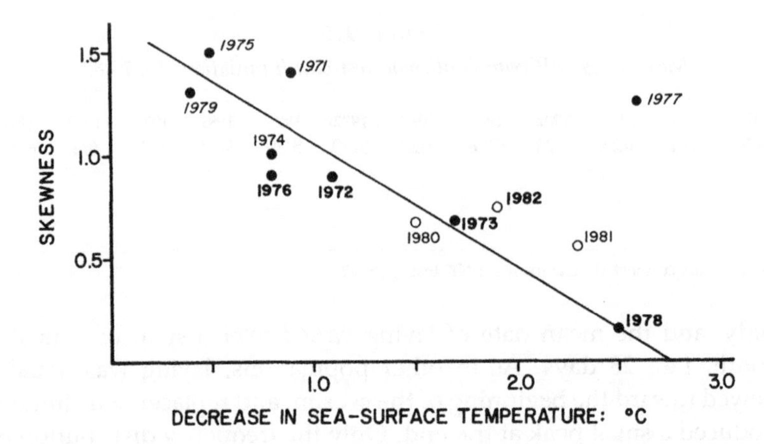

Figure 9.5. Relationship between guillemot laying synchrony, measured by skewness, and intensity of the spring transition, 1971–82; open circles indicate estimates from egg data collected once per week rather than every two days as in other years; without the 1977 value, $y = -0.49x + 1.5$, $r = -.900$ (with 1977, $r = -.676$).

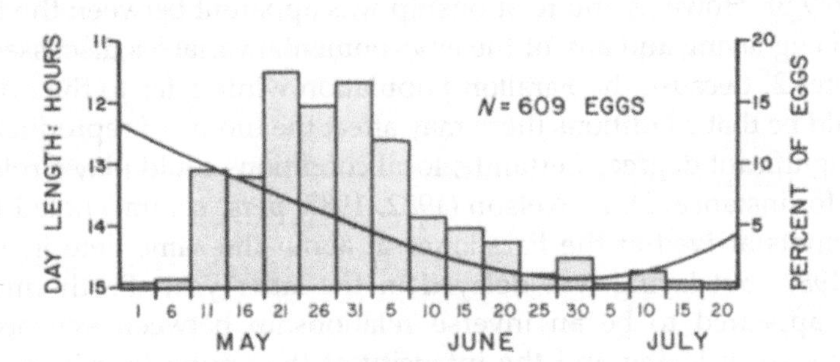

Figure 9.6. Percentages of guillemot eggs laid in five-day periods, all years combined, 1971–82, compared to day length (curved line).

darte Island, British Columbia (cf. Drent et al. 1964, figure 17; see also Chapter 12, Figure 12.1). Further comparison indicates that the cessation of egg laying also occurred at the same time at both sites and corresponded with the shortening of day length in early July. Photoperiod is known to have a major effect on reproductive activities of most birds that live outside the equatorial region. We would guess that wherever information on the Pigeon Guillemot's egg laying is collected, the cessation of laying will be similar, as we found to be true for various species of auklets (Chapter 10). Indeed, even in Denmark at 57° N, in Hudson Bay at 62° N, or in the Barents Sea at about

68° N (Farallones at 37° N, Mandarte at 49° N), guillemots laid few eggs after the first week of July (Asbirk 1979, Cairns 1984, Belopol'skii 1961). On the other hand, at Prince Leopold Island at 74° N, where the persistence of pack ice is a severe constraint on seabird ecology, guillemots did not begin to lay until late June and continued to late July (A. Gaston and D. Nettleship, unpubl. data).

Clutch size

The maximum size of a guillemot's clutch is two eggs. Occasionally three or four eggs appear in a nest, but such a clutch results from the activities of two females. A small proportion of pairs produce only one egg, and in many cases such pairs consist of young birds (W. Preston 1968, Asbirk 1979). The proportion of one-egg clutches reported in other studies is relatively small, with some variation, from 8% at Mikelskaren, Finland (Bergman 1971), to 28% at Cooper Island and 30% at Naked Island, Alaska (Divoky, Watson, and Bartonek 1974, Kuletz 1983, M. P. Harris and Birkhead 1985; Table 9.6). In contrast, the proportion of one-egg clutches at the Farallones ranged much more widely, from 10% in 1971 and 1979 to 42% in 1976 and 83% in 1978 (Appendix 9.2)! As a result of this variability, mean clutch size ranged from 1.2 to 1.9 eggs (Table 9.7), although the 12-year average (1.8) is consistent with clutch sizes measured over shorter terms elsewhere (M. P. Harris and Birkhead 1985; Table 9.6). The most unusual years were 1976 and 1978, when many one-egg clutches were produced, and 1971, 1977, and 1979, when many two-egg clutches were produced. Clutches averaged smaller during all ENSO/warm-water years except 1973. Such variation at the Farallones could be due to fluctuation in the proportion of young breeders or to interannual differences in the ease with which females could find food. The especially large mean clutch sizes during 1971, 1977, and 1979 might well have been due to younger birds finding food easier to come by and thus being able to lay larger clutches. Indeed, chicks were fed at higher rates during 1979 than during 1980 and adults spent more time at the colony, an indication that food was easier to find in 1979 (D. Nelson, unpubl. data).

The proportion of nests in which eggs were laid varied little (Appendix 9.2). All sites checked by us were used at least once during the 13 years of study, and we take the percentage used in a given year as an indirect measure of the proportion of pairs attempting to breed. These proportions in absolute terms are underestimates because, as noted above, some pairs may control more than one site. Exceptional

TABLE 9.6

Pigeon Guillemot nesting characteristics at various localities

Site (years studied)	Breeding (%)	One-egg clutches (%)	Clutch size	HS (%)	FS (%)	Chicks per pair	Fledging weight (gm)	Source
Prince William Sound (4)	55–70%	30%	1.7	70%	71%	1.1	417–504	Kuletz 1983 & unpubl.
Mandarte Island (5)	80	9	1.9	62	74	1.2	411–13	Drent et al. 1964, Drent 1965, Koelink 1972
Puget Sound (1)	95	21	1.8	52	86	0.8	438	Thoresen & Booth 1958
Farallones (12)	80	24	1.8	76	67	1.0	398	This study

NOTE: *FS*, fledging success; *HS*, hatching success. No eggs laid in one of the 13 years at the Farallones (1983).

TABLE 9.7

Mean clutch size, brood size, and fledglings per pair, Pigeon Guillemots, 1971–82

Clutch size

1971	1973	1977	1979	1981	1974	1975	1980	1982	1972	1976	1978
1.9	1.9	1.9	1.9	1.9	1.8	1.8	1.8	1.8	1.7	1.5	1.2

Brood size

1977	1979	1971	1981	1974	1975	1972	1973	1980	1982	1976	1978
1.7	1.7	1.6	1.6	1.5	1.5	1.4	1.4	1.4	1.4	1.2	0.4

Fledglings per pair

1977	1971	1981	1974	1975	1979	1972	1980	1973	1976	1982	1978
1.6	1.5	1.3	1.2	1.1	1.1	0.9	0.8	0.6	0.6	0.6	0.2

NOTE: Lines connect similar means; SNK test, $p > .05$.

were the years 1978 and 1983, when the proportion of cavities occupied, or of birds attempting to breed, was especially low. This pattern of relatively constant site use, coupled with the wide fluctuation in the proportion of one-egg clutches, indicates that Farallon guillemots, given the chance, almost always made an attempt at breeding, even if they produced clutches of reduced size. On the other hand, Kuletz (1983) detected a twofold difference in the proportion of guillemots breeding but almost no year-to-year variability in clutch size at Naked Island, Alaska. Disturbance as well as natural environmental factors may have been involved. Similarly, Cairns (1984) noted that disturbance in the form of a predator's presence reduced the proportion of guillemots attempting to breed at colonies at the Nuvuk Islands, Hudson Bay.

If the first clutch is lost as a result of such phenomena as courtship squabbles or contests for nest sites, guillemots will lay a replacement clutch (Figure 9.4). As with all late clutches, replacement clutches tend to be smaller, 1.5 ± 0.5 (n = 26) compared to 1.8 ± 0.4 (n = 868) eggs (see also W. Preston 1968). In one case during 1971, a pair produced a third egg when the first was lost before the second was laid. Replacement clutches appeared only during the six years when the mean laying date preceded 30 May.

Incubation period

The usual interval between laying of the first and second eggs of a guillemot clutch is three days (74%), although the interval ranges from one to four days (Drent 1965), or longer in some cases. At the Farallones, where we checked nests every other day, the laying interval ranged correspondingly from two to six days; had we checked daily, we would presumably have encountered an interval of one to five or six days. No between-year differences in the interval existed (Appendix 9.3). Using temperature probes, Drent (1965) discovered that the first eggs at Mandarte Island were incubated only during the day, and that steady incubation begins one to three days after the second egg is laid. Such an incubation rhythm results in different "incubation" (laying to hatching) periods for the two eggs.

Compared with other alcids, guillemots possess the third shortest incubation period, which is consistent with an egg size that is proportionately small, i.e., only 12% of body weight (Sealy 1972). Drent et al. (1964) determined that the first-laid egg hatches in 32 days and the second-laid egg hatches in 30 days after laying. We, too, found a two-day interval on average between the hatching of the two eggs

TABLE 9.8

Mean incubation and nestling periods for single and first eggs and chicks, and for second eggs and chicks, Pigeon Guillemots, 1971–79

(*Days*)

Incubation period, single and first eggs

1973	1974	*1975*	*1977*	**1972**	**1976**	**1978**
29	29	29	29	29	30	32

Incubation period, second eggs

1974	**1973**	*1977*	**1972**	*1975*	**1976**	**1978**
26	27	27	27	27	28	32

Nestling period, single and first chicks

1977	*1979*	*1971*	*1975*	1974	**1976**	**1973**	**1972**	**1978**
35	36	37	37	39	39	40	41	41

Nestling period, second chicks

1977	*1979*	*1971*	1974	*1975*	**1973**	**1972**	**1976**	**1978**
37	38	38	41	41	45	46	—	—

NOTE: Lines connect similar means; SNK test, $p > .05$.

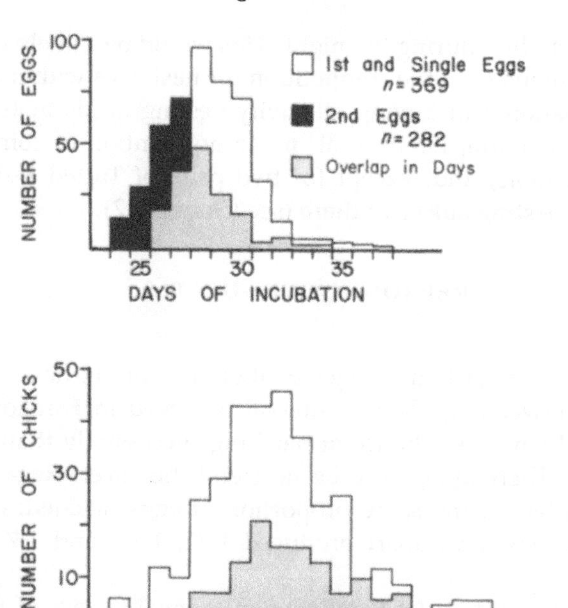

Figure 9.7. Frequency distributions of incubation and nestling periods in Pigeon Guillemots; all years combined, 1971–79.

(71% of cases). First and single eggs hatched in 29.1 ± 1.8 days (n = 369) and second eggs hatched in 27.2 ± 1.9 days (n = 282; $t = 13.04$, $p < .05$). We also found that the apparent mean incubation period varied among years, ranging from 29 to 32 days in first eggs and from 26 to 32 days in second eggs (Table 9.8). Single eggs hatched in the same amount of time as first eggs of two-egg clutches (Figure 9.7). The longest incubation periods occurred during the warm-water years of 1976 and 1978. Food shortage during those years (see Chapter 3) probably contributed to parents' inattentiveness. That is, they may have left the egg or eggs unattended for short periods when sufficient food was particularly hard to find. Guillemots can seemingly afford to do this because they lay their eggs in a cavity protected from predators.

The average incubation periods of eggs at the Farallones being shorter than those at Mandarte Island (Drent et al. 1964), 29 vs. 32 days, respectively, for first eggs and 27 vs. 30 days, respectively, for second eggs, indicates that Farallon guillemots may be more attentive to their eggs. This could indicate that they occupy their nests more regularly, including during the night. This would be a likely response to the high intensity of the competition for nest sites within the guillemot population and among all cavity-nesting alcids at the Farallones. For Mandarte, Drent et al. made no mention of competition among guillemots, and, except for two pairs of Tufted Puffins, no other cavity-nesting auks live there (see Chapter 12).

REPRODUCTIVE SUCCESS

Hatching success

One might expect that an egg unattended for six days, i.e., the maximum between-egg laying interval observed in Farallon guillemots, would have less chance of hatching successfully than one neglected less. There appears to be no trend, however. Regardless of the laying interval, the same proportion of eggs hatched; eggs laid two, four, or six days apart produced 1.71, 1.73, and 1.71 chicks, respectively.

Hatching success at the Farallones over the 13 years of intensive study averaged higher than in the great majority of guillemot populations studied elsewhere (Divoky, Watson, and Bartonek 1974, M. P. Harris and Birkhead 1985; Table 9.6). Significant between-year variability nevertheless occurred, with the range in values exceeding that observed among all other sites (Appendix 9.2). Hatching success was particularly low during the warm-water years of 1973, 1976, and 1978. Initial brood size, which results from the combination of clutch size and hatching success, varied accordingly (Table 9.7). Average brood sizes in all warm-water years beginning with 1972 were at the low end of the scale.

We checked nests every day once the expected hatching date drew near. The interval between the hatching of the two eggs of a clutch ranged from zero to four or more days (Appendix 9.3). Unlike laying interval, hatching interval varied significantly between years. The interval was particularly short during 1971, 1977, and 1979, all years of exceptional food availability (Chapter 3). The proportion of eggs hatched was highest during those years as well (Appendix 9.2).

Chick development

Single chicks and the older members of two-chick broods fledged on average in 38.0 ± 3.8 days (n = 368), and younger members of two-chick broods fledged in 39.3 ± 4.1 days (n = 140). These differences are significant ($t = 3.37$, $p < .05$) and are based on a schedule of daily nest checks. Significant between-year differences also existed in these periods (Table 9.8); nestling periods were longest during warm-water years.

Growth data were available only for 1971–77; fledging weights were available for all years except 1983, when no chicks fledged. In the average logistic growth curves (Figure 9.8), no difference existed between single chicks and the older members of two-chick broods (ANOVA: hatching and asymptotic weight, $p > .40$; growth constant, $p > .09$). Thus, data for these individuals were combined when the growth curves were drawn. For these chicks, hatching weight, asymptotic weight, and the growth constant (K) averaged 38.9 gm, 401.5 gm, and 0.1756 per day, respectively; for younger siblings, the values were 37.2 gm, 380.1 gm, and 0.1793. For all three parameters, there was a highly significant intercept ($p < .01$), indicating a real difference between the older and the younger chick in a nest. During most years, the older chick weighed more at hatching and at asymptote but reached asymptote at a slower rate (Appendix 9.4).

Maximum growth occurred between the ages of nine and 12 days and reached 14.0 and 19.1 gm per day for younger chicks and single or older chicks, respectively. The average was 16.5 gm per day, a

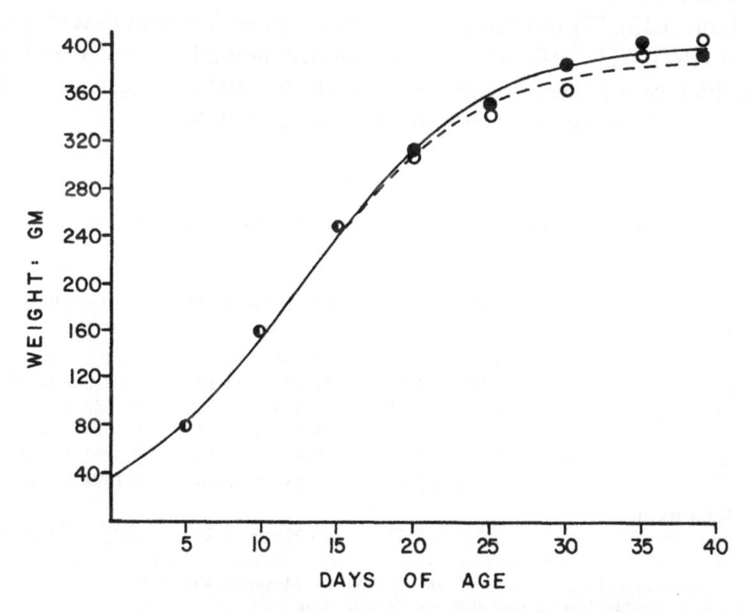

Figure 9.8. Logistic growth curves of Pigeon Guillemot chicks (ANOVA; n = 169 chick pairs). Dots represent mean values for single and older chicks; open circles (and dashed line) represent the younger members of two-chick broods; all years combined, 1971–77.

value similar to those measured by Koelink (1972) at Mandarte Island (15.9 gm/day) and by Kuletz (1983) at Naked Island (15.8 gm/day). For the Black Guillemot in Denmark, Asbirk (1979) determined a value of K ranging from 0.170 to 0.192, depending on year.

The magnitude of differences between the older and younger chicks in two-chick broods changed from year to year (Table 9.9). Years accounted for 11% of variation in hatching-weight differences, 20% of variation in asymptote differences, and 50% of variation in growth-rate differences. During no year did the actual magnitude of differences between siblings in hatching and asymptotic weights become statistically less or greater than in any other year, probably because of the small sample size, but the relative difference for the two weight parameters held for all years. On the other hand, the difference in growth constant between the two chicks ranged from positive to negative, with an average value of −0.022 (younger chick reached its lower asymptote faster on average than the older chick), and the disparity changed significantly in certain years. During 1973, a warm-water year when rockfish were rare in the guillemot diet (Chapter 3), there was greater disparity than in years when rockfish were abundant; however, there was no absolute correlation of growth and diet (cf. Table 3.13). Within the sample of years in which growth was measured (1971–77), 1976 was similar environmentally to 1973, but no two-chick broods were available to us during 1976. Thus, one might consider 1973 patterns as indicative of those in 1976.

TABLE 9.9

Yearly variation in Pigeon Guillemot mean growth parameters between younger and older chicks, 1971–77

Year[a]	Pairs	Hatching weight[b] (gm)	Asymptotic weight[c] (gm)	Growth constant (K)
1971	12	−1.323 ± 2.44	−11.834 ± 6.31	−0.1355 ± 0.008
1972	3	5.667 ± 4.31	21.496 ± 11.18	−0.0227 ± 0.013
1973	5	4.988 ± 3.44	10.806 ± 8.92	−0.7587 ± 0.011[d]
1974	20	−1.488 ± 2.06	−5.316 ± 5.33	0.1083 ± 0.006
1975	14	−2.501 ± 2.37	3.496 ± 6.13	0.3292 ± 0.007[d]
1977	15	−5.340 ± 2.65	−18.648 ± 7.20	−0.1894 ± 0.007
Total mean difference	69	−1.722 ± 2.52	−5.226 ± 6.58	−0.0225 ± 0.007

NOTE: Comparison of regression coefficients by ANOVA. Means are shown ± SE.

[a] Too few two-chick broods were available for analysis in 1976.

[b] No values are statistically significant departures from the total difference, which is significant between chicks ($t = 4.72$, DF = 63, $p < .001$).

[c] No values are statistically significant departures from the total difference, which is significant between chicks ($t = 11.58$, DF = 63, $p < .001$).

[d] Statistically significant departures from the total difference ($t > 4.45$, $p < .01$); total value also significant ($t = 3.01$, DF = 63, $p < .01$).

TABLE 9.10

Between-year trends in Pigeon Guillemot growth patterns, 1971–77

Parameter	Single/older chicks			Younger chicks		
	%	*p*	*F*	%	*p*	*F*
Hatching weight (gm)	7.8%	< .02	2.83	41.3%	< .001	8.99
Asymptotic weight (gm)	27.9	< .001	12.86	10.5	.20	1.50
Growth constant (*K*)	6.9	< .04	2.49	56.0	< .001	16.32
DF		5,166			5,64	

NOTE: Comparison by ANOVA. Percentage of variation (and significance level) explained by year is indicated. Data for 1976 not included because of too few two-chick broods.

Single, older, and younger chicks showed significant yearly variation in all growth parameters except for asymptotic weight, in which younger chicks did not vary much (Table 9.10). Hatching and asymptotic weights were much lower during 1973 and 1976, whereas growth rates were higher (Appendix 9.4). The different result for younger chicks is curious. Among 62 younger chicks in two-chick broods, only two ever attained weights greater than 460 gm, although 50% reached between 420 and 460 gm. In contrast, a quarter of older siblings reached asymptotic weights exceeding 460 gm. The seeming 460-gm upper boundary to the weight of younger chicks in two-chick broods may have been a result of competition with their older siblings.

Within years, asymptotic weight was the only parameter that varied significantly (Table 9.11): later-hatching chicks reached a lower asymptotic weight than did earlier ones. This was especially true for single chicks, but many single-egg clutches were likely laid by young females (noted above; W. Preston 1968, Asbirk 1979), so such a result is logical: the lower asymptotic weights of chicks could be a result of parental inexperience, especially in food gathering. Much of the within-season trend was contributed by chick weights during 1973. For example, for the seasonal trend in the weights of single chicks during 1973, $t = 6.89$, but for other years, $t < 2.77$. We excluded 1976 from this analysis because no chick pairs were available, but that being a warm-water year also, one would expect the same phenomenon.

Fledging weights were determined every year, regardless of whether other growth parameters were investigated. During the 1971–77 period, fledging weights fluctuated consistently with asymptotic weights (cf. Appendix 9.4, Table 9.12). For example, within that period, both asymptotic weights and fledging weights were low in 1973 and 1976 and high in 1977. Therefore, fledging weight offers insight into growth characteristics, particularly relative year-to-year differences.

TABLE 9.11

Significance level and percentage of variation for Pigeon Guillemot growth parameters, explained by hatching date, 1971–77

		Percentage of variation (significance)		
Chick sample	*n*	Hatching weight (gm)	Asymptotic weight (gm)	Growth constant (*K*)
Single chick	105	11.8 (> .5)	40.0 (< .001)	9.5 (> .5)
Older of two	69	20.8 (> .4)	57.4 (< .02)	23.8 (> .2)
Younger of two	69	34.9 (> .5)	61.4 (< .1)	52.1 (< .1)

NOTE: Comparison by ANOVA; data from 1976 not included because of too few two-chick broods.

TABLE 9.12

Mean fledging weights, Pigeon Guillemots, 1971–82

(gm)

1979	*1977*	*1971*	*1975*	1980	1974	1981	**1972**	**1976**	**1973**	**1978**	**1982**
450	443	426	422	420	419	409	383	382	344	339	336

NOTE: Lines connect similar means; SNK test, $p > .05$.

One would expect from our analysis of growth parameters that within two-chick broods, younger chicks would fledge weighing less than older ones. In some years, particularly those of warm ocean temperatures, this was true. On the other hand, all years combined, no significant differences existed between the average fledging weights of chicks from one-chick broods and those of chicks from two-chick broods. In most years, the weights of single chicks ranged only from 4 to 18 gm more than those of chicks from two-chick broods, but in two years, 1971 and 1981, the reverse was true, with chicks from two-chick broods averaging 27 gm *more* than their single counterparts. Thus, to cope with these complexities, we combined the data from all chicks when we looked at annual variation in fledging weight. Chicks fledged at lowest weights during all ENSO and warm-water years; the heaviest chicks fledged during years of high rockfish availability, especially 1977 and 1979 (Table 9.12).

As with asymptotic weight, notable within-season variability existed in fledging weight. In a number of years, late-hatching chicks fledged at lower weights than did earlier-hatching individuals (Figure 9.9). This was especially true during 1971, 1974, 1975, 1977, 1980,

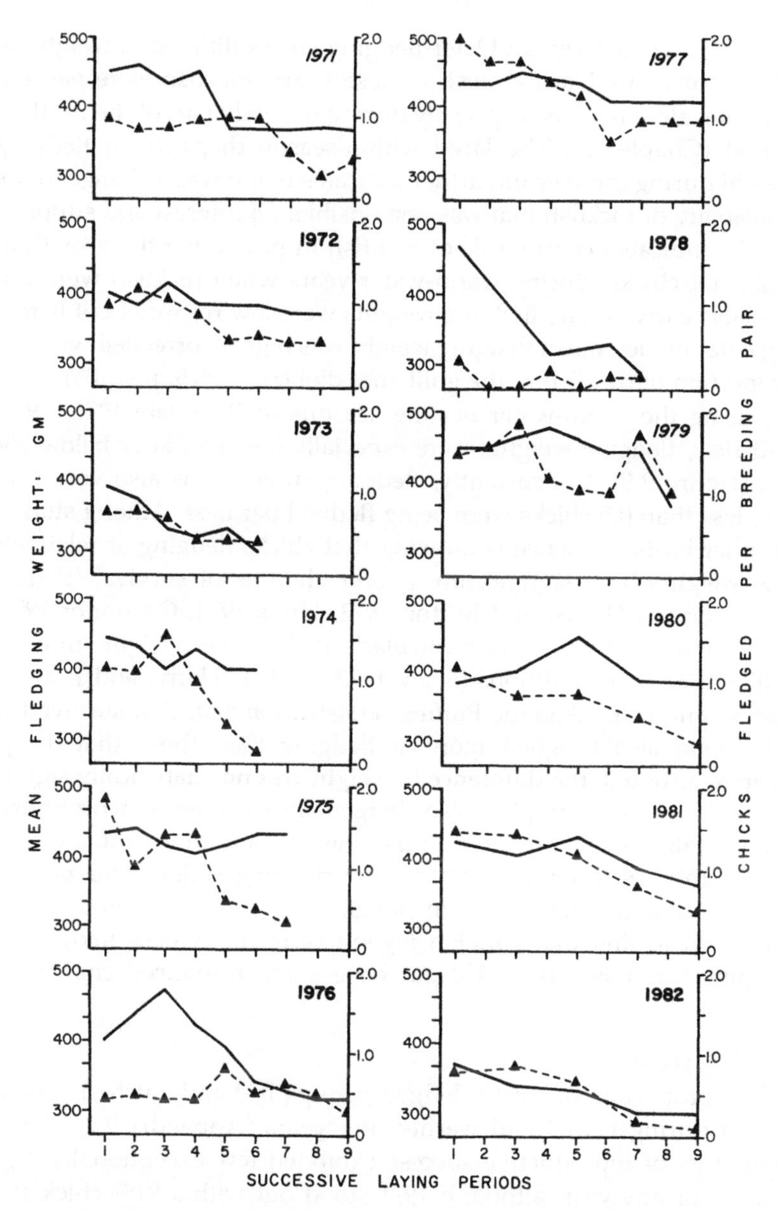

Figure 9.9. Within-year variability in breeding success (solid lines) and fledging weight (dashed lines) in Pigeon Guillemots, 1971–82. Data in five-day periods 1971–79; seven-day periods thereafter.

and 1981, when early- and later-fledging chicks differed in weight by 100 gm or more. During most of these years, rockfish were particularly prevalent in diets, especially during the early part of the nestling period (Chapter 3). The large within-season disparity in fledging weight during these years further indicates that it was a change in the availability of rockfish that was responsible. Of interest and supporting the indicated central role of rockfish in promoting the growth of guillemot chicks, during warm-water years when rockfish were unavailable early or late, fledging weights were low regardless of hatching date. In fact, mean fledging weight was highly correlated with the proportion of rockfish in the guillemot diet ($r_S = .825$, $p < .05$).

During the warm-water nesting seasons of 1973, late 1976, 1978, and 1982, fledging weights were especially low, i.e., at or below 350 gm (Figure 9.9). Concurrently, fledging success was also very low, i.e., less than 0.5 chicks were being fledged per nest. Among studies of other birds, evidence is growing that chicks fledging at relatively low weights have significantly poorer chances of survival (Perrins 1965, Perrins, Harris, and Britton 1973, Jarvis 1974, O'Connor 1976, Coulson and Porter 1985), particularly if the lower weights indicate difficult feeding conditions (Sealy 1973b). M. P. Harris and Rothery (1985) found that Atlantic Puffins, *Fratercula arctica*, that survived to adulthood also weighed more at fledging than those that disappeared, although the difference in weight was not statistically significant; the puffins sampled also fledged much more synchronously than Farallon guillemots. Given the seasonal and interannual coincidence and consistency of patterns of fledging weight and nesting success and the wide range in fledging weights, we would not be surprised to find that Farallon Pigeon Guillemot chicks fledging at weights less than about 350 gm have a much-reduced chance for survival.

Breeding success

The proportion of chicks fledged per egg laid and hatched was reduced during ENSO and warm-water years (Appendix 9.2). These indicators of reproductive success exhibited few exceptionally high values for any year, although 1971 stood out with a 91% chick survival. Years of low egg survival also tended to be years of low chick survival; exceptions were 1978, when egg survival was extremely low but chick survival was fairly high, and 1982, when egg survival was high but chick survival was very low. In 1971, three of the 61 nests failed because of interference from puffins. All three nests were in deep cavities located high on the ridge, and in two of the cases we suspect that prospecting pairs of puffins were involved. We also suspect that our presence in the area eventually discouraged the puffins from prospecting there after that year. In the case of the third cavity, the puffins continued to use it during at least the next few years, and we deleted that nest from the guillemot sample.

The number of chicks fledged per breeding pair, i.e., those pairs that produced at least one egg, is a product of clutch size, hatching success, and fledging success (Appendix 9.2, Table 9.7). The latter three factors accounted for 91% of the annual variability in the number of chicks fledged per pair ($F = 12.708$, DF $= 3,8$, $p < .003$). A comparison of chicks fledged per pair per year against clutch size, hatching success, and fledging success indicated that the latter two factors may be the most important of the three ($r_S = .505, .862$, and $.811$; $p < .05, .001$, and $.001$, respectively; DF $= 11$). If year-to-year differences in the number of chicks fledged per pair are considered, the warm-water years stand out as those in which reproduction was severely reduced; exceptionally productive years were 1971 and 1977 (Table 9.7). The productivity of guillemots was strongly correlated with the proportion of juvenile rockfish in the diet, 1972–83 ($r_S = .618$, $p < .05$). As discussed above, fledging weights tended to be low in the nonproductive years as well, except for 1980, when the few chicks fledged weighed more than in all but four other years (Appendix 9.2, Table 9.12). The latter pattern suggests that food became available rather abruptly late in the 1980 season (see above discussion of within-season differences in fledging weight).

Unlike the lack of a relationship between laying interval and hatching success, hatching interval had a strong bearing on fledging success. In nests where both chicks hatched within a day of one another, 1.6 chicks on average fledged; in nests where the chicks hatched within two days of one another, 1.3 chicks on the average fledged. When chicks hatched three or more days apart, the younger was more likely to die because the older chick had too much competitive advantage. Accordingly, in nests where chicks hatched three or more days apart, 1.1 chicks on average fledged.

As did fledging weight, the number of chicks fledged per breeding pair showed significant within-year variation (Figure 9.9). Especially during warm-water years, eggs laid late in the laying period tended to produce fewer young. As noted above, such fledglings also tended to be lighter and, therefore, would be expected to have low chances of survival, particularly during or preceding periods of warm water, when all indications point toward difficult feeding conditions (Chapter 3). This expectation is supported by the recovery rates of banded chicks (i.e., Table 9.4). Chicks fledging at a heavy weight during those years would have more reserves to sustain them while they learned, on their own, to forage successfully. During 1974, 1975, 1977, and, to a slightly lesser extent, 1979, fledging success remained high throughout the reproductive period. These were all years when rockfish were particularly prevalent in guillemot diets, and diet did not diversify much toward the end of the breeding season (Chapter 3).

SENSITIVITY OF GUILLEMOTS TO DISTURBANCE

Fledging success for guillemots on the Farallones during the entire 13-year span of the study averaged 0.9 chick per breeding pair (c/p) of adults per year. Not including the exceptional year of 1983, when no young were produced, the rate was 1.0 c/p (Figure 9.9). Either rate was equaled or surpassed in only five of 12 other guillemot studies, all of the five being relatively short-term (Table 9.6). Higher rates for Pigeon Guillemots were detected by Kuletz (1983, unpubl. data) in a four-year study at Naked Island, Alaska (1.1 c/p), and by Drent et al. (1964) in a four-year study at Mandarte Island, British Columbia (1.2 c/p); an equivalent rate (1.0 c/p) was detected for Black Guillemots by Cairns (1981) during one season at one colony in the Gulf of St. Lawrence. At a nearby colony that year, Cairns detected a lower reproductive rate. Cairns (1984) listed unpublished information on Black Guillemots from one other North American, presumably one-season, study, as well as information from a study in Iceland, where reproductive success reached 1.2 and 1.3 chicks per pair, respectively. During two years, 1971 and 1977, reproductive success at the Farallones exceeded all other published rates for either Pigeon or Black guillemots.

Two phenomena must be considered when the validity of comparing reproductive parameters among the many guillemot studies is addressed: (1) the effect of different degrees of observer interference and (2) the tremendous annual variability in reproductive success that can occur in this species, as demonstrated in the Farallon data. Cairns (1980, 1984) felt that the critical periods for disturbance were during egg laying and the first few days of the nestling period, when chicks are still brooded; during the remainder of the nestling period adults spend much of their time elsewhere (Slater and Slater 1972). Cairns found a noticeable difference in reproductive success between a colony where nests were checked daily, which is the interval used in most studies, and a nearby colony where they were checked every four days. Many researchers have attempted to band adult guillemots, which Cairns apparently also did not do in the colonies he was comparing; such activity is likely to be especially disruptive and may make guillemots even more sensitive to mere visits by researchers to the colony (Drent et al. 1964, Kuletz 1983; see also Spear 1988b). At what level observer activity becomes interference is not known. Observer activity in the present study was somewhere between the two visitation regimes tested by Cairns, perhaps closer to his lower level of disturbance. Although we made visits every two days during egg laying, visits were terminated once a clutch was completed, and daily visits did not begin until chicks hatched. Furthermore, by not marking eggs (after 1972), we could move rapidly through the study plot, in some cases being able to peer into nests from five or more meters away. Comparing results from nests receiving our "intensive" level of effort with those from a sample of nests visited only twice during the entire season indicated no appreciable differences in reproductive success. It seems then that to obtain realistic estimates of reproductive success and perhaps other parameters, one has to reduce observer activity to a minimum to avoid disruption. Daily visits and handling of adult guillemots early in the season are particularly disruptive. Reducing such activities, however, means sacrificing opportunities to study other aspects of reproductive success and behavior. This is a drawback in small populations of guillemots, because a sufficient sample of nests is not available to allow a variety of research protocols. Similar problems are often faced in studies of other seabirds sensitive to observer activity (see Chapter 4).

THE GUILLEMOT BREEDING STRATEGY

Annual variability is a difficult problem to address because of the long-term effort required, and unfortunately most researchers in avian ecology do not have the luxury of time. Shorter periods of study can sometimes lead to wrong conclusions about certain subjects and can reduce insight into the range of characteristics of a species. Such a plight was admitted by Ainley, LeResche, and Sladen (1983) when they compared their conclusions about penguin biology from a longer-term data set with those from some "preliminary" reports based on shorter-term results. The variability in overall reproductive success observed in the present study was far broader than that observed in other, mostly short-term, studies (see Cairns 1984, M. P. Harris and Birkhead 1985; Table 9.6). The other studies suggest that guillemots are the archetypical seabird because they seem to exhibit a conservative, low-amplitude response to environmental variability, and against this mass of information the Farallon population appears to be atypical in its variability. In their four-year study, Drent et al. (1964) experienced one season when reproductive success was so much lower than during the other three (but high compared to some of the years at the Farallones) that they considered it atypical of guillemots and treated it separately from the other three.

Another long-term study of guillemot reproduction is that by Bergman (1971, 1978) on Black Guillemots between 1963 and 1977. During 15 years, Bergman noted two years of virtually total reproductive failure. Otherwise, he presented detailed information on the extent of variability in reproductive success only for the eight-year period 1963–70. Reproduction during that span ranged from 0 to 1.3 chicks per breeding pair. Bergman felt that the situation he was studying corresponded "poorly to the normal requirements of the species," i.e., he thought it to be suboptimal, based largely on the years of reproductive failure. The extent of variability, however, was equivalent to that at the Farallones, which we view as being a situation far from suboptimal for the species. The high productivity, large population size, and intense competition for nesting sites indicate that the Farallones provide a highly favorable, if not optimal, breeding situation. Virtually all other studies of guillemots have been conducted at sites where populations are small, usually less than a few hundred pairs. Nevertheless, other large guillemot populations do occur (see Sowls, Hatch, and Lensink 1978, Nettleship and Evans 1985), and a study conducted at one or more of those sites would be interesting for comparison with the Farallon situation.

A number of factors, such as population size, proportion of birds nesting, clutch size, chick survival, and chick growth, indicated years of unusually favorable conditions for guillemots breeding at the Farallones, as in 1971, 1977, and 1979, as well as years of unfavorable conditions, as in the warm-water years of 1973, 1976, 1978, 1982, and 1983. This variability can be explained by what we learned about guillemot feeding ecology during the nesting season (Chapter 3) and by certain connections between feeding ecology and breeding biology that are better understood in guillemots than in most seabirds (see also studies of the Great Tit, *Parus major*, e.g., Perrins 1979). Drent (1965) and Slater and Slater (1972) noted that individual guillemots carrying fish to their young tend to specialize on certain species. Cairns (1984) noted that they specialize for varying lengths of time depending on previous success; in other words, individuals return to the same habitat or feeding site as long as they are successful. Kuletz (1983) was able to show that the degree of specialization affected both the timing and success of reproduction. That is, breeding was earlier and more successful in one of three years when, because of the superabundance of one prey species, individual specialization was less evident. Within another season, she also found that individuals that habitually fed on one type of fish tended to breed later or earlier, as the case might be, than individuals that habitually fed on other species. Thus, it would seem that as more individuals become specialized in feeding, and as the number of specialties increases, the more asynchronous breeding might be.

We did not determine the diet of individual guillemots at the Farallones, although in making the collections reported by Follett and Ainley (1976; most fish and fish species collected in 1973) we too noticed that certain prey were more likely to be found in some nests than in others. More important to our present discussion, we found that diet diversity within the population varied widely among years (Chapter 3). In some years (1971, 1975, 1977, and 1979, especially), rockfish were predominant in diets and diet diversity was low; thus, it is likely that few individuals specialized on other prey in those years. Indeed, as noted in Chapter 3, guillemots fed in flocks, usually with other seabird species; in those years, not only was egg laying more synchronous (i.e., skewed and peaked; Figures 9.4 and 9.5), but fledging weights and reproductive success were also high (Appendix 9.2; Tables 9.7, 9.12). The years when diets were diverse and individual specialization likely was prevalent were those in which breeding was asynchronous and unsuccessful. These were years when rockfish were not prevalent. Bergman (1971, 1978), too, noted that his guillemot population failed to reproduce during years when an important prey species failed to appear. Had the Farallon data been available to him, perhaps he would not have concluded that such variability was evidence for suboptimal nesting conditions. We will return to the relationship between feeding ecology and breeding biology in Chapter 12.

10

Cassin's Auklet

David G. Ainley, Robert J. Boekelheide, Stephen H. Morrell, and Craig S. Strong

Five species of auklets are confined to the North Pacific, and a sixth is confined largely to the North Atlantic. Considering various facets of the group's zoogeography, Udvardy (1963) and Bédard (1985) traced the evolutionary roots of these birds to the Pacific Basin. Four of the extant species are now confined to the Bering Sea, but one, the Dovekie, *Plautus alle,* occurs in the North Atlantic, and another, Cassin's Auklet, occurs in subarctic waters along the western boundary of North America. Of the six, only Cassin's Auklet frequents waters that are not covered or influenced by winter pack ice. Only the Dovekie can be considered a true pack-ice species; all the other high-latitude species wait for the ice to retreat before occupying their summer waters.

Until recently, little detailed information was available on the natural history of auklets. During the last 20 years, however, owing to the efforts of a rather small group of researchers, mainly M. D. F. Udvardy, A. C. Thoresen, J. Bédard, S. G. Sealy, S. Speich, D. A. Manuwal, M. S. W. Bradstreet, and K. Vermeer, much information has been gathered. When Sealy began graduate research on three species of auklets on St. Lawrence Island in 1966, the overall breeding strategies of auklets could not be constructed, but by the time he finished in 1972, so much work had been completed by the above researchers that he was able to summarize effectively and compare the distinctive and common features of the breeding biology of the group. Based largely on Sealy's (1972) Ph.D. dissertation, the following is a sketch of auklet breeding ecology. Information on Cassin's Auklet comes from the work of Thoresen, Manuwal, Speich, and Vermeer.

Auklets are small, compact seabirds. The Cassin's, about midway in a rather limited size range (100–190 gm), weighs about 170 gm as an adult at the start of the breeding season and decreases slightly in weight, to 160–165 gm, by the end. All auklets are rather clumsy on land because their legs are placed far back on the body, an adaptation for better swimming. Sealy characterized auklets as being offshore feeders. All have throat pouches that they fill with food before returning to the colony to feed chicks. This morphological characteristic affords them the opportunity to travel long distances for food if necessary. They can dive 30–50 m deep by using their wings for propulsion (see Chapter 3).

Auklet plumage is not particularly remarkable, except for various tufts and crests on the head. The bill and feet are usually brightly colored. Cassin's Auklet is perhaps the most conservative in plumage; besides its blue feet, its only other notable feature is a white eyebrow that flashes when the bird blinks. Otherwise, the Cassin's Auklet is steel gray above with a white belly. Auklets are not noticeably sexually dimorphic, although males have larger bills on the average (D. A. Nelson 1981, Knudsen 1976).

By and large, the sexes share equally the duties of incubation and chick rearing. The male, however, takes the most active role in construction of the burrow, which perhaps compensates the female for the energy she spends forming the egg. Auklets dig burrows by using their strong legs and sharp claws as well as their beak, or they may occupy natural crevices and hollows in talus slopes. According to Manuwal (1974a,b), about 40% of Cassin's Auklets at the Farallones nest in natural cavities and the remainder nest in burrows; highest densities, reaching one burrow entrance per square meter of surface area, occur where a layer of deep soil is protected by thick annual vegetation (i.e., Farallon Weed). Burrows may be up to about two meters long, but in talus and rock the birds use any suitable cavity. As indicated in Chapter 4, auklets on the Farallones usurp the nesting cavities of the smaller storm-petrels when they have the chance, but in their turn Cassin's Auklets are evicted by Pigeon Guillemots and puffins from cavities of larger size. Auklets retain the same nesting cavity for several seasons. Most auklet species, including Cassin's, typically are densely colonial in their nesting.

Most auklets are diurnal in their visits to the nesting colony, with the population reaching peak attendance during mid-morning. One exception is Cassin's Auklet, which is strictly nocturnal. Sealy argued that colony visitation patterns evolved largely in response to daily cycles in the availability of suitable prey and secondarily in response to the habits of predators.

Although auklets have two incubation patches, they lay but one unmarked white egg. Cassin's Auklets will readily replace a lost egg and will even under certain circumstances lay a second egg after fledging the first chick. In other auklet species, replacement laying is rare and second clutches are nonexistent. Auklet eggs are relatively large, equaling between 13 and 19% of body weight, with a relationship inverse to body size among the six species (Cassin's about 17%). In Cassin's Auklet, egg formation requires 12 to 13 days (Roudybush et al. 1979, Ainley et al. 1981). Members of a pair usually switch incubation duties once per day, and they alternately guard the newly hatched chick for its first three to five days.

Among auklets, the incubation period lasts 24 to 39 days, and the period's length is proportional to the bird's body size. Exceptional is Cassin's Auklet, which is midway in size but has the longest incubation period. The nestling period is proportional to incubation period, ranging between 24 and 41 days (again, Cassin's longest). The longer incubation and nestling periods of the Cassin's Auklet are likely the result of a slower rate of chick growth and food demand. The slower rate of growth is required because, visiting the island nocturnally, parents can each bring only one food load to the chick per day. Chicks hatch with a covering of down and are semiprecocial; they can move about fairly well and are able to thermoregulate within a few days. By the time they fledge they have attained 80 to 90% of adult weight. When they fledge, chicks are independent of adults and must learn to find food and forage efficiently on their own. Cassin's Auklets begin breeding when two to four years old and probably live ten to 20 years (PRBO, unpubl. data).

It should be obvious from this discussion that the Cassin's Auklet is the most "outlying" member of the auklet group. It does not nest in the Arctic, it is nocturnal on its breeding grounds, its incubation and nestling periods are unusually long, and it readily lays replacement or even second clutches. Near the end of this chapter, we will return once again to the unusual character of this auklet.

METHODS

Our major effort in this study was to record the nesting success and phenology of auklets residing in marked burrows. About half of the "burrows" were actually nest boxes with openings larger than those of storm-petrel boxes (see Figure 4.1). These boxes were almost entirely buried in the ground. We followed 66 to 84 nests in the same four plots each year. Two of these plots were among the ten used by Manuwal in his studies. As long as burrows or boxes remained intact, we studied the same sites year after year; there was, however, an annual turnover in study cavities of about 10% due to deterioration during the nonbreeding season. Each spring, well before egg laying began (February), we visited the plots and made sure that holes we excavated from the surface down to the nesting chambers in natural burrows would allow our easy access. We also checked the condition of boxes. Access holes (our access) were covered by a tightly fitted board, which in turn was weighted down with a rock. If we discovered that a burrow studied the previous year had caved in, we found another one suitable for study to replace it.

Beginning on 1 March each year, we began to visit sites at two-day intervals until the first egg was found. From 1972 to 1979, we then checked all in which laying had not yet occurred every other day to determine egg dates; from 1980 to 1983, the interval was once every seven days. In the analyses that follow, data are grouped in five-day intervals; the data for 1983, what few there were (due to ENSO), were extrapolated to fit a five-day interval. After an egg was discovered, we did not normally recheck the nest site until 38 days later, when we checked to discover whether the egg had hatched. If it had not, we continued to check daily or every other day until it did. We then usually left the parent and chick undisturbed for five days, after which we began a schedule of weighing the chick. Coulter and Higbee (ms), in 1974, found that following their nightly feeds, chicks lost weight rapidly during the day until about noon, after which weight loss was negligible. Because of this pattern, we weighed chicks in the early afternoon every year. From 1972 to 1979 (except 1976), we weighed all chicks daily until fledging; in 1976 and from 1980 to 1983, we weighed chicks from 35 days of age until they fledged; thus we determined only fledging weight. Using these procedures, we attempted to keep disturbance of adults at the lowest possible level.

We placed treadles and microswitches in ten burrows to monitor auklet arrivals and departures between 1973 and 1979. When a switch was tripped by an auklet, the event was logged by an Esterline-Angus recorder (see Chapter 1 for more details). We analyzed the resulting charts to determine visitation patterns and compared patterns on overcast (> 6/10 cover) vs. clear nights (≤ 6/10 cover) as well as by moon phase. For the latter, months were divided three ways: by the five days on either side of the full moon, half moons, and new moon.

FARALLON POPULATION

Cassin's Auklet is currently the most abundant seabird nesting on the South Farallon Islands. Individuals can be found nesting anywhere there is sufficient cover, be it a board, a rock, a cormorant nest platform, or soil. The densest concentrations occur where the soil is deep: at the base and on the lower portions of Lighthouse Hill on Southeast Farallon and at the base of Maintop on West End (Figures 1.3 and 10.1). Manuwal (1972) estimated a breeding population of 105,000 birds on Southeast Farallon at any given time during the nesting season. This he called the "estimated breeding population," which has

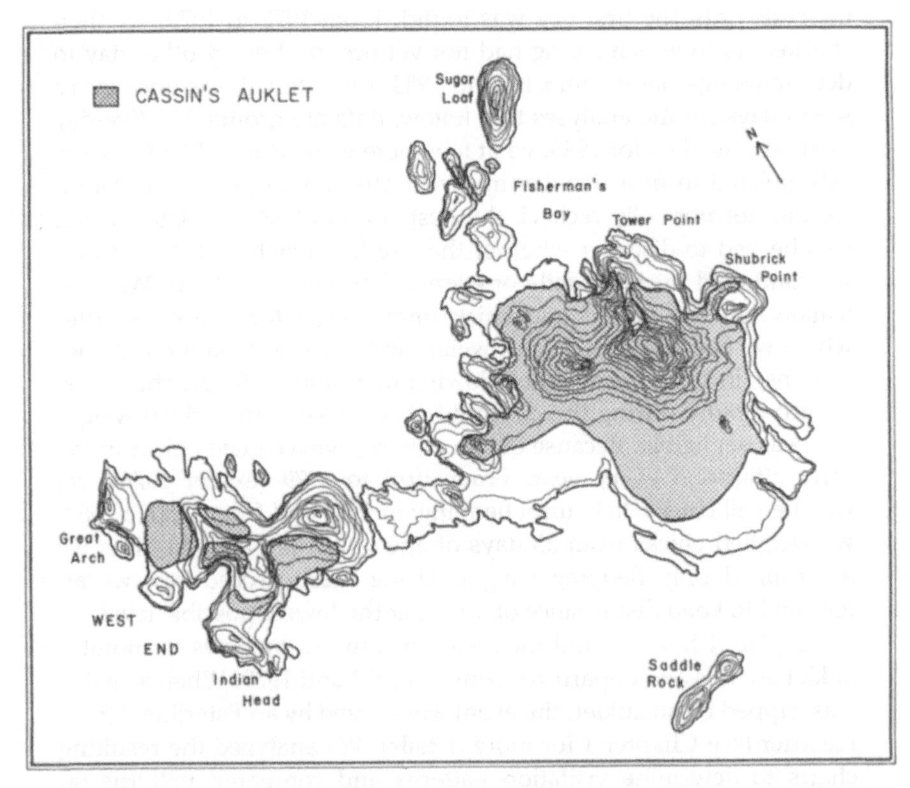

Figure 10.1. Nesting distribution of Cassin's Auklets on Southeast Farallon, 1971–82.

been quoted ever since by him and others. The estimate has also been used as the auklet population of the entire Farallon chain (Sowls et al. 1980). Two interesting facts, though, if considered, indicate that this population estimate is low. First, Manuwal's figure is for Southeast Farallon only and does not include West End. Considering the size of West End and the type of habitat available, and using Manuwal's (1974b) estimates of burrow densities by habitat type, we would guess that about 30,000 auklets nest on West End. This brings the size of the population on the South Farallones to about 135,000 birds; the number breeding on the North Farallones is probably too small to worry about, given the lack of suitable habitat and the large statistical error in the estimate for the South Farallones. Second, Manuwal (1972, 1974b) convincingly demonstrated that if you remove the original occupants of a burrow, a new pair will take up residence and about 70% of these pairs from the "floater" population will breed. Speich and Manuwal (1974) estimated about 40,000 birds in excess of the 105,000 that breed on Southeast Farallon; 70% of the surplus, or about 28,000 individuals, could be considered a special part of the nonbreeding population. If similar ratios are used for West End, the estimated total breeding population, including special nonbreeders, on the Farallon Islands should be about 171,000 birds, although about 36,000 are prevented from breeding by lack of a territory (i.e., nest cavity). These "floaters" socialize en masse in particular areas and visit burrows to determine if they are occupied. Manuwal (1972, 1974b) was among the first researchers to demonstrate the existence of a "floating population" in wild birds (see Glossary). That this should occur in a marine bird is perhaps not so surprising, because a limited number of islands offer suitable breeding habitat for any given species. In this case, it is even less surprising, because the finite number of nest cavities further limits the chances for a pair to nest. Other populations of seabirds are likely to have subpopulations of floaters as well, particularly burrow nesters but also tropical species that nest in bushes or trees, which are often in limited supply on coral atolls. It requires a special effort, however, to demonstrate the existence of such floating populations.

We have little indication of changes in the size of the population of auklets breeding from one year to the next. During ENSO 1982–83, however, we did note increased turnover in the banded occupants of a sample of burrows we observed between 1978 and 1985. Thus, mortality of burrow occupants had apparently increased (PRBO, unpubl. data).

TABLE 10.1
Estimated numbers of Cassin's Auklets nesting on the North American Pacific coast, 1965–81

Area	Birds	Sites	Source
Alaska	600,000	21	Sowls, Hatch & Lensink 1978
British Columbia	1,063,200	6+	Vermeer et al. 1983
Washington	70,000	4	Manuwal, Wahl & Speich 1979, Varoujean 1979
Oregon	480	7	Varoujean 1979
California	237,170[a]	10	Sowls et al. 1980, Manuwal 1972
Mexico	10,500	2	DeLong & Crossin ms
Total	1,981,350	50+	

NOTE: Most recent data.
[a] The Farallon population (145,000 birds) constitutes 61% of this figure (but see text).

The population of Cassin's Auklet on the South Farallones represents 87% of the breeding population of this species in California and 10% of the world total (Table 10.1). Thus, because of its size, the Farallon population is an important one. The largest single population occurs at Triangle Island, British Columbia, where about 720,000 auklets nest (Vermeer et al. 1979). About 54% of the total world population nests in British Columbia. These figures do not reflect the still-incomplete inventory of Cassin's Auklet populations in Alaska (Sowls, Hatch, and Lensink 1978), and they do not reflect the possibility that floating populations occur at sites other than the Farallones.

OCCURRENCE PATTERNS

Distribution at sea

When we have explored the ocean in the vicinity of the Farallones, we have frequently discovered dense concentrations of auklets in waters above the continental slope. The slope stretches north to south a few miles to the west of the Farallones. Persons making trips to the Cordell Bank, 25 km northwest of Southeast Farallon, have reported large flocks of auklets in that area also. Farther north, few auklets have been encountered. It appears, therefore, that the Farallon auklets regularly use waters over the continental slope as much as 50 km away, especially to the northwest, when not on the island (see Chapter 3).

We banded about 10,000 auklets between 1968 and 1983, and recaptures at the Farallones number in the thousands. From farther afield, however, we have received reports of recoveries of only three birds, all young and all from Monterey Bay, about 80 km to the south. This low recovery rate, the fact that dense concentrations of auklets occur over slope waters adjacent to the island throughout the year, and the fact that auklets visit the island in large numbers throughout the year except during late summer and early fall (when they are molting; Payne 1965; see Figure 12.1) indicate that auklets of the Farallon population are localized and sedentary. They seem to prefer waters on the coastal side of the California Current, especially those most strongly influenced by coastal upwelling (Briggs et al. 1987). Upwelling renders these waters the coldest in the region. Sea-surface temperatures in offshore coastal waters from central California to Oregon, in fact, are colder during summer than those from Washington to British Columbia and southeast Alaska. During the spring and summer upwelling period, sea-surface temperatures from central California south to northern Baja California are similar to those off the Pacific Northwest during summer. Thus, oceanographically, it is probably incorrect to say that the Cassin's Auklet nests from subarctic to subtropical waters (e.g., Ashmole 1971, Manuwal 1984). In effect, if latitude is disregarded, the species is subarctic throughout its breeding range. The Cassin's Auklet is largely confined to the "upwelling domain" of the eastern North Pacific (Favorite, Dodimead, and Nasu 1976).

Colony attendance

Auklets visited burrows essentially all year, although the majority of fall visitors may be subadults (PRBO, unpubl. data; Figure 12.1). During April 1973, and especially 1983, heavy rains flooded burrows on the marine terrace, delaying occupation of a large proportion of burrows (Chapter 2). Otherwise, two annual patterns of visitation were apparent, depending on whether egg laying was late (1973, 1977, 1978) or early (1974, 1975, 1976, 1979; see Egg Laying). During years when laying began in March (early), breeding auklets visited their burrows almost every night from January through June (Figure 10.2). During the prelaying period, January to March, about 20% fewer individuals visited on nights of a full moon, especially ones with clear skies, than on nights of a new moon. Once egg laying began, however, moon phase and sky conditions had little bearing on the proportion of breeding auklets making visits. Visitation decreased substantially in July, and continued to decline through December. During the fall, more auklets again visited on overcast, dark nights than on those of clear skies and full moon.

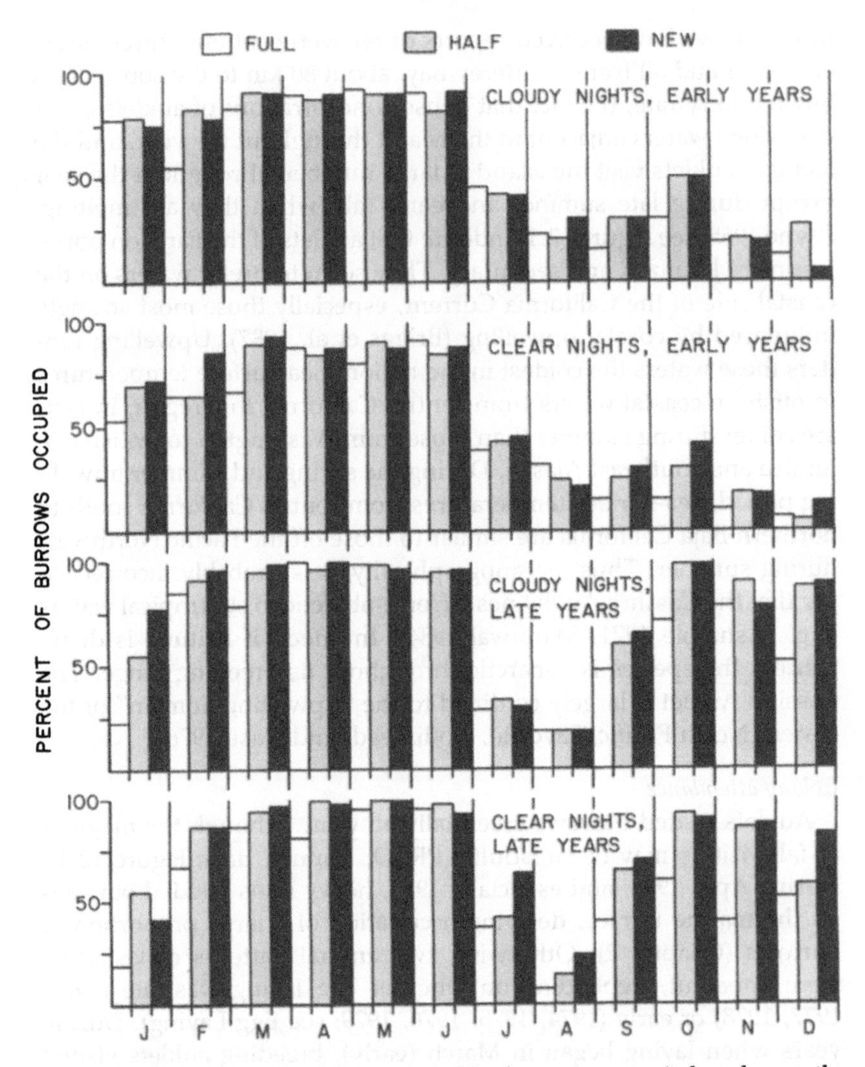

Figure 10.2. Percentage of Cassin's Auklet burrows occupied each month, comparing years of early (1974–76, 1979) and late (1973, 1977–78) laying, as well as the effect of moon phase and sky conditions.

During years when laying commenced in April (late), so no replacement or second clutches were laid (see below), visitation was less frequent during the preceding January but soon reached a peak that was carried through June. In contrast to the other set of years, visitation increased during September, and large numbers continued to visit their burrows through the fall and winter. The effects of moon and sky conditions were the same as in the years of early laying.

Why the two annual patterns of visitation existed is a difficult question to answer in the absence of information on at-sea distribution and prey availability. The dip in visitation frequency during July and August corresponds to the period of heavy molt. In years of early laying, when Cassin's Auklets are able to lay and tend to replacement eggs and/or second clutches, many find themselves overlapping nesting duties with molt. In years of late laying, very little such overlap occurs. Overlap of molt and breeding may bring about the observed patterns.

The fact that fewer auklets visit their burrows, and the island, on clear nights with a full moon is evidence supporting the hypothesis that nocturnal visitation is an adaptation to reduce predation. It is interesting, however, that once breeding begins, auklets visit regardless of light levels. This pattern suggests that breeding has its risks. On the other hand, by spring and summer, gulls—the prime auklet predator at the Farallones—usually have plenty of more easily available prey, and thus the risk is perhaps reduced. In support of this hypothesis, as indicated in Chapter 3, gulls do not eat measurable numbers of auklets except in summers of extreme food shortage.

On full-moon nights, auklets arrive at their burrows about an hour later and depart about 45 minutes earlier than on (dark) new-moon nights (Table 10.2). As a result, they spend almost two more hours on the island on dark nights than on light ones. On dark evenings and mornings, much of their post-arrival and pre-exodus time is spent in bouts of singing outside, at the burrow's entrance, interspersed with returns to the burrow. On light nights, auklets spend almost all their time inside the burrow. These responses to light levels are also likely induced by predation pressure.

Besides sky conditions, the total number of hours spent on the island during the night is a function of night length, stage of breeding,

TABLE 10.2

Mean times of auklet arrival and departure in relation to sky conditions and moon phase, Farallones, 1972–80

	Cloudy skies[a]		Clear skies[b]	
Time of:	Full moon	New moon	Full moon	New moon
Arrival	2225	2108	2236	2129
Departure	0354	0439	0344	0422
Total (hours)	5.5	7.5	5.1	6.9

[a] 3,541 data entries from 936 nights during years of late egg laying.
[b] 4,407 data entries from 1,027 nights during years of late egg laying.

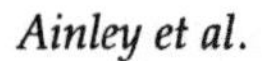

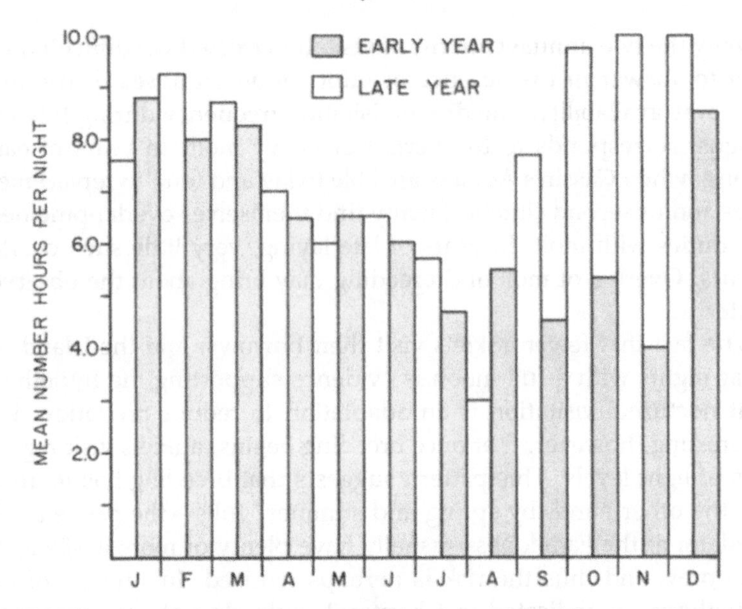

Figure 10.3. Amount of time per night that Cassin's Auklets remained at their burrows, by month, 1973–79; the data summarize elapsed time between first arrival and last departure.

and year (Figure 10.3). Maximum time is spent ashore during non-breeding periods, September through March, except in early-laying years, when relatively little time is spent ashore September through November. The amount of time spent ashore is least from April through August, with a gradual decline in the number of hours throughout this period. Not only are nights short then, but adults are incubating eggs and feeding chicks as well. Less time is probably spent advertising and defending territory, especially since at least one bird (adult or chick) is occupying the burrow at all times.

EGG LAYING

During most years, Cassin's Auklets began to excavate and repair burrows in earnest during January and February. In the process of maintaining burrows, the birds dig vigorously and forcefully kick soil from the burrow entrance. After nights when many thousands of birds dig and fling soil from burrow entrances, nesting areas resemble prairie dog towns, pock-marked with mounds of excavated dirt. It may take an auklet pair several weeks to complete their burrow. A nesting chamber is made, depending on the workability of the soil, at the end of the burrow and is large enough barely to allow two adults and a chick to maneuver. During most years, all but a few (less than 5%) of the auklets that possessed burrows eventually laid eggs (Appendix 10.1). During the consecutive years for which we have data, only 1983 stands out with regard to burrow use. In that year, auklets laid eggs in only 47.6% of burrows.

In our sample of 1,273 egg dates compiled from 1970 to 1983, laying ranged from the second week of March to the end of the second week of July (Figures 10.4, 12.1). Whereas the initiation of laying ranged over a three-month period (Figure 10.5), termination was much more regular in timing. The last egg was laid during the first week of July during six of the 14 years, and during four others it was laid during the week preceding that. We know of only two eggs laid later than the first week of July, one in 1971 during the third week (Manuwal 1979) and one in 1976 during the second week. In contrast to the constancy in termination of laying, first-egg dates ranged from the second week of March to the last week of May. The average dates of clutch initiation ranged from 24 March to 29 May, a range of two months, and the mean laying date in any given year was statistically similar to mean laying dates in no more than three of the 13 other years (Table 10.3). In only one year, 1983, was the mean laying date different from (later than) that in *all* other years. Egg laying was late during years that began with sea-surface temperatures warmer than average (Chapter 2).

Although photoperiod probably stimulates auklets to come into reproductive condition or readiness (Sealy 1968), egg laying itself in Cassin's Auklet is triggered obviously by an environmental factor that

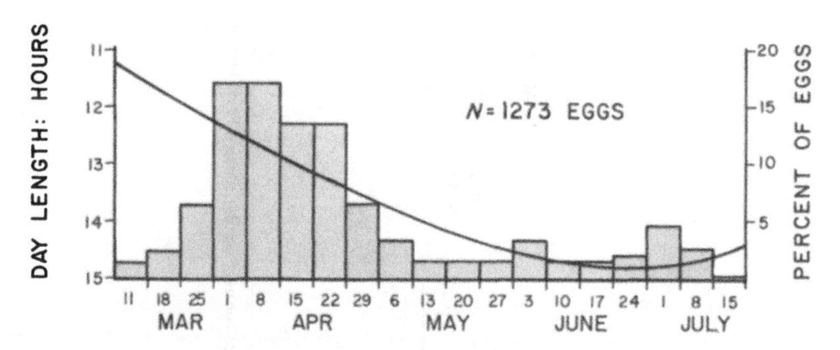

Figure 10.4. Percentages of Cassin's Auklet eggs laid in seven-day periods, all years combined, 1972–83, compared to day length (curved line).

TABLE 10.3

Mean annual laying dates for first-clutch eggs, Cassin's Auklets, 1970–83

1971	1974	**1976**	*1979*	*1975*	1981	**1982**	**1973**	**1972**	*1977*	1970	1980	**1978**	**1983**
3/24	3/28	3/30	3/31	4/4	4/5	4/8	4/8	4/12	4/16	4/18	4/19	4/24	5/29

NOTE: Lines connect similar means; SNK test, $p > .05$.

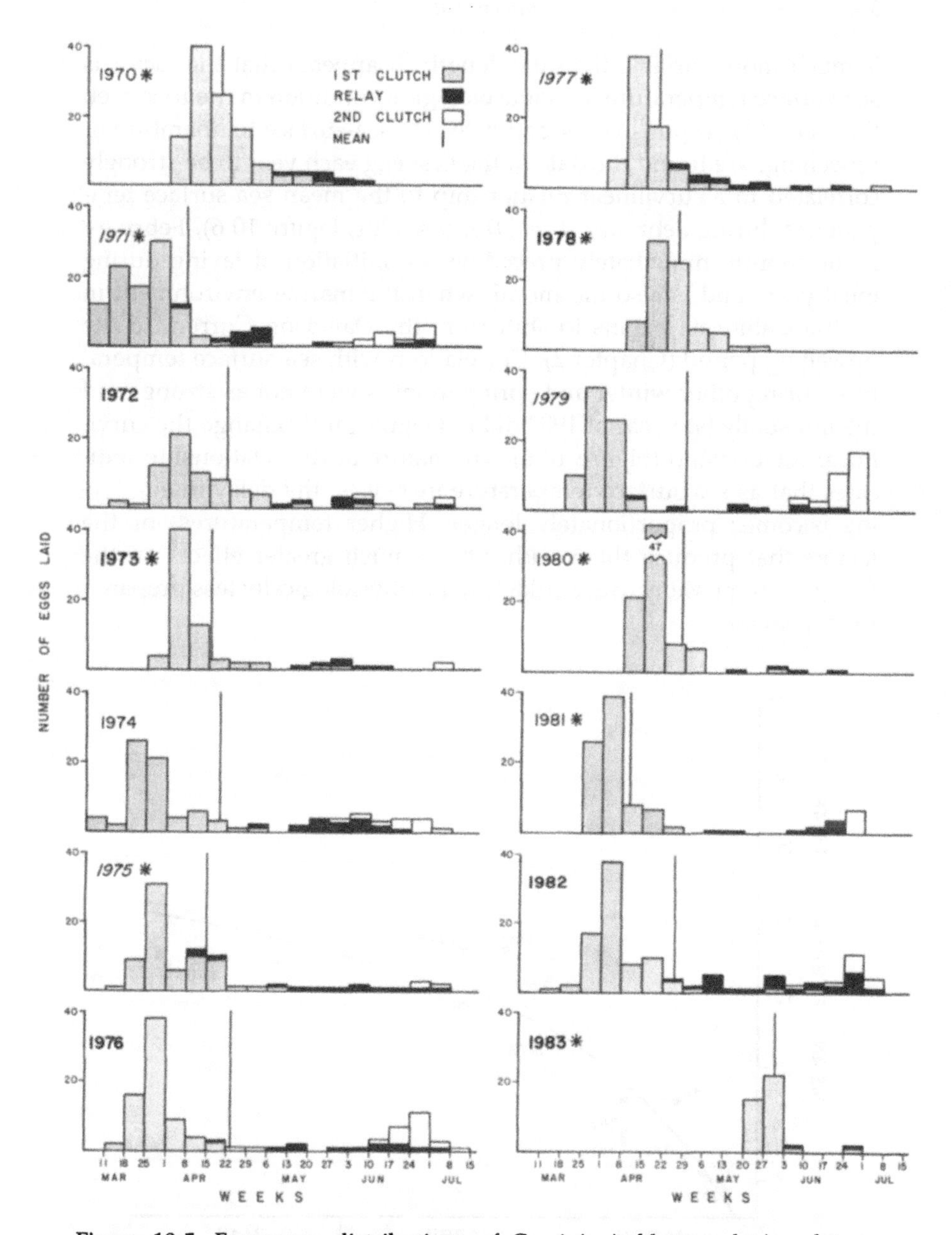

Figure 10.5. Frequency distributions of Cassin's Auklet egg-laying dates, 1970–83; in years marked by an asterisk (*), laying was significantly peaked and skewed (t test, $p < .05$).

is much more variable than day length. It appears that the factor is sea-surface temperature, or some biological condition in the food web that is highly responsive (or correlated) to sea-surface temperature or upwelling. We found the date of the first egg each year to be strongly correlated in a curvilinear relationship to the mean sea-surface temperature during February ($r = .915$, $p < .001$; Figure 10.6). February is the month immediately preceding the initiation of laying during most years and is also the month when the marine environment in central California begins to shift from the Davidson Current to the upwelling period (Chapter 2). Correlations with sea-surface temperature during other winter and spring months were not as strong, and the unusually late year of 1983 did not significantly change the curvilinear relationship (Figure 10.6). The nature of the relationship indicates that as sea-surface temperature increases, the delay in egg laying becomes proportionately longer. Higher temperatures, or the factors that produce them, either have much greater effects on the food web or possibly cause auklets to be physiologically less prepared for egg laying.

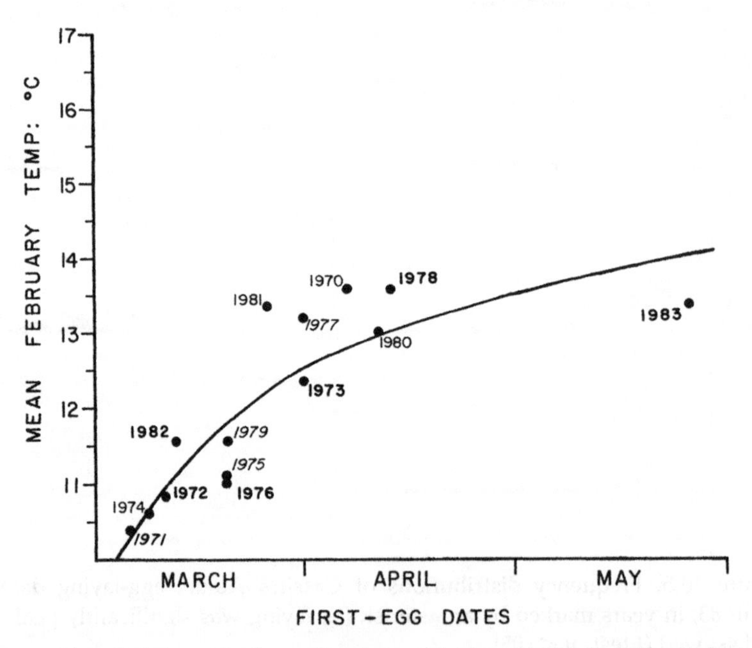

Figure 10.6. Relationship between date of laying of the first Cassin's Auklet egg and mean sea-surface temperature during February, 1970–83; $x = 2.18y^{0.38}$, $r = .915$, for the curved line.

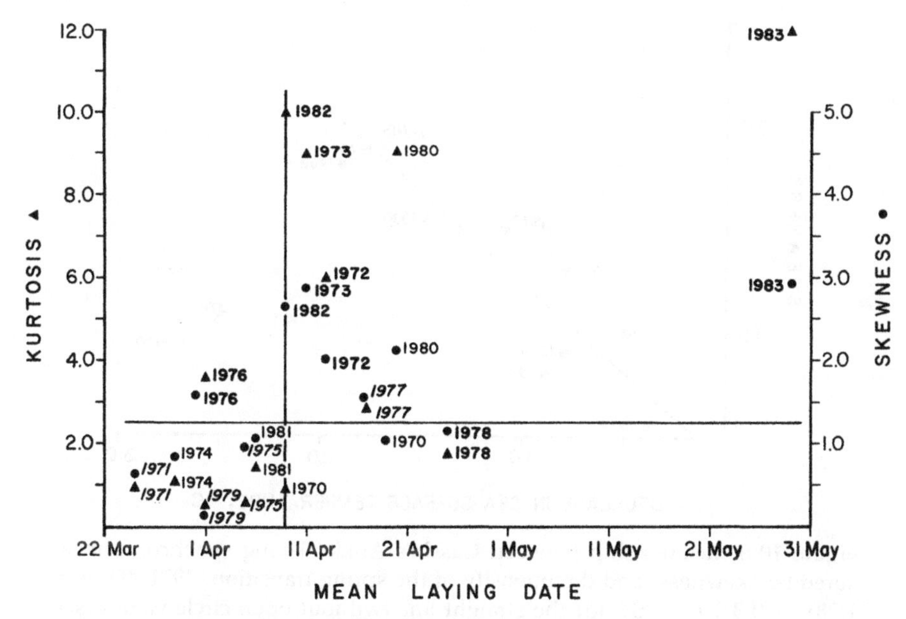

Figure 10.7. Scatter plot showing the relationship between mean laying date and the clumping of laying, as measured by skewness and kurtosis, for the Cassin's Auklet laying distribution, 1970–83.

During every year but 1979, egg laying was significantly skewed toward the early part of the laying period (Figure 10.5). The degrees of both skewness and clumping (kurtosis) were related to the mean laying date (Figure 10.7). In most years of later laying dates, such as 1972, 1973, 1980, 1982, and 1983, the laying period became more skewed and more synchronous (1970, 1977, 1978, and 1981 were exceptional; see below). As with the initiation of egg laying, the shape of the laying curve for first eggs was strongly affected by a variable in the marine environment, in this case the sharpness of the transition between the Davidson Current and the upwelling periods.

Skewness, as a measure of the clumping of laying, was related directly to the rate at which sea-surface temperature declined during the month when it dropped most, as long as the drop was no more than 2.0° C (Figure 10.8). In 1971, 1974, and 1979, all years of less skewed or unsynchronized laying, the water around the Farallones was cold through the entire winter, and the temperature changed little as the laying period began (Chapter 2). In contrast, years with strongly skewed and synchronous laying dates (1972, 1973, 1980, and 1982) had anomalously warm winter sea temperatures followed by

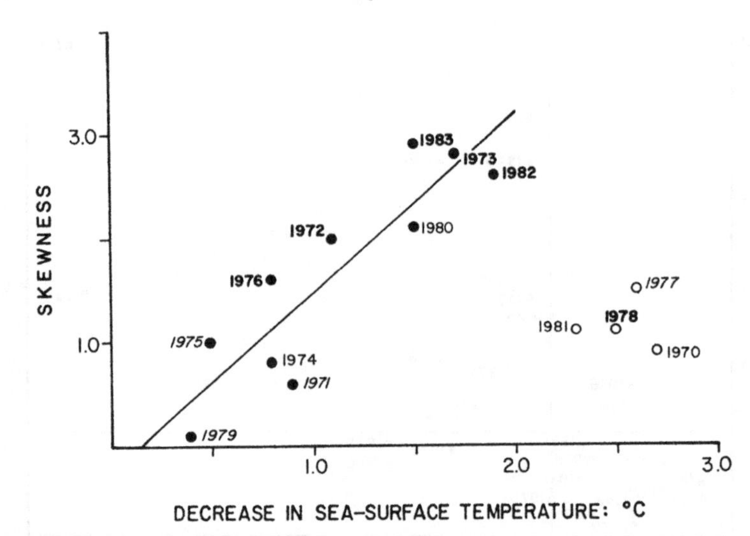

Figure 10.8. Relationship between Cassin's Auklet laying synchrony, measured by skewness, and the intensity of the spring transition, 1971–83; $y = 1.78x - 0.31$, $r = .901$, for the straight line (without open-circle values; see text).

moderate upwelling episodes and a rapid decline in sea-surface temperature during March or April.

Diverging from this relationship were four years, 1970, 1977, 1978, and 1981, when the transition exceeded 2.0° C. These years did not fit the linear relationship just described (Figure 10.8); the relationship could prove to be curvilinear, but more data are needed. These four years were also the ones in which sea-surface temperatures changed unusually abruptly during the critical transition period, from a positive anomaly exceeding +1.0° C to a negative one exceeding −0.6° C. It seemed as if such an abrupt change in the marine environment caught the auklets "off guard," and thus rather than egg laying being even more clumped it was just the opposite. Egg laying was late during these four years, and in that case, as indicated above, it should have been greatly clumped.

Replacement eggs

As noted in the introduction to this chapter, Cassin's Auklet is an unusual auklet in that, in some years, it will readily replace an egg that is lost or that fails to hatch, and in many years it will lay a second clutch after the first chick fledges. After an egg or chick is lost or deserted, Cassin's Auklets are able to lay a replacement egg on average in 17.5 ± 8.0 days (range 9–37 days; n = 18, including four between-egg intervals published by Manuwal 1979). After the first chick fledges, auklets on average lay a second clutch (one egg) in 14.5 ± 3.0 days (range 8–24, n = 75). Ainley et al. (1981) found that auklets can form an egg in 12 to 13 days. Thus, an auklet's physiological response to loss of an egg or chick, or to the fledging of its first chick, is practically immediate.

As one would expect, combining replacement eggs and second clutches with first clutches makes the mean date of egg laying significantly later during most years. Only during warm-water years, when few eggs other than first clutches are laid, was mean laying date not altered by this method of calculation.

Incubation period

Manuwal (1974a) calculated an incubation period of 37.8 days (n = 86 eggs), but his calculations did not include the first day that an egg was in a burrow. Perhaps he considered that most eggs are not incubated until one night after laying. On the other hand, he showed that eggs not incubated on the first night have a much lower chance of being incubated subsequently and thus have a much lower chance of hatching. Including that first day, we calculated a mean incubation period of 39.0 ± 2.0 days (range 37–57, n = 493). The majority of eggs (58%) hatched at 38 days (Figure 10.9). In other auklet species, the male takes the first turn at incubation, and the two members of the pair then switch turns every 24 hours (Sealy 1968, 1972). This appears to be so in Cassin's Auklet as well (Manuwal 1974a).

Whereas storm-petrels take advantage of the safety that nesting in a burrow brings and thus leave their egg unattended for short periods when need be (Chapter 4), the low variability in the auklet incubation period (Figure 10.9) indicates that Cassin's Auklets rarely do this. Like that of the storm-petrels, however, the length of the auklet incubation period varied within the breeding season. Incubation was longest among earliest-laid eggs; it then dropped to a minimum among eggs laid during the laying peak, and thereafter increased slightly (Table 10.4). Even though we are considering a range in the mean length of the incubation period of only 1.3 days at most, these differences are statistically significant. The pattern of decrease to a minimum and then gradual increase is also consistent year after year. This latter trend alone indicates that the phenomenon is not an artifact of sampling.

The longer incubation period of early eggs could result from in-

TABLE 10.4

Mean incubation and nestling periods, and hatching and fledging success, Cassin's Auklets, 1972–79

Variable	Breeding-season period[a]				
	1	2	3	4	Total
Incubation period (d ± SD)[b]	39.8 ± 2.4	38.5 ± 1.4	39.0 ± 2.1	39.4 ± 2.6	39.0 ± 2.0
n	70	211	139	102	493
Hatching success[c]	84.2%	84.4%	85.3%	42.8%	72.0%
n	76	250	163	208	697
Nestling period (d ± SD)[d]	43.0 ± 3.0	42.0 ± 3.3	42.0 ± 3.2	41.6 ± 3.2	42.2 ± 3.2
n	60	185	122	46	413
Fledging success[e]	85.9%	88.2%	88.5%	45.9%	79.2%
n	64	211	139	109	523

[a] Breeding season divided into four periods: *1*, eggs laid during the ten days before the five-day period of peak laying; *2*, eggs laid during the peak; *3*, eggs laid during the ten days immediately after the peak; *4*, eggs laid during the remainder of the breeding season.

[b] All pairs of sample means are statistically different from one another (SNK test, $p < .01$) except samples 3 and 4.

[c] Success during period 4 significantly lower than during the three earlier periods (SNK test, $p < .01$).

[d] Sample mean during period 1 significantly higher than means of periods 2–4 (SNK test, $p < .01$).

[e] Fledging success = percentage of hatched chicks fledging.

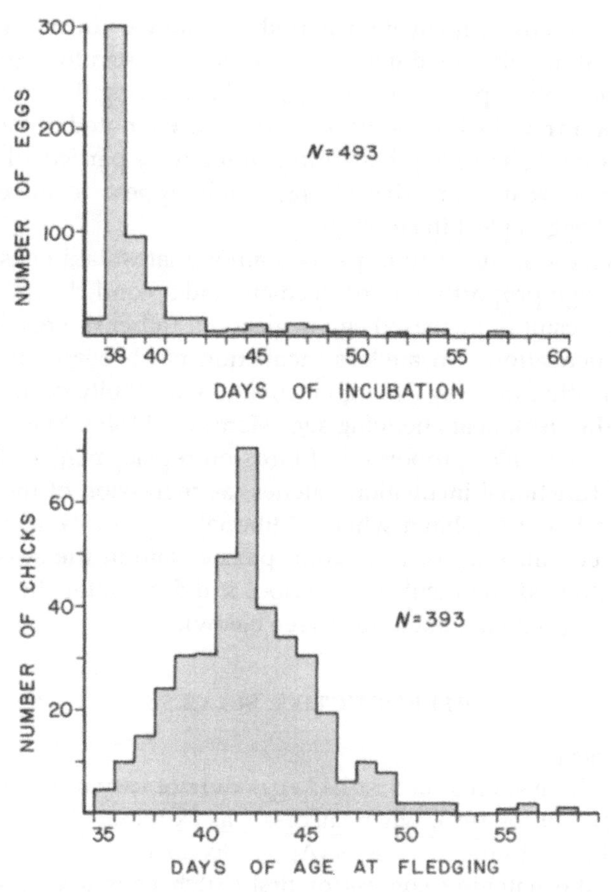

Figure 10.9. Frequency distributions of Cassin's Auklet incubation and nestling periods; all years combined, 1972–82.

creased short-term (1–2 days) egg neglect. We found this to be true of early storm-petrel eggs (Chapter 4), and, as in storm-petrels, it could be the result of the more inclement weather early in the breeding period (Chapter 2). On the other hand, inclement spring weather at the Farallones is quite similar to the typical weather experienced by Cassin's Auklets in the Pacific Northwest and by other auklet species in the Bering Sea. Unfortunately, there is no information on within-season variation of the incubation period for these other auklet populations. If one or two days of neglect is the cause of the longer mean incubation period among early eggs, it did not lower hatching success (Appendix 10.1).

At the Farallones, like that of the Ashy Storm-Petrel, the mean length of the auklet incubation period was longer during the years 1974–76, although the difference was not statistically significant. Among incubation periods exceeding 43 days during the years 1972–79, 77% occurred during the 1974–76 period. We noted in the storm-petrel chapter (Chapter 4) that spring winds were particularly strong during these years, and that strong winds appear to increase the amount of egg neglect in storm-petrels.

The increase in incubation periods among latest-laid eggs, which include a high proportion of replacement and second clutches, might not be the result of increased egg neglect but rather the result of less efficient incubation. An auklet's incubation patch begins to regress and to refeather midway through incubation, and fully disappears by the time the chick nears fledging age (Manuwal 1974c). Manuwal also found that a smaller proportion of birds on replacement and second eggs had functional incubation patches, as regression of the incubation patch is not inhibited when additional eggs are laid. Although the reduced efficiency of incubation patches late in the season may have lengthened the incubation period, it did not alter the hatching success of eggs incubated to term (see below).

REPRODUCTIVE SUCCESS

Hatching success

The hatching success of first-laid eggs did not vary much from year to year, except for 1983, when success was extremely low because of increased abandonment of eggs (Appendix 10.1). During the period 1970–82, the hatching success of first-clutch eggs averaged 78.9% (n = 1,484 eggs), a rate slightly higher than that observed by Vermeer and Lemon (1986) over two seasons in the Queen Charlotte Islands. At the Farallones, the hatching rate of latest-laid eggs was much lower than that of eggs laid during the earlier two-thirds of the laying period. No doubt this is due to the progressively lower success of replacement and second clutches, a pattern that in turn is probably related to a seasonal change in food availability (Chapter 3). Hatching success in replacement clutches averaged 65.8% and in second clutches averaged 47.1% (Appendix 10.1). The increased failure to hatch among the three successive types of clutches was coincident with and related to an increasingly greater desertion rate, reaching 11% in first clutches, 21% in replacement clutches, and 54% in second clutches. Desertion rates were highest during warm-water years. In contrast, it appears that the proportion of eggs that were addled, i.e., never hatched even though incubated to term, remained constant (10–12%) regardless of which clutch was involved. Manuwal (1979) noted that some eggs—as many as 8% in 1970 (a warm-water year), but none in 1971—were never incubated.

Chick development

Cassin's Auklet chicks remained in the nest an average 42.2 ± 3.2 days (n = 413), a figure one day longer than that calculated by Manuwal (1974a) and two days shorter than that calculated by Thoresen (1964). The length of the nestling period declined from an average 43.0 days among chicks from early-laid eggs to 41.6 days among those from latest-laid eggs (Table 10.4). Sealy (1968) also noted that late chicks fledged more quickly in Least, Crested, and Parakeet auklets (*Aethia pusilla*, *A. cristatella*, and *Cyclorrhynchus psittacula*).

On the Farallones, the length of Cassin's Auklet nestling periods varied much less than those of Ashy Storm-Petrels (compare Figures 4.10 and 10.9). Accordingly, the storm-petrels and Cassin's Auklet differ in that no relationship between fledging date and the phase of the moon was evident for Cassin's Auklet chicks (storm-petrel chicks extend their occupancy of the nest until the sky is favorably dark; Chapter 4). During the years 1972 to 1979, 47.2% of auklets fledged during the dark half of the lunar cycle, compared with 52.8% during the light half. However, early in the summer, when most auklets fledge, it is foggy or overcast on the vast majority of nights. In contrast, during the fall, the time of year when storm-petrels fledge, skies are often clear. Thus, if it is important for auklet chicks to fledge when nights are dark (to avoid predation), normally overcast conditions may preclude the need to fledge in synchrony with moon phase.

The shape of the auklet's growth curve, calculated with the logistic model, is typical of many seabirds' (Figure 10.10). Using analysis of variance, we estimated the average hatching weight, the asymptotic weight, and the growth-rate constant (K) to be 21.1 gm (range of annual averages 18.7–26.4), 163.8 gm (range 160.7–169.0), and 0.1367 per day (range 0.107–0.159), respectively, with little between-year variability (Tukey-Kramen Studentized Range test, $p > .4$). Chicks hatched at a slightly heavier weight in 1978, and the growth-rate constant was accordingly lower that year than in any of the other years. Consistent with these patterns, asymptotic weight showed no significant between-year differences ($F = 0.48$, DF = 5,133, $p > .5$).

Within years, only asymptotic weight was affected by hatching date: early chicks became heavier than late ones, although a great deal

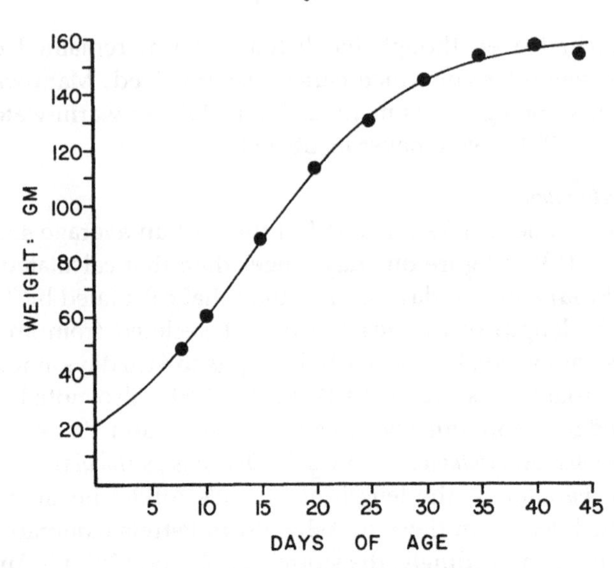

Figure 10.10. Logistic growth curve of Cassin's Auklet chicks (ANOVA; n = 138). Dots are mean values; all years combined, 1972–79.

of variability existed (Figure 10.11). No such trends were evident for hatching weight or growth-rate constant (ANOVA, DF = 133; hatching weight, $t = 3.58$, $p > .5$, growth rate, $t = 6.27$, $p > .3$).

We also analyzed actual weights at fledging. These are not necessarily asymptotic weights, because in many instances chicks lost weight just prior to departure. Auklet chicks from first-clutch eggs fledged at 151.6 gm, and unlike that in asymptotic weight, the annual variation in fledging weight was significant (Appendix 10.2). Mean fledging weight ranged from a low of 144.6 gm in 1983 to a high of 158.0 gm in 1971. During all warm-water years except 1972–73, fledging weight averaged less than 150 gm. In 1959, another warm-water year, A. Thoresen (unpubl. data) determined a mean fledging weight of 139.3 gm (n = 20 chicks).

There was significant within-year variation in mean fledging weights, as there was in asymptotic weights. Except during 1981 and 1982, average fledging weight declined sharply as the season progressed, especially in the latter half of the fledging period (Figure 10.12). Latest chicks, in fact, fledged at weights so much lighter than those of earlier chicks that it is doubtful they survived their first few days at sea. At Triangle Island, British Columbia, in 1978, Vermeer (1981) recorded a mean fledging weight of 159.2 ± 11.7 gm (n = 41),

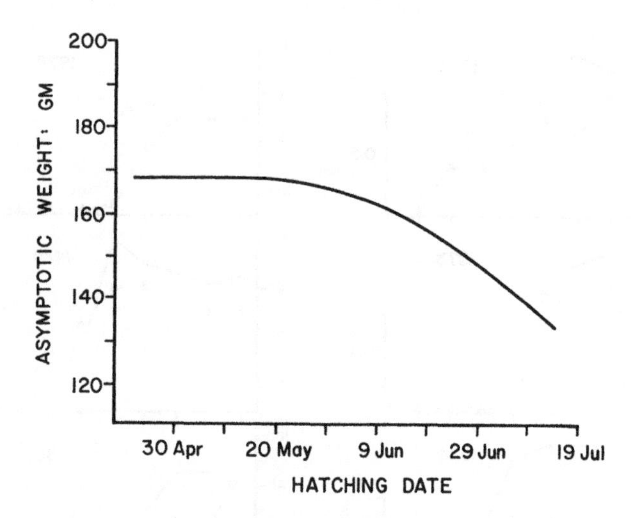

Figure 10.11. Variation in asymptotic weight as a function of hatching date among Cassin's Auklet chicks, 1972–79 (ANOVA, $r = .404$, DF = 133, $t = 17.65$, $p < .001$).

which is not much different from the highest mean weight recorded at the Farallones. Comparison, however, is difficult, because it is not known if Vermeer's sample was from early or late in the fledging period.

Chicks fledge at about 90% of adult weight, but growth after fledging must be much slower than it was before. Even several months after fledging, young auklets are still noticeably smaller than adults (Ainley pers. obs.).

Breeding success

As in hatching success, little annual variation existed in chick survival (Appendix 10.1). Among chicks hatched from first-clutch eggs, survival ranged from a low of 71.6% in 1981 to a high of 92.9% in 1977. Even in 1983, when hatching success was exceedingly low, nestling survival was within the limits set during other years. Over the entire 14-year span, survival among chicks hatched from first-clutch eggs was 82.9% (n = 1,183), a rate slightly lower than that observed during two seasons by Vermeer and Lemon (1986) in the Queen Charlotte Islands.

The pattern was much different among chicks hatched from eggs laid later. Survival among nestlings hatched from replacement eggs declined to 58%, and among those hatched from second clutches sur-

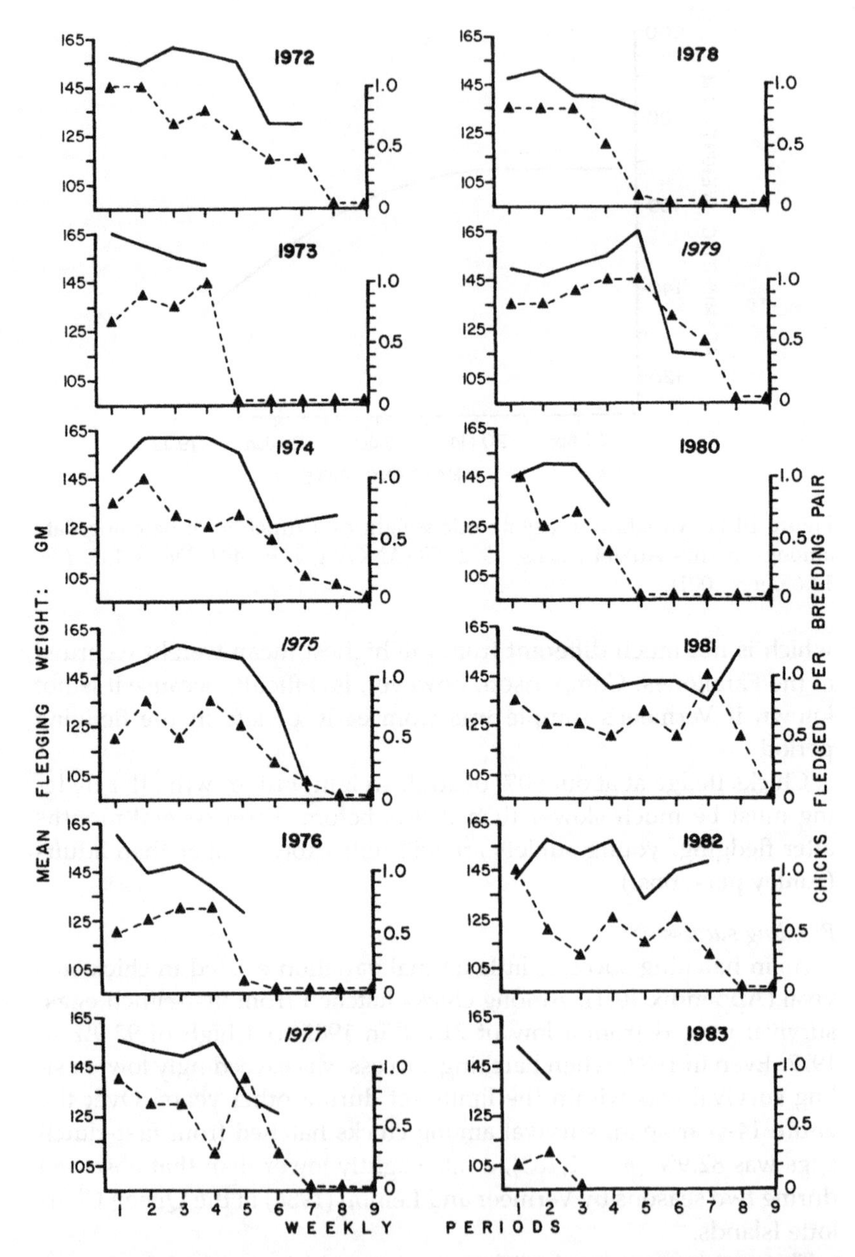

Figure 10.12. Within-year variability in breeding success (dashed lines) and fledging weights (solid lines) in Cassin's Auklets, 1972–83. Data are in weekly periods from date of first egg laying.

vival was only 24%. In these chicks, survival varied greatly among years, ranging from 17 to 100% from replacement clutches and 12 to 100% from second clutches. Overall, only 46% of chicks from eggs laid during the last third of egg laying survived, compared to 87% from eggs laid earlier.

Within first clutches, the number of chicks fledged per nesting pair over the 14-year span of the study averaged 0.6, a rate similar to that observed during two years by Vermeer and Lemon (1986) in the Queen Charlotte Islands. At the Farallones, except for unusually high output from first clutches in 1973 and 1979 and unusually low output in 1983, productivity varied little from year to year (Appendix 10.1). Among replacement eggs, productivity averaged over the 14 years fell to 0.3 chick per pair, and among second clutches to 0.1 chick per pair. In addition to being lower, the productivity of later eggs was more variable from year to year than that of first clutches. The productivity of replacement eggs, calculated as chicks produced per egg laid, ranged from 0.0 to 0.7, and the productivity of second clutches ranged from 0.0 to 0.4 chick per pair. Relatively high productivity in replacement eggs did not necessarily correlate with high success in second clutches. On the other hand, high productivity in second clutches followed good success in replacement clutches. Chicks were successfully fledged from second-clutch eggs only during 1971, 1974, 1979, and 1981 (Appendix 10.1), all cold-water years of early breeding that followed ENSO or warm-water years (Chapter 2).

If annual rates of productivity of first-clutch eggs are compared with corresponding figures for hatching success ($r = .601$) and chick survival ($r = .273$), it appears that the year-to-year fluctuation in productivity was determined largely by hatching success (multiple rank correlation, $p < .05$). The data in Table 10.4 indicate that hatching success is affected strongly by the desertion or loss of eggs rather than failure to hatch following complete incubation.

Within breeding seasons, productivity varied considerably but in a consistent pattern (Figure 10.12). During all years except 1983, productivity from earliest-laid eggs was at or above 0.8 chick per pair; it remained near that level for eggs laid during the next ten days, but then began to decline. At about the same time that chick productivity declined, fledging weights also dropped rapidly from about 145 gm on average to 135 gm or below. Thus, within each season there was a definite transition from high productivity and high fledging weights to low productivity and low fledging weights. The period of egg laying that leads to high productivity and high fledging weights might be called the "effective breeding season." If a pair laid an egg, be it a first, replacement, or second clutch, within the effective breeding season, they had a high chance for success. Otherwise, prospects were poor. If we define the end of the effective breeding season as the point where the sharp decline in success and fledging weights occurs, then it is evident that lengths of the effective season varied widely, from ten days in 1983 to 80 days in 1981 (Table 10.5). Short effective breeding seasons tended to occur during years when egg laying was late and/or during years influenced by ENSO or warm sea-surface temperatures. Low overall fledging weights and overall fledging success also occurred during years when the effective season was short, but the opposite was not necessarily true (i.e., short seasons leading to low productivity and weights). Although the duration of the effective breeding season ranged between ten and 80 days, a factor of eight, the proportion of birds laying during that critical period ranged only from 78 to 98% of breeders, a factor of less than one (Figure 10.13). This is not really surprising given our discovery discussed earlier that the clumping of egg laying varied widely, too, and that laying was more synchronous during late (warm-water) years.

The proportion of birds breeding within effective breeding seasons

TABLE 10.5

Nesting and environmental variability, Cassin's Auklets, 1970–83

	Nesting[a]				Environment[a]			
Year	Timing of season	Season length (d)	Chicks per pair	Mean fledging weight (gm)	Early SST	UW index	Sea level	Rain
1970[b]	L	20?	0.7	?	H	L	H	H
1971[b]	E	50	0.8	150	L	H	L	—
1972	—	35	0.7	153	L	—	L	—
1973	—	20	0.8	157	H	L	H	H
1974	E	35	0.8	156	L	—	—	L
1975	E	35	0.6	155	L	—	L	L
1976	E	20	0.7	148	L	H	L	L
1977	L	35	0.8	153	H	H	L	—
1978	L	20	0.7	145	H	L	H	H
1979	E	35	0.9	151	L	—	H	—
1980	L	15	0.6	147	H	—	?	—
1981	—	80	0.7	155	—	—	?	—
1982	—	50	0.7	150	—	—	?	H
1983	L	10	0.2	145	L	L	H	H

[a] E, early; *H*, high; *L*, late or low; *SST*, sea-surface temperature; *UW*, upwelling.
[b] Data from Manuwal 1972.

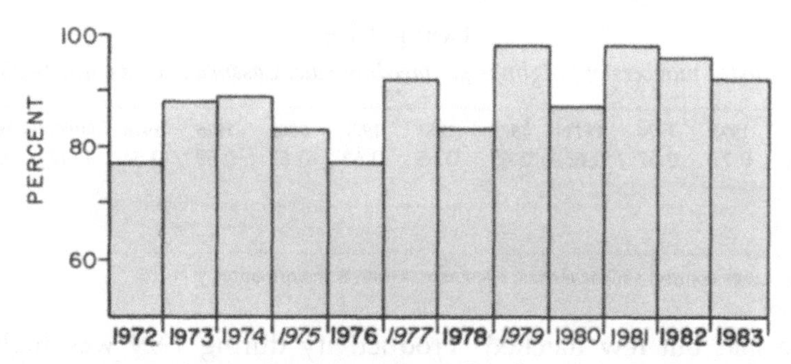

Figure 10.13. Proportion of Cassin's Auklets laying eggs within the "effective breeding season," by year, 1972–83.

(Figure 10.13) increased during the study period ($r_S = .629$, $p < .05$). This is hard to explain but is coincident with some other trends: an increase in average annual sea-surface temperature (Chapter 2), with a concomitant trend toward more synchronized laying, a decrease in the length of the effective breeding season (Table 10.5), and a decrease in average fledging weights (Appendix 10.2).

The population's overall productivity during any given year was determined largely by the success of first clutches (Appendix 10.3). During the 14 years of study, replacement clutches increased productivity on average by only 0.06 chick, and second clutches increased it by only 0.02 chick per pair. The utility of continued laying appears to be even less if chicks that fledge at weights lighter than 135 gm have little chance for survival, which is likely (see arguments, end of Chapter 9), and are subsequently removed from calculations of breeding productivity (most of these light chicks are from the latest-laid eggs). The reproductive benefit of replacement clutches then falls to 0.03 chick and that of second clutches becomes less than 0.01 chick per original breeding pair.

Within the 12 years for which we have data, only during 1981 did second clutches make any measurable contribution to the population's overall productivity (Appendix 10.3). On the other hand, the value of second clutches must be measured on the basis of benefit to the individual pairs. Clearly, even from that perspective, however, the benefit appears slight, particularly if breeding increases stress or increases chances of predation. We will return to this subject below.

Reproduction during 1981 was only average, in spite of an effective breeding season much longer than in any other year (80 days). By far, productivity was highest during 1979, when many second clutches

TABLE 10.6

Adjusted numbers of fledglings per breeding pair, Cassin's Auklets, 1972–83

1979	**1973**	1974	**1972**	*1975*	1981	*1977*	**1982**	**1976**	**1978**	1980	**1983**
0.81	0.77	0.67	0.65	0.65	0.65	0.63	0.61	0.59	0.54	0.47	0.18

NOTE: Lines connect similar means; *t* test after arcsin transformation, $p > .05$.

were laid but few hatched. Productivity during 1979 was higher, whereas during 1983 it was lower than in other years (Table 10.6). Otherwise, productivity tended to be high during early, cool-water breeding seasons and low during late, warm-water seasons, when food was scarce (1972–73 was the exception). An analogous pattern has been observed in the Great Tit by Perrins (1979).

THE CASSIN'S AUKLET BREEDING STRATEGY

The phenology of reproduction in Cassin's Auklets on the Farallon Islands exhibited several notable features: (1) initiation of egg laying in different years ranged over a three-month period and was correlated to sea-surface temperatures just before the onset of the upwelling period; (2) the mean date of laying for first-clutch eggs varied over a two-month period, and, even within the 14-year span of our data, the average date of egg laying in any given year was similar to that in few other years; (3) as the mean date of laying grew later, egg laying became more clumped and more skewed toward the early part of the laying period as a function of the intensity of the transition between winter and upwelling conditions; (4) regardless of the degree of clumping, the same proportion of eggs was laid during the period when the chances of success were highest, but as a result of annual variation in the degree of laying synchrony, that effective period varied in length from ten to 80 days; and (5) the termination of egg laying varied little, ending rather precisely during the first week of July.

We can construct a hypothetical scenario to interrelate these phenomena. Cool sea-surface temperatures during winter and early spring indicate a weak California Countercurrent and a strong California Current. Correspondingly, there are increased nutrients in surface waters and increased levels of zooplankton biomass (Chapter 2). Under these conditions, when changing photoperiod stimulates the auklets to reproductive readiness, the auklets are also ready for signs that the full upwelling period has begun. The latter is signaled by a large drop in sea-surface temperatures. Such an event in turn causes a dramatic change in the behavior of the euphausiid *Thysanoessa spinifera* (the auklets' virtually exclusive prey during upwelling years; Chapter 3), which begins to swarm at or near the surface even during the day, making it easily available to auklets. Beyond that, little is known about between-year variability in the range of euphausiid responses to upwelling. In such years of cool winters, if the onset of the upwelling period is late and the transition to upwelling conditions is relatively sharp, then it appears that an even greater proportion of auklets are primed for breeding when the transition finally occurs, because laying is even more synchronous. A variation of this scenario occurs in years when winter sea-surface temperatures are warm, the onset of upwelling is late, and the transition to upwelling is exceedingly sharp. Then it appears that the auklets are caught "off guard," i.e., few are in condition to breed, and laying becomes rather asynchronous.

Validation of this scenario, of course, requires much more information on annual and seasonal variation in the availability of suitable prey. It was constructed, however, with the knowledge that Sealy (1968) and Bédard (1967) found that the *chick period* of auklets in the Bering Sea corresponded closely with an increase in the abundance of certain zooplankters and an abrupt switch in diet by the auklets to these organisms. The same pattern of prey switching occurred both during seasons in which laying was early and during seasons in which it was late. However, on the basis of current (meager) information, a potentially important difference between the Farallon and Bering Sea patterns is that the key switch of prey among Farallon auklets occurs well before chicks hatch. We now believe, on the basis of our 1985–88 work on the year-round diet of auklets (unpubl. data), that breeding patterns observed in Cassin's Auklets as a response to upwelling phenomena are indeed mediated by prey responses to upwelling, but four years of study are not satisfactory for understanding the range of variability we have observed in auklet breeding.

Egg laying each year terminated approximately at the summer solstice, or, more precisely, about two weeks later than when the sun sets at its latest time in the entire year (Figure 10.4). The sun sets latest from 25 to 30 June, but by 10 July it still sets only three minutes earlier. Thereafter, it sets earlier at an increasing rate of change. Thus, it appears that egg laying ends when the sun begins to set earlier each day, and, therefore, it seems likely that photoperiod affects both the onset (Sealy 1968) and the termination of reproduction in this species regardless of food availability (consider 1982, and especially 1981, when auklets lost interest in reproduction in spite of no late-season decrease in chick fledging weights; Figure 10.12).

Interestingly, in spite of a 37° span in latitude, Cassin's Auklet in British Columbia and the three species of auklets that nest on St. Lawrence Island in the northern Bering Sea also terminate egg laying during the first or second week of July (Sealy 1968, Vermeer 1981). Any latitudinal differences in the timing of reproduction in auklets appear to manifest themselves in the onset of laying, which becomes progressively later with latitude, rather than in its termination.

ECOLOGICAL SIMILARITIES TO HIGH-LATITUDE AUKLETS

In the introductory comments, we indicated that Cassin's Auklet breeds in central California only because of the subarctic conditions that upwelling brings to the West Coast of North America. The scenario presented above indicates that within this southern extension of range Cassin's Auklet reproduction is keyed to, or is perhaps at the mercy of, annual vagaries of ocean conditions. The occurrence of Cassin's Auklets in subtropical latitudes is thus analogous to the northward extension of the range of diving petrels, *Pelecanoides* spp., from subpolar to subtropical latitudes in the eastern South Pacific, an extension made possible by intense coastal upwelling in the Peru Current. Many other similarities in the behavior and morphology of Cassin's Auklet and diving petrels have also been found (Storer 1960, Thoresen 1969, Watson 1968).

We also indicated that Cassin's Auklets share many characteristics of breeding ecology with other auklet species but that various species possess some interesting differences. The similarities confirm the close affinity of Cassin's Auklet with its arctic and subarctic brethren, but the differences along with some other ecological characteristics lend support to Udvardy's (1963) idea that Cassin's Auklet has occupied the southern extension of its present range rather recently. First, in all auklets, reproductive condition appears to be brought on by increasing day length and to be terminated when day length no longer increases. Among auklets of high latitudes, contraction of the reproductive season, which Sealy (1968) considered to be an adaptation to the short summer, is accomplished by restrictions to the onset of reproduction more than to its termination; high-latitude auklets apparently breed as soon as the melting snow makes nest sites available. Although the termination of laying during early July in the Arctic seems the right choice for auklets there, it seems to be rather premature for auklets in California, where winter does not arrive until November. This shortening of the reproductive season at its terminus is one "ecological strategy" that appears to link Cassin's Auklet closely to higher-latitude species. Second, in all auklets, the incubation patch develops when the first egg is laid and regresses by the time the chick is about ready to fledge. That pattern coincides well with the reproductive phenology of arctic auklets but becomes incongruous in Cassin's Auklet because Cassin's, unlike the arctic species, readily lays replacement eggs and even second clutches, although the incubation patch has become nonfunctional (Manuwal 1974c). This then is another characteristic that links Cassin's Auklet to a rather recent closer association with a higher-latitude existence. Third, the later eggs laid by Cassin's Auklets almost never increase reproductive output. Thus, Cassin's has no higher reproductive potential than auklet species that lay but one clutch. The situation almost appears to be one in which the low-latitude auklets (Cassin's), having begun laying quite early in the year, find themselves still in prime reproductive condition after fledging their first chick. With day length still on the increase, they continue laying, and they continue the reproductive effort until decreasing day length "turns them off." In Cassin's Auklet, this continued but wasted reproductive effort must be neutral in an evolutionary sense, because, if production and care of second clutches entail additional energetic costs, selection pressures should otherwise have rapidly phased out any costly efforts of no useful consequence. Alternatively, Cassin's Auklets moved from higher latitudes relatively recently and an evolutionary solution to wasted reproductive effort has not yet been accomplished.

There is one important complication in trying to explain the apparent wastage of late-season reproductive effort as an artifact of relatively recent zoogeographic invasion. As indicated by Manuwal (1974b), competition for nesting sites is so intense that many pairs are prevented from breeding because they are unable to secure a site. It could be that the auklets' readiness to continue laying has important competitive value, because by remaining intensely territorial they are able to hold a burrow until potential competitors pass from reproductive condition and lose interest in securing a site. For auklets or pairs of auklets that do not possess a site, it is important to remain in physiological condition for breeding as long as possible (i.e., until decreasing day length "causes" them to pass from condition) in the event that a site becomes available owing to the death of its owner. For Cassin's Auklets with and without breeding sites, then, it is important to come into peak breeding condition as early as possible or to remain in that condition as late as possible, respectively, in the first case to take advantage of conditions favorable for breeding (e.g., food availability) and in the second to be able to breed should a burrow become vacant in time to do so successfully. As a result, those birds that hold territories continue to "waste" breeding effort. Such effort, however, is not a waste if it ensures reproduction in the next year as well.

We have speculated that certain elements of Cassin's Auklet life history may be evidence for recent movement of the species from higher latitudes. We have seen how the present life history of this interesting seabird is a collage of adaptations having possible roots in its ancestry as well as in recent conditions. If anything, the suite of reproductive adaptations possessed by Cassin's Auklet probably allowed it to cope well with the several advances and retreats of ice sheets and the corresponding changes in marine conditions that have occurred in the eastern North Pacific over recent millennia. Rather than necessarily occupying a "refuge" during the last great glacial period (Wisconsin glaciation), as suggested by Udvardy (1963), the Cassin's Auklet could have fared well without taking refuge. At present the species appears to be "preadapted" for existence at higher latitudes mainly because, in an evolutionary sense, it has not "forgotten" its strong ecological and behavioral affinities with other members of the auklet group (see also Chapter 12).

11

Rhinoceros Auklet and Tufted Puffin

David G. Ainley, Stephen H. Morrell, and Robert J. Boekelheide

The Rhinoceros Auklet and Tufted Puffin are both "puffins" (Storer 1945). Both have a large, deep, laterally compressed bill with a pronounced cere at its base, and the structure of their pelvis and hind limbs allows them to walk upright with much greater facility than the nonpuffin members of the family Alcidae. The Rhinoceros Auklet is sometimes referred to as the Horn-billed Puffin (American Ornithologists' Union 1983). The Tufted Puffin is striking in appearance—jet black with a red bill and feet, green cere, white face, and large white head tufts. In comparison, the Rhinoceros Auklet is relatively drab. Its bill is a dull yellow, its legs are dark, and it has thin white plumes above each eye. The "Rhino's" name comes from the hornlike projection rising vertically from its cere, at the base of the upper mandible. Both species are among the larger alcids, the Rhino weighing about 520 gm and the Tufted Puffin about 780 gm (Figure 1.3).

The Rhinoceros Auklet is entirely sympatric with the puffin; both nest from central California north and west around the Pacific rim to northern Japan. Only the Tufted Puffin, however, nests north of the Aleutians, and it does so as far as the Bering Strait (Udvardy 1963). The puffin is most abundant in the central and northern part of Alaska, where the auklet is absent. The latter is more concentrated from southern Alaska south to Washington (Sowls, Hatch, and Lensink 1978, Vermeer 1979; Table 11.1).

The puffin and Rhinoceros Auklet each lay a single white egg, which is 12 and 15%, respectively, of adult body weight (Sealy 1972). Like the smaller auklets, puffins have two incubation patches, and they hold the egg against one or the other by using a wing. The Tufted Puffin has the longest incubation period of any alcid, about 45 days, and the Rhinoceros Auklet is tied with the Horned Puffin for

TABLE 11.1

Estimated numbers of Rhinoceros Auklets and Tufted Puffins nesting in North America, 1973–81

	Breeding pairs		Sites		
Area	Auklet	Puffin	Auklet	Puffin	Source
Alaska	100,000	2,000,000	12	502	Sowls, Hatch & Lensink 1978
British Columbia	104,000	31,000	13	17	Vermeer 1979 & unpubl., Vermeer et al. 1983
Washington	28,800	12,580	6	16	Manuwal, Wahl & Speich 1979, Varoujean 1979
Oregon	50	260	3	10	Varoujean 1979, J. M. Scott et al. 1974
California	256[a]	125[b]	8	11	Sowls et al. 1980
Total	233,100	2,044,000	42	556	

NOTE: Most recent data.
[a] The Farallon population constitutes 41% of this total.
[b] The Farallon population constitutes 40% of this total.

the next longest at 42 days. Agreeing with Lack (1968), Sealy (1972) surmised that the puffins' long incubation period is consistent with their long nestling period. The extended length of these periods is the result of slow development, affecting both embryo and nestling. Among alcids, the Rhinoceros Auklet has the longest nestling period, 51 days, and the Tufted Puffin the next longest, 43 days (Vermeer and Cullen 1982). Presumably the slower rate of chick development is an evolutionary response to the parents' reduced provisioning capabilities. Puffin chicks fledge at 64 to 68% of adult weight (Sealy 1972).

One of the major contrasts between the two species is their daily pattern of onshore activity. The Tufted Puffin comes and goes from its burrow during the day, whereas throughout most of its range the Rhinoceros Auklet does so in the evening or at night. The different patterns may not be a function of feeding schedules (Chapter 3). The difference in coloration, i.e., the puffin being gaudy and the Rhino drab, likely relates to this contrast in colony-attendance cycles. We will return to this subject of colony attendance below.

FARALLON POPULATIONS

The Farallon breeding populations are in the southernmost parts of the two species' ranges in the eastern Pacific. The Rhino nests at two other sites in central California farther south than the Farallones. The Farallon populations of both species are small compared to the majority of populations farther north. Nevertheless, Farallon populations are important, because neither species breeds in large numbers anywhere south of Washington State. Populations at the Farallones contribute 40 and 41%, respectively, of the numbers of Tufted Puffins and Rhinoceros Auklets in California (Table 11.1).

Tufted Puffins were much more numerous at the Farallones during the nineteenth century than they are now. Once in the thousands, numbers declined drastically because of oil pollution, the introduction of the European hare, and, possibly, the disappearance of sardines (Ainley and Lewis 1974; Chapter 1). Oil pollution has been reduced significantly in the Gulf of the Farallones, and in 1974 we eliminated rabbits, which occupied all but the most precipitously situated cavities suitable for puffin nesting. Since the early 1970's, when PRBO first estimated puffin population size, the number of breeding pairs has grown by about 7% per year, from about 27 pairs in 1972 to about 50 pairs in 1982 (Appendix 11.1). As a result of ENSO, few birds appeared during 1983, and after the 1982–83 event numbers were lower than before (Figure 11.1). Thus, with the exception of a decline due to increased mortality in 1982–83, population growth has been slow but steady.

No information exists to indicate how large the Farallon Rhinoceros Auklet population was before modern human society arrived at the

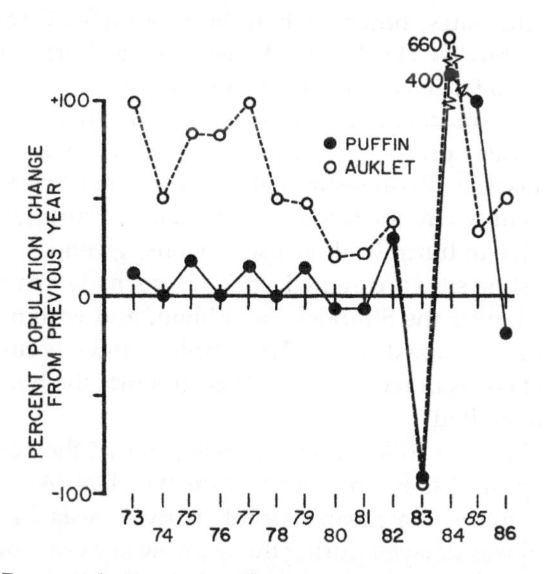

Figure 11.1. Percent change in breeding populations of Rhinoceros Auklets and Tufted Puffins on the South Farallon Islands, by year, 1972–86 (see Appendix 11.1 for number of pairs).

islands. The species was not present from the 1860's, when several were collected for scientific specimens, until 1972. In the latter year, two pairs may have bred. Since then, on the basis of counts of birds rafting at Fisherman's Bay during the evening, as well as tallies of known nesting sites, numbers have grown rapidly. The population doubled every year from 1973 to 1977, after which it continued to grow but at a slower rate (Appendix 11.1; Figure 11.1). Some mortality appears to have resulted from the ENSO of 1982–83.

Many burrows now occupied by Rhinoceros Auklets are known to have been prime rabbit warrens. Thus, removal of rabbits has likely contributed to the Rhino's recovery, and no doubt to the growth of the Tufted Puffin population as well. The auklet's particularly rapid population growth must also have been aided by immigration. The source of immigrants, however, is not known, although several Rhinoceros Auklets banded in British Columbia have been recovered in California (A. Gaston pers. comm.). The species is surprisingly abundant year round in California waters seaward of the continental slope, including those west of the Gulf of the Farallones (Briggs et al. 1987; Chapter 3). It seems likely that Rhinos should be colonizing other sites along the Pacific coast as well, which they indeed are (see J. M. Scott et al. 1974, Sowls et al. 1980).

As far as we know, neither puffin digs nesting burrows on the Farallones; at other sites, however, both burrow extensively (e.g., Richardson 1961, Sowls, Hatch, and Lensink 1978, Vermeer 1979). All nests located so far have been in natural cavities or clefts in the rock. These cavities are deep, usually deeper than an average person's reach. Puffins also prefer some slope or altitude to their sites, perhaps to aid their takeoffs. Because the soil on the Farallones is shallow, and virtually absent where the terrain slopes, it is not surprising that puffins do not live in burrows. It is also obvious, given the Farallon terrain, that nesting sites are limited. In 1973, we made a cavity from tile pipe and rock near the Shubrick Point blind, and within two years a pair of puffins occupied it. Similarly, Rhinos have occupied several artificial nest boxes in recent years. This supports the hypothesis that nest sites are limited.

Tufted Puffins are so few on the Farallones that they escape notice in the spring until they are observed on the cliffs (Appendix 11.1). During this study, the mean date of "arrival" was 24 March (±7 days); arrival was delayed during the warm-water years of 1973, 1978, and 1983, but not in 1976. Arrival at the Farallones was almost two months earlier than in British Columbia (see Drent et al. 1964). Counts of birds standing on perches indicated an increase in numbers

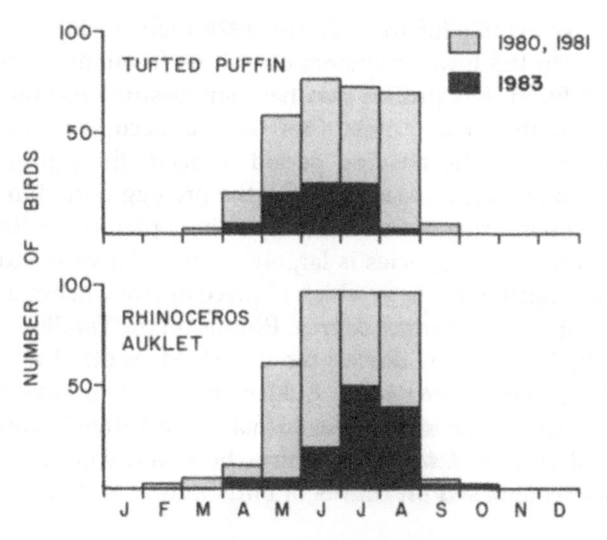

Figure 11.2. Seasonal change in numbers of puffins observed at the South Farallon Islands, comparing two sets of years, 1980–83.

through May, a plateau, and then a sudden drop after August (Figure 11.2). No puffins have been observed after September (Figure 12.1). An appreciable proportion of this species' nest sites and perches are located on the higher parts of the island (Figures 1.3 and 11.3).

During late April, before eggs were laid, Tufted Puffins often perched on outcrops near nest-site entrances. In 1977, we conducted 15 all-day watches and counted the number of visible puffins hourly. From these counts we found that they frequented perches from early morning until late afternoon and thereafter were absent (Figure 11.4). This pattern was continued by the nonincubating member of pairs through the egg stage of nesting, approximately May to mid-June. Once chicks hatched, the behavior of adults changed. No longer did they stand about much during the day, but they did so a great deal during the evening, usually after delivering the day's last load of fish to the chick. The pattern differed slightly during 1978, when we conducted ten all-day watches, though none during the pre-egg stage of nesting (Figure 11.4). During the egg stage that year, few adults stood about past noon. During the nestling stage, many more stood about in the morning and early afternoon than did so during 1977, and few were present during the evening. These differences are likely related to feeding conditions. As detailed in Chapter 3 and other chapters, 1977 was a very productive year for all Farallon seabirds and 1978 was one of the poorest for the entire study period. Puffins likely spent more time searching for food during 1978 than during 1977, which would explain the lower numbers of puffins frequenting perches in 1978. In addition, few puffins may have successfully produced chicks in 1978, and thus many nest sites were unoccupied during what should have been the nestling period. Indeed, the puffins' loafing pattern became quite similar to that of the pre-egg period in 1977.

We have no information on the daily activity patterns of Rhinoceros Auklets, because the species is largely crepuscular or nocturnal and we had no available nests in which to place microswitches and treadles (see Chapter 1). To some degree, Rhinos on the Farallones are also active at their nest sites during the day (J. M. Scott et al. 1974; see below). Seasonally, Rhinoceros Auklets appeared to have a cycle of island occupation that was similar to that of the Tufted Puffin (Figures 11.2 and 12.1). Our data for the Rhino, however, were less exact because they were based on counts of birds rafting on the water in the

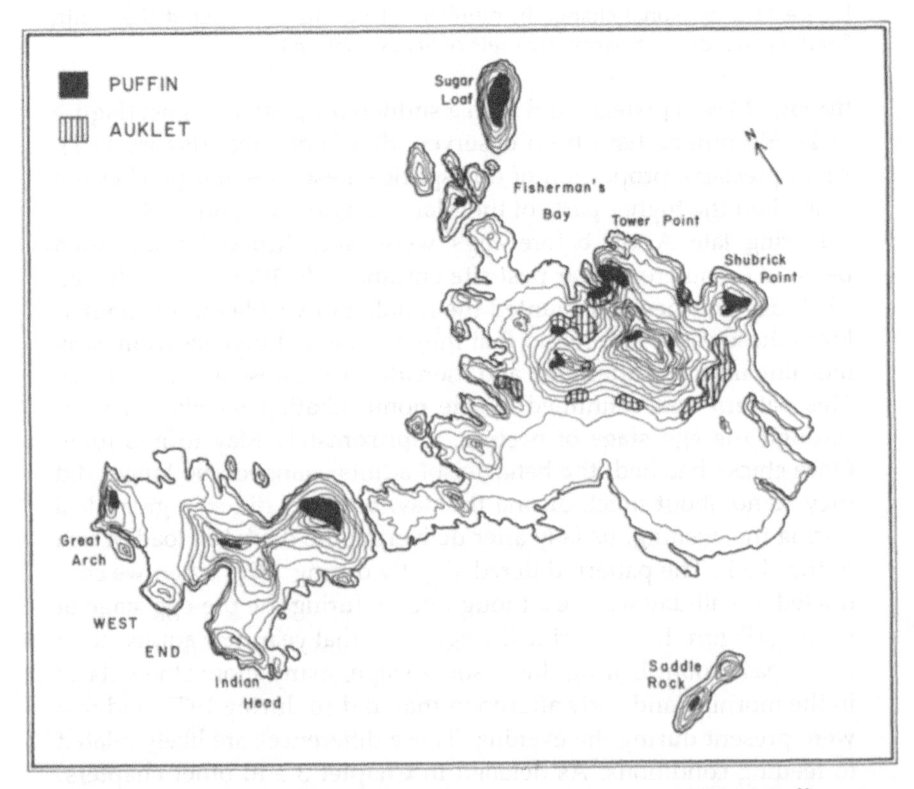

Figure 11.3. Major nesting areas of Rhinoceros Auklets and Tufted Puffins at the South Farallon Islands as of 1982.

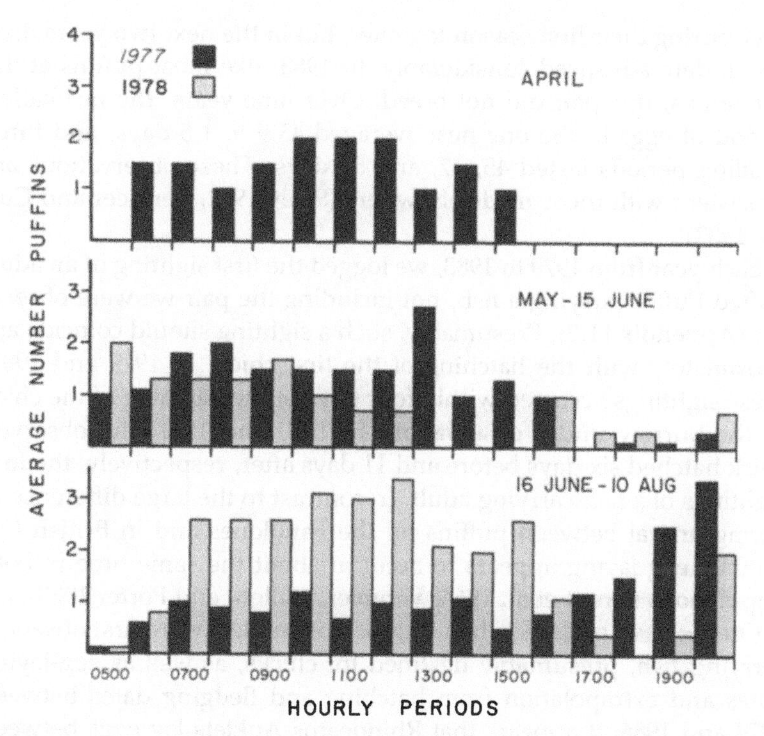

Figure 11.4. Daily pattern in counts of Tufted Puffins seen at burrow entrances throughout the day; data averaged by month, 1977–78.

evening, rather than on counts of birds on the island. Unlike Tufted Puffins, Rhinoceros Auklets are seen in low numbers year round near the island, but whether these are Farallon breeders is not known.

REPRODUCTION

Because of the inaccessibility of nests and the small populations present, we have had little opportunity to investigate the breeding biology of these two species. Appendix 11.2 summarizes information gleaned from the only puffin nest site we were able to observe. We became so familiar with the occupants that we could identify them by bill and feet characteristics. Laying dates were earlier during the first four years than during the later ones, a trend consistent with the change in egg-laying patterns of other species (see other chapters). In 1980, one member of the pair we had been observing for eight years failed to appear, and the surviving puffin acquired a new mate. As happens in long-lived birds, the new combination of mates laid much later during their first season together, but in the next two years their laying date advanced considerably. In 1983, like most puffins at the Farallones, this pair did not breed. Over nine years, the incubation period of eggs in the one nest averaged 43.9 ± 1.5 days, and three nestling periods lasted 45, 47, and 55 days. These observations are consistent with those made elsewhere (Sealy 1972, Vermeer and Cullen 1982).

Each year from 1979 to 1983, we logged the first sighting of an adult Tufted Puffin carrying a fish, not including the pair we were observing (Appendix 11.2). Presumably, such a sighting should coincide approximately with the hatching of the first chick. In 1979 and 1982, these sightings occurred within four days of the hatching of the chick in the burrow under observation. In 1980 and 1981, the observed chick hatched six days before and 11 days after, respectively, the first sightings of a fish-carrying adult. In contrast to the large difference in spring arrival between puffins on the Farallones and in British Columbia, egg laying appears to occur at about the same time in both populations (Drent et al. 1964, Vermeer, Cullen, and Porter 1979).

On the basis of dates when Rhinoceros Auklets were first observed carrying fish, presumably destined for chicks, as well as egg-laying dates and extrapolation from hatching and fledging dates between 1974 and 1986, it appears that Rhinoceros Auklets lay eggs between mid-April and mid-June (Sander 1986). Breeding in the two puffin species thus overlaps broadly at the Farallones (Figure 12.1).

INTERACTIONS BETWEEN PUFFIN SPECIES

Puffins come and go from their nesting burrows during daylight, as well as at dusk, when it is almost too dark for us to see. In contrast, in most of their range Rhinoceros Auklets come and go from their nest sites mainly between sundown and sunup. At the southern end of their range, however, the auklet brings fish to its young at dusk and moves about by burrow entrances during the day (J. M. Scott et al. 1974, Thoresen 1983). Such a departure from the norm opens to question the advantages of nocturnality to the Rhinoceros Auklet. Most nocturnal alcids are small, burrow-dwelling species (Figure 1.1), and the proposal that nocturnality is an anti-predator strategy for them seems logical (Sealy 1972). Summers (1970) proposed that nocturnality in the Rhinoceros Auklet is also a response to predation. If so, this relationship would likely be weak, because Rhinos are larger, and hence less susceptible to predation, than at least three other diurnal cavity-dwelling alcids. Even if nocturnal activity at the breeding island were a response to such predators as eagles and falcons, which certainly exist in fair numbers farther north in the Rhino's range (and did so not long ago at the Farallones; DeSante and Ainley 1980), why guillemots and other puffins would not also become nocturnal is puzzling. It is possible that nocturnality reduces kleptoparasitism by gulls attracted to the large and conspicuous loads of food that Rhinos bring to their chicks, presumably one load per adult (J. M. Scott et al. 1974). Other puffin species, however, bring equally large loads, and are subject to kleptoparasitism (Nettleship 1972, Pierotti 1983), but yet remain diurnal. Farallon observations indicate that in spite of diurnal activity, Tufted Puffins also usually bring only one, sometimes two, loads per adult per day.

The Rhinoceros Auklets' daylight activities on the Farallones have afforded us the opportunity to observe their interactions with Tufted Puffins, and these observations may provide insight into the problem of differences in daily cycle. Tufted Puffins go out of their way to respond to any Rhinoceros Auklet they observe on shore. The two species display to one another, often using the "ritualized walk" illustrated by Birkhead (1985), and the puffins usually advance on the auklets until the latter depart (the Tufted Puffin is much larger; Figure 1.1). It may be that Rhinoceros Auklets are nocturnal to avoid interference by Tufted Puffins. The scarcity of Rhinoceros Auklets within the main portion of the Tufted Puffin's breeding range supports this idea. Whittam and Siegel-Causey (1981) analyzed assemblages of nesting seabird species in Alaska, attempting to determine whether any species tended to be absent where others nested. If such a pattern occurred, it would be evidence of competitive exclusion of one species by another. For unknown reasons, they excluded Rhinoceros Auklets from their analysis, which for our purposes is unfortunate.

The reason why the two species interact may relate to competition for similar nesting habitat and sites. Both species nest in deep cavities on high or sloping terrain. The Rhino nests under forest canopies on some islands in British Columbia and Alaska (Sowls, Hatch, and Lensink 1978, A. Gaston pers. comm.), whereas the puffin apparently rarely does so; both species nest in the open on many islands. At many locales, the birds dig cavities, but at the Farallones only natural ones are available. The deep cavities are limited in supply and their availability is likely to limit both species' populations. Intense competition for sites exists between the two puffins and between them and Pigeon Guillemots. During the 1800's, several ornithological accounts mentioned that puffins nested around the islands' periphery, a habitat from which they are now absent and the area that Rhinoceros Auklets have been recently colonizing. D. A. Nelson (1982, 1987) documents several instances of Rhinos evicting guillemots from deep cavities at the Farallones, and he describes a guillemot warning call specifically directed at Tufted Puffins. In Chapter 9, we mentioned several instances of puffins evicting guillemots. It is possible that with continued growth of the two puffin populations, reduction of the guillemot population may result (see Chapters 9 and 12).

We are excited about the return of the Rhinoceros Auklet and encouraged by the continued population growth of both it and the Tufted Puffin. These trends, no doubt, are related to the demise of the feral rabbits. Few places exist where both puffin species nest together, and it may be that with continued population growth, we will have an opportunity to investigate the biological differences and conflicts between these two interesting and closely related species.

12

Farallon Seabirds: Patterns at the Community Level

David G. Ainley

Thus far, we have concentrated on the patterns of feeding and breeding among the component populations of the Farallon marine-bird nesting community. Here, we will attempt to draw together some of our findings by considering certain patterns from the community perspective.

In the various chapters we have detailed notable within- and between-season variability in most of the aspects of breeding ecology examined. When the project began in 1971, we were resolved to continue until we encountered a breeding season in which patterns fairly well repeated those of a previous one. In the context of the myriad shorter-term studies available then (and now), most of which showed relatively little variability, we fully expected our project to last maybe five or six years. We were, therefore, psychologically geared to undergo the stress and strain required to study 11 species simultaneously. Thirteen years later, we decided to call it quits. This decision stemmed partly from mental and physical exhaustion, but mostly from a feeling that we had finally glimpsed the spectrum of variation that occurs in major aspects of the breeding and nesting-season feeding ecology of Farallon seabirds. In no small way, we were aided by Mother Nature, who arranged an array of environmental conditions that in slightly more than a decade marked the extremes experienced on the West Coast during this and the previous century. As described in Chapter 2, not only did we experience the remarkable 1982–83 ENSO, as well as the "lesser" warm-water events of the 1970's, but our study period also included the sets of environmental conditions that brought to California both the most severe drought and the heaviest precipitation since the initiation of modern records (more than 100 years). Rain and drought are a function of marine conditions, and, as was made abundantly clear in Chapters 4 through 11, as marine conditions change, so do the responses of Farallon seabirds.

If, through all the environmental instability, the Farallon birds had mechanistically persevered, disappointed we surely would have been. On the contrary, in spite of our exhaustion we were pleased with the exciting "natural experiments" that the marked variability presented to us. Only now, however, with the analysis of data presented in these pages, have patterns emerged from the seeming chaos as one supposedly "atypical" year gave way to the next. Though many of these patterns are discussed at the species level in the preceding chapters, certain patterns bear on some of the larger questions that seabird biologists and other ecologists have been discussing for a long time.

TIMING OF BREEDING

Several Farallon species terminated their breeding season at the same time that they changed their diet, from one in which one or a few prey species predominated to another far more diverse (Chapter 3). This change, in conjunction with the disappearance of certain of these predominant prey species (euphausiids, juvenile rockfish; Chapter 2), suggests that reproduction was timed to take advantage of abundant prey. A similar pattern of diet diversification toward the end of the reproductive season was observed by Belopol'skii (1961) for several species in the Barents Sea and by Bédard (1969a) for auklets in the Bering Sea. Bédard (1969a), in particular, was able to gather the same type of data as did we (but without the independent confirmation of prey availability) and came to the same conclusion regarding the link between the timing of reproduction and the availability of prey. This conclusion was consistent with a hypothesis that had been offered earlier by Lack (1954), who noticed the temporal correspondence between breeding seasons and seasonal spurts in potential prey populations and general oceanic productivity (see also Sealy 1968, Pearson 1968, M. P. Harris 1969b, Boersma 1978, Croxall 1984). Other patterns at the Farallones that further support the idea are the late-season decreases in breeding success observed for most species and the late-season decreases in growth rates and fledging weights observed for Ashy Storm-Petrel, Pigeon Guillemot, and Cassin's Auklet chicks. These decreases occurred at the same time as the diversification of diet; in some years, however, diet did not diversify and late-nesting birds did as well as early ones.

Evidence indicates that other important physiological factors, not just the proximate one affected by food availability, have operated to terminate reproduction. In several species, including the Western Gull, Pigeon Guillemot, Common Murre, and Cassin's Auklet, no eggs were laid after about the first week of July (Table 12.1), the time when the seasonal decrease in day length begins to accelerate noticeably (see Chapter 10). The startling thing about this was that egg laying terminates at the same time in virtually all other alcid and gull populations for which data are available, including several genera and species. Such a pattern strongly indicates a link between photoperiod and gonadal state for these birds (see Farner 1965, Wolfson 1965). This response to photoperiod would help to explain why Farallon seabirds cease laying even in years when prey availability presumably remains high, as indicated by the fact that late nesters are equally as or even more successful than early ones. It also perhaps offers the grounds to explain at least in part why breeding synchrony is high in many Farallon species during years when the initiation of laying is late. The hypothesis is as follows: lacking the proximate trigger of food availability because the important breeding-season prey base has yet to develop, more and more individuals are physiologically primed to breed but do not lay eggs. Then, when prey appear, egg laying begins en masse. Some indirect evidence supporting this idea is that late egg-laying seasons are shorter than early ones, rather than length being independent of timing (see especially Chapter 10).

An artifact of the termination of egg laying is its effect on the incidence of replacement and second clutches. Although second clutches, in the case of Cassin's Auklets, and replacement clutches, in the case of other species, occurred most frequently in years when the high

TABLE 12.1

Egg-laying seasons of Farallon seabirds, 1971–83

Species	Overall mean laying date (± SD, in days)	Range of laying dates	Length of laying season (days)
Ashy Storm-Petrel	9 June ± 20	1 May–16 Aug.	107
Double-crested Cormorant	1 May ± 10	1 Apr.–12 June	72
Brandt's Cormorant[a]	11 May ± 10	18 Apr.– 8 July	81
Pelagic Cormorant	30 May ± 10	8 May–25 July	78
Western Gull	10 May ± 12	20 Apr.–10 June	51
Common Murre	17 May ± 7	4 May–26 June	53
Pigeon Guillemot	28 May ± 8	1 May–11 July	71
Cassin's Auklet	23 Apr. ± 24	8 Mar.–12 July	126

[a] Colony I only.

level of breeding success indicated abundant food, few replacement clutches were laid during years when nesting began late, including years such as 1977, which was one of the most productive in our study. This pattern appears to be a general one, exhibited by most Farallon species, and is also related to the effect of photoperiod on reproductive condition.

Why seabirds with northern zoogeographic affinities terminate egg laying at the same time each year, regardless of food availability, is another question. Sealy (1968) grappled with the question in regard to auklets of the Bering Sea, but could not pin down a clear solution. Somehow the answer must lie in the deterioration of environmental conditions that usually occurs during the early fall, particularly in the Arctic: storms increase in frequency and severity during July and August, the upper layers of the ocean become less stratified, the thermocline deepens, and consequently prey are "diluted" or less concentrated and, therefore, become less available (G. Hunt pers. comm.). Individuals that terminate breeding then and vacate northern waters must have higher survival rates over the long term. In the California Current, the oceanic period, bringing water depleted of nutrients, arrives in August and September. Many visiting seabird species, however, reach their annual peaks in abundance at that time (Ainley 1976, Briggs et al. 1987). Therefore, food does not yet appear to be limiting, at least through the fall. It well may be that Farallon seabirds carry the same genetically controlled behaviors as their northern brethren.

Two exceptions exist to the apparent (approximate) summer solstice-mediated termination of egg laying. One is the Ashy Storm-Petrel (Table 12.1), which continues to lay well into August; in fact, some individuals lay even much later (Chapter 4). The other is the Brandt's Cormorant, although this is not obvious in Table 12.1. As indicated in Chapter 5, egg laying in some populations has been reported during the winter, and in some Brandt's Cormorant colonies at the Farallones (but not our study colonies), egg laying occasionally continued into August. During years when abundant food persisted (1979 especially), many parents still fed chicks during September and October. Late laying also occurred during 1978. Interestingly, among Farallon species, these two species have the most southerly zoogeographic affinities, having no close ties to subarctic or arctic species, and, in fact, they are the two most characteristically endemic seabirds of the California Current system. The Western Gull, which might also be considered a California Current endemic, interbreeds extensively with a subarctic gull "species" to the north (Glaucous-winged Gull; J. M. Scott 1971, Hoffman, Wiens, and Scott 1978).

Thus far, by considering the termination of laying, we have touched on the ultimate and proximate factors that may affect the timing of reproduction in Farallon seabirds. This may seem a backwards way to address the problem, but we lack much direct information on the initiation of laying, mainly because we lack data on diet and prey availability during the prebreeding period. Presumably, photoperiod and its effect on gonadal development bring Farallon seabirds into breeding condition, but the extreme variability in the onset of laying described for most Farallon species indicates linkage to a more variable environmental factor that finally triggers actual laying.

For the Western Gull and Cassin's Auklet (Chapters 7 and 10), two species that feed heavily on euphausiids during spring, the laying phenology correlated with a drop in sea-surface temperatures a month or so prior to the breeding season. Changes in sea-surface temperatures in turn indicate the timing of the spring transition (i.e., initiation of upwelling), to which the surface swarming of local euphausiid species is linked (Chapter 2). Though the trends were evident, a statistically significant relationship between sea-surface temperature and egg laying was not apparent in another early-season euphausiid predator, the murre, but, as explained in Chapter 8, such a relationship could easily be masked at present by factors related to the rapidly changing age structure of the murre population. On the other hand, there was correlation between the murre's and the auklet's average date of laying ($r_S = .622$, $p < .05$); that is, on a relative basis, early and late laying by these species tended to correspond. Such a correspondence was not evident among many other Farallon species (see below). This correspondence suggests that a similar factor was affecting the onset of reproduction in both the auklet and the murre. The weak correlation between mean laying dates of the auklet and Pelagic Cormorant ($r_S = .570$, $p < .05$) is less easy to explain, although both species feed more on crustaceans than do many other Farallon species.

Neither did a clear relationship exist between onset of upwelling and breeding in species that feed during spring at the next highest trophic level (piscivores), though a relationship was evident to some degree in Brandt's and Pelagic cormorants. On one hand, a higher trophic level adds a layer of complexity to interactions within the food web. On the other, the availability of suitable fish prey in the Gulf of the Farallones is related not just to the seasonal spurt in productivity brought by upwelling but also to the survival of fish larvae, which is first contingent on the pattern of upwelling, and only later on the level of productivity (Chapter 2). In fact, intense upwelling too early can be detrimental to fish availability (Bakun and Parrish 1982). Of interest, as indicated by annual means of laying dates, the onset of egg laying by the Pelagic Cormorant and Pigeon Guillemot, two species whose feeding ecology overlapped more than that of any other pair, was highly correlated from 1971 to 1983 (Chapter 3; $r_S = .864$, $p < .001$). Onset of laying by the Western Gull and Brandt's Cormorant was weakly correlated during the same 13-year period ($r_S = .519$, $p < .05$). Both species tend to feed heavily inshore along the mainland coast during the prelaying period. To proceed further with unraveling the mystery of what proximate factors affect breeding phenology, we will have to learn more about populations of prey fish and diets of avian piscivores during the January–April prebreeding period.

According to Cody (1973, 1974) and others (see also Croxall and Prince 1980, Croxall 1984), another factor that could affect the timing of breeding in seabirds is competition for a resource, especially food. The underlying assumption is that cropping of the prey base by one species could reduce food availability for another, so the latter should nest earlier or later, when food for its chicks would be more easily gathered. At the Farallones, we found no evidence that breeding seasons are timed to reduce competition for food. Indeed, breeding seasons appeared to be timed so that competition for food, if it occurs, is assured (Table 12.2, Figure 12.1)! It is evident that many species pairs

TABLE 12.2

Overlap in egg-laying seasons of Farallon seabirds, 1971–83

(*Percent*)

Species	ASP	BC	PC	DCC	WG	CM	PG
Ashy Storm-Petrel (ASP)	—						
Brandt's Cormorant (BC)	*67%*						
Pelagic Cormorant (PC)	*74*	*61%*					
Double-crested Cormorant (DCC)	44	*65*	29%				
Western Gull (WG)	33	*63*	34	59%			
Common Murre (CM)	50	*66*	41	44	56%		
Pigeon Guillemot (PG)	*66*	*81*	*74*	39	49	*61%*	
Cassin's Auklet	45	*65*	46	57	40	40	54%

NOTE: Laying seasons measured between earliest and latest clutch-initiation date; values of overlap greater than 60% are italicized for emphasis.

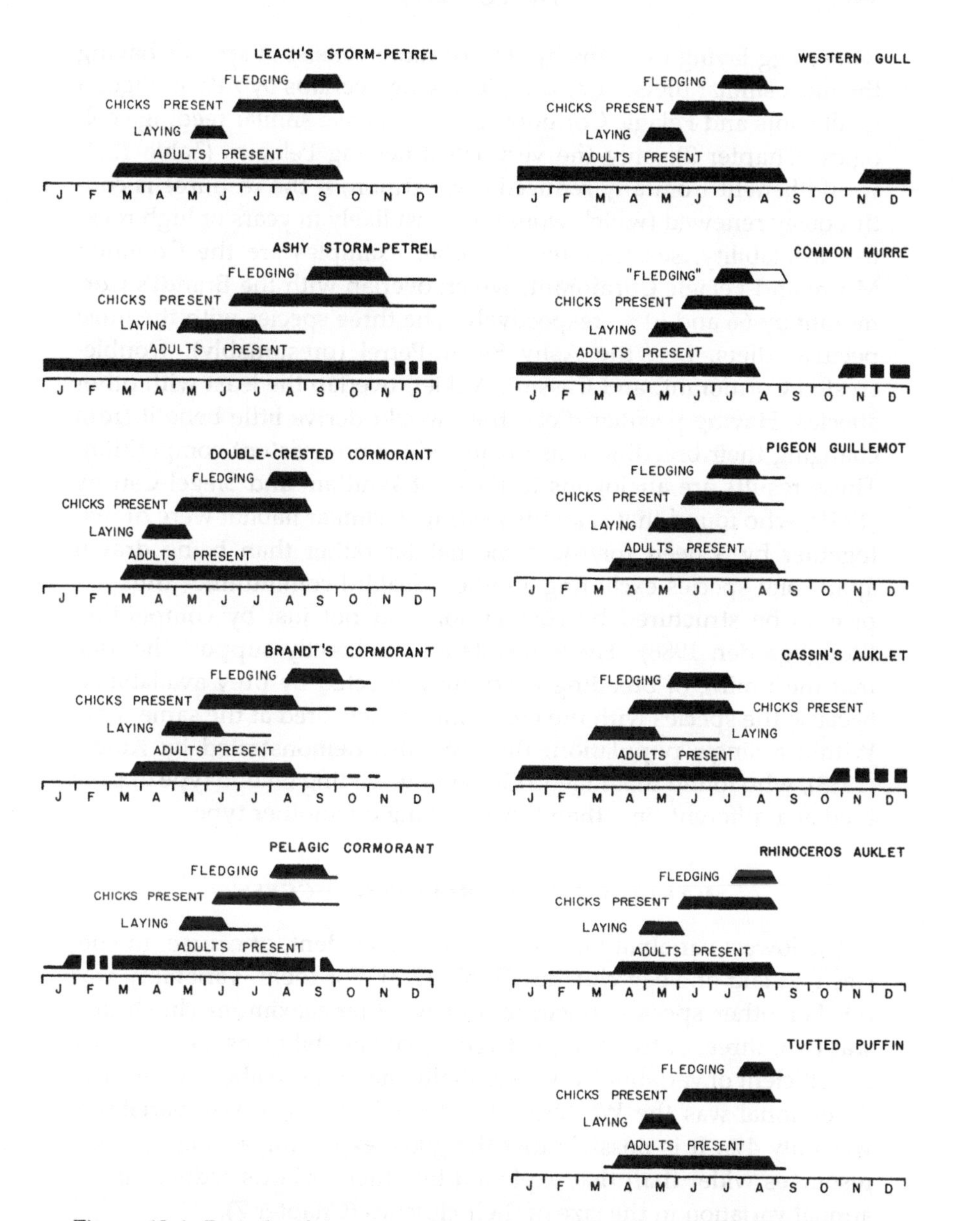

Figure 12.1. Periods of the year when various components of breeding populations are present on the Farallones. Dashed bars show periods of less regular attendance; thin bars show periods when less than the entire population is present. For murres, the dark "fledging" bar shows when chicks jump from cliffs; the light part of the bar estimates when chicks attain independence from parents.

whose egg laying overlaps by 60% or more consist of species having the most similar diets. For example, laying overlaps by 74% in Pigeon Guillemots and Pelagic Cormorants, which have similar feeding ecologies (Chapter 3); only the very latest-nesting Pelagics (Table 12.1) would benefit from any reduced competition, if the resource is continuously renewed (which would be most likely in years of high rockfish availability; see Chapter 3). Other examples are the Common Murre and Pelagic Cormorant, which overlap with the Brandt's Cormorant by 66 and 61%, respectively. The three species with the most peculiar diets, i.e., the Ashy Storm-Petrel (presumably), Double-crested Cormorant, and Cassin's Auklet, overlap the least with other species. Having peculiar diets, they would derive little benefit from changing their breeding seasons to reduce nonexistent competition. These results are analogous to those of Whittam and Siegel-Causey (1981), who found that seabirds nesting in similar habitat were drawn together by a need for the same habitat rather than being drawn apart, one species excluding another. Seabird communities thus appear to be structured by coevolution and not just by competition (Roughgarden 1986). These results also indirectly support the idea that the timing of breeding is strongly affected by prey availability, because the species with the most similar diets bred at the same time. Within a single population, this was also demonstrated by Kuletz (1983), who found that Pigeon Guillemots feeding on one type of prey bred at a different time than those feeding on another type.

VARIABILITY IN BREEDING SUCCESS

The lowest variability in clutch size was evident, of course, in species that laid one-egg clutches (Table 12.3). For them, variability was nil. For other species, regardless of whether maximum clutch size was two, three, or five eggs, between-year variability, as measured by a coefficient of variation, was essentially the same, at about 10 or 11%. Exceptional was the Western Gull, for which clutch-size variability was only 4%. It is possible that the gulls' exploiting a range of food resources wider than that exploited by other seabirds reduces inter-annual variation in the size of their clutches (Chapter 7).

Hatching success and fledging success (proportion of chicks fledged of those that hatched) were both inversely proportional to clutch size (Table 12.3). In other words, species that laid fewer eggs suffered lower mortality of eggs and chicks than those that laid more eggs. The overall result was that the number of chicks fledged per breeding

TABLE 12.3

Reproductive success of Farallon seabirds, 1971–83

Species	Mean clutch size[a]	CV,[b] clutch size (%)	HS (%)	FS (%)	Chicks per pair	Decrease, eggs to chicks (%)	CV,[b] chicks per pair (%)
Ashy Storm-Petrel	1.00	—	78%	87%	0.7 ± 0.1	32%	11.5%
Common Murre	1.00	—	86	95	0.8 ± 0.2	23	29.3
Cassin's Auklet	1.00	—	78	83	0.6 ± 0.2	38	24.3
Pigeon Guillemot	1.8 ± 0.2	11.4%	73	67	1.0 ± 0.4	44	41.6
Western Gull	2.7 ± 0.1	4.1	76	71	1.4 ± 0.4	47	24.5
Brandt's Cormorant	3.1 ± 0.4	11.8	60	77	1.3 ± 0.7	51	56.1
Pelagic Cormorant	3.4 ± 0.4	10.3	48	61	1.0 ± 0.9	76	85.3
Double-crested Cormorant	3+	?	?	?	1.6 ± 0.6	?	> 38.2

NOTE: *FS*, overall fledging success; *HS*, overall hatching success. All data extrapolated from chapters on individual species; means are shown ± SD.

[a] Does not include years in which no complete clutches were laid: 1983 for Pigeon Guillemot, 1978 and 1983 for both Pelagic and Brandt's cormorants.

[b] Coefficient of variation in annual breeding performance.

pair varied by a factor of two to three, from 0.6 chick in Cassin's Auklet to 1.6 in the Double-crested Cormorant, whereas clutch size varied by a factor of three to four, from 1.0 egg in three species to 3.4 eggs in the Pelagic Cormorant. Reproductively conservative species, which laid only one egg, ultimately fledged chicks from greater than 65% of their eggs, whereas species that laid the most eggs ultimately fledged chicks from less than 55% of theirs. This pattern appears to be an interspecific example of the pattern Boyce and Perrins (1987) have noted in the Great Tit: the "bad years effect," namely, in years in which the survival of young is poor, the effect is much more pronounced in species (or individuals) that lay large clutches than in those that lay small clutches. The existence of such a pattern suggests that species that lay fewer eggs have developed behavior to take better care of their eggs than have species that can "afford" to lose a few eggs.

The relatively similar fecundities among Farallon species also suggest that although nesting strategies, as represented by clutch size, may differ widely among seabirds, natural-history strategies that pertain to life stages after fledging may not differ much. For one thing, with such a low interspecific range in fecundity, we might expect Farallon seabirds' demographic parameters to be similar. Investigation of that idea, however, remains for the future (we have begun).

The between-year variability in fledgling production, as expressed by the coefficient of variation (CV; Table 12.3), ranged from 11% in the Ashy Storm-Petrel to an astounding 85% in the Pelagic Cormorant. A direct relationship to clutch size was evident, except for the Western Gull, in which variability was low. This may be the result of the gull's extreme trophic plasticity and the availability of alternate, unnatural food resources such as garbage and fish offal. For the gull's relative in the Bering Sea, the Black-legged Kittiwake, which does not have such a large supply of alternate food but like the Western Gull is restricted to feeding at the sea surface, years of complete breeding failure occur regularly (Hunt, Burgeson, and Sanger 1981, Hunt, Eppley, and Drury 1981). In addition, during two of ten years, A. Gaston (pers. comm.) observed complete reproductive failure in the Glaucous Gull, *Larus hyperboreus*, at Prince Leopold Island. It would have been instructive, but alas is impossible now, for the biology of large larids to have been observed prior to the advent of modern civilization.

That variability in seabird reproduction is related to food resources is suggested by the very high correspondence between years of high

TABLE 12.4

Spearman correlation coefficients for annual productivity values of various Farallon species, 1971–83

Species	BC	PC	WG	CM	PG
Brandt's Cormorant (BC)	—				
Pelagic Cormorant (PC)	.467				
Western Gull (WG)	.357	.254			
Common Murre (CM)	.617[a]	.208	.677[a]		
Pigeon Guillemot (PG)	.624[a]	.801[b]	.420	.572[a]	
Cassin's Auklet	.449	.587[a]	.582[a]	.355	.499

[a] $p < .05$. [b] $p < .001$.

and low production in the Pelagic Cormorant and Pigeon Guillemot (Table 12.4). As noted several times, the feeding ecology of these two species exhibits virtually complete overlap in all respects. Correspondence was less strong among other species and was nonexistent for storm-petrels (not shown), which exhibited practically no variability in nesting success (Chapter 4).

The high annual variability in fledgling production among species that lay large clutches, i.e., overall chick production relative to clutch size, suggests that laying a large clutch may be an evolutionarily derived mechanism by which certain seabird species can better cope with or take advantage of widely fluctuating food availability (see discussion of "brood reduction" theory in Chapters 5 and 7). If this is true, it is not surprising that cormorants, which lay the largest clutches among seabirds, numerically dominate the avifaunas of the world's major upwelling systems, the Peru, Benguela, and California currents. Such systems are known for their high but unpredictable productivity (see Chapter 2, and Population Stability, below). Conversely, the murres and auklets, which lay but one egg, are the numerically dominant members of avifaunas in the Arctic, where productivity is lower and perhaps more predictable.

CHICK GROWTH: WITHIN- AND BETWEEN-YEAR PATTERNS

Much has been written about patterns of growth in birds, including seabirds, as indicated in the review by Ricklefs (1983). On an interspecific basis, patterns of chick growth have been described for a wide variety of seabirds, and these patterns have been related to various other aspects of natural history (e.g., Sealy 1973a, Vermeer and Cullen 1982, Pennycuick, Croxall, and Prince 1984). In the forms of the growth curves we constructed for Farallon seabirds there were no surprises. A point made by Ricklefs (1983) is that information on within- and, especially, between-year variability in the growth parameters of seabirds is sparse. It is to this gap in our knowledge that the data on growth at the Farallones make their greatest contribution.

Within-season variability in fledging weights, in all cases later-hatching chicks being lighter at fledging, has been detected in a number of alcids—murres (Hedgren 1979, Birkhead and Nettleship 1981), Razorbills (Lloyd 1979), auklets (Manuwal 1974b, Vermeer 1981, Sealy 1981), and puffins (Vermeer and Cullen 1982, M. P. Harris and Rothery 1985)—as well as in the Glaucous-winged Gull (Hunt and Hunt 1976b). The Farallon data expand this species pool by demonstrating within-season variability, not just in fledging weights, but also in asymptotic weights and the growth constant for the Ashy Storm-Petrel, Western Gull, and Pigeon Guillemot. Birkhead and Nettleship (1981) found within-season differences in the growth rate of murres in the Canadian Arctic, but Pearson (1968) found no such trends in any of the several larids, alcids, and cormorants that he studied in Wales. More importantly, the Farallon data showed that within-season variability was evident in some years but not in others. Years in which late-hatching chicks grew more slowly and reached lower asymptotes were those in which prey were less available (see Chapter 2). One of the few other studies that detected between-year differences in the within-year variability of chick growth was that of Thick-billed Murres by Gaston, Chapdelaine, and Noble (1983).

Between-year variability in growth parameters has apparently been demonstrated in still fewer seabirds (Ricklefs 1983). For murres, such variability has been detected by Gaston, Chapdelaine, and Noble (1983) but not by Hedgren (1979), although the latter investigator compared fledging weights over five years, and Pearson (1968) detected no variability in any of the species that he investigated for two years. Substantial between-year variability in fledging weights has been demonstrated for Razorbills (Lloyd 1979), guillemots (Kuletz 1983), and puffins (Summers and Drent 1979). Such variability was also evident in asymptotic weight, growth rate, and fledging weight for all four of the Farallon species in which we studied those characteristics. Trends were similar among species across years and were correlated with food availability. The relationship between food availability and growth parameters has been demonstrated for sulids (J. B. Nelson 1978, Ricklefs, Duffy, and Coulter 1984), terns (Dunn 1975), gulls (Hunt and Butler 1980), and puffins (M. P. Harris 1984), but mainly through indirect inferences about food availability and, except for the gull study, mainly through between-locality comparisons rather than through between-year comparisons at the same locality.

The marked variability in the growth parameters of Farallon seabird chicks, both within and between years, like the variability we observed in other aspects of reproduction, warn against the dangers of inter- and intraspecific comparisons among seabirds using a meager sample of years. Koelink (1972), for example, found that at Mandarte Island guillemots could successfully raise three chicks in a brood during some years but not others: what a difference years would make in conclusions drawn about the provisioning capabilities of parent guillemots! More importantly, the great variability in the growth parameters of Farallon seabirds and their correspondence with fluctuations in breeding success and diet demonstrate their utility as powerful indicators of food availability (Hunt and Butler 1980, Ricklefs, Duffy, and Coulter 1984, M. P. Harris 1984).

ECOLOGICAL STRUCTURE

In Chapter 1, we mentioned three obvious ecological guilds represented among Farallon seabirds. These are the cavity-nesting, sea-surface gleaners of zooplankton and micronekton (the two storm-petrels), the cavity-nesting, diving piscivores (two puffins and the guillemot), and the surface-nesting, diving piscivores (three cormorants and the murre). Classically, the composition of guilds, as defined by Root (1967) and as applied to seabirds, has been considered along only one niche dimension, for instance, nesting habitat (e.g., Whittam and Siegel-Causey 1981) or feeding ecology (e.g., Croxall and Prince 1980). This, of course, does not preclude guild composition from being defined along more than one dimension, but doing so reduces the number of species per grouping and increases the complexity of interactions we are trying to understand. This defeats the purpose of delineating guilds, namely, to identify the basic unit in which interactions between similar species can be viewed. J. Diamond (1986) noted that only when closely related species are completely sympatric do they begin to segregate along more than one niche dimension. We have only two groups of congeners at the Farallones (storm-petrels, cormorants), and within each, species are only partly sympatric.

If one looked at segregation within a co-occurring group of species along all niche dimensions, the result would not be guilds, but, by ecological definition, just the opposite, a collection of individual species (Hutchinson 1959). We used the grouping in Chapter 1 mainly to make the points that several assemblages of ecologically similar, even closely related species reside at the Farallones and that the potential for interaction was high and thus attracted our interest. At this point, it is best to return to the usual concept of a guild.

Nesting niches

Among Farallon cavity-nesters, breeding space is allocated by body size and aggressive ability, which is also somewhat a function of size (Figure 1.1). Little soil exists in which birds can burrow, and thus most species nest in cavities in the talus and in cavities wrought by the rotting of granite, wave erosion, and other effects of the marine environment. Competition among the Farallon cavity-nesters is intense; each chapter presents details of larger species supplanting smaller ones from cavities into which the larger ones can fit. This system of dividing space resources by body size has been discussed by Bédard (1969c) and Sealy (1968).

The only Farallon species that burrows extensively in the limited soil is Cassin's Auklet. These birds are much heavier and more aggressive than storm-petrels, so it is not surprising that the latter (Leach's especially), known to burrow on many islands elsewhere, are confined entirely to the small cavities in talus or in the rock walls of old building foundations. Larger cavities in talus are taken by auklets. Tufted Puffins and Rhinoceros Auklets, which also burrow elsewhere, do not do so on the Farallones, because they prefer some altitude to their nesting habitat and the only deep soil available is on the flat marine terrace. They, too, along with Pigeon Guillemots, nest in the larger crevices offered by rock rubble, moving Cassin's Auklets out at every opportunity. In their turn, guillemots are displaced by Rhinos, and both are displaced by Tufted Puffins. If coevolution is indeed the process that structures seabird communities, as suggested above, it appears that coexistence of species is greatly advanced through adjustment of body size (see below for further discussion).

Allocation of nesting space among Farallon surface-nesters is a function not only of size but of timing and nesting density. Only two surface-nesting species select obviously different habitats: Pelagic Cormorants choose single, isolated ledges on which to cement their nests, and Double-crested Cormorants prefer large outcrops or the shoulders of hills. Among the remaining three surface-nesting species broad overlap occurs. In general, murres and Brandt's Cormorants nest on broad, gently sloping ledges and terraces on the windy portions of the island. Murres also nest on the ledges of steep slopes and even cliffs. Gulls, though they prefer flat but not featureless terrain, nest anywhere they can find an unoccupied site.

These descriptions of nesting habitat are consistent with those summarized by Whittam and Siegel-Causey (1981), who classified 20 species of Alaskan seabirds according to seven nesting guilds: cliff, ledge, talus, talus and burrow, flat ground, boulder rubble, and burrow and rock crevice. For six Farallon species, this classification suffices. The Pelagic Cormorant is under the "cliff" category, Ashy Storm-Petrel is under "talus," Leach's Storm-Petrel and Cassin's Auklet are under "talus and burrow," Western Gull as the large larid is under "flat ground/gentle slope," and Pigeon Guillemot is under "boulder rubble." Classifying the remaining five species by the Alaskan system is problematic. The Double-crested Cormorant would move from "cliff" to a new category, "steep slope," whereas Brandt's Cormorant would be assigned to "gentle slope." The Common Murre would occupy a "ledge/gentle slope" category, indicating its use of ledges, as defined by Whittam and Siegel-Causey (see also Squibb and Hunt 1983) and its sharing of moderate slopes with other surface-nesting Farallon species. The two puffins (Tufted and Rhinoceros) would fall in the Alaskan "burrow/rock crevice" category, except that little opportunity exists for them to burrow on the Farallones. More properly they should reside with the Pigeon Guillemot in a "rubble/rock crevice" category. Why Whittam and Siegel-Causey did not include the Rhinoceros Auklet in their analysis is not clear and most unfortunate from our perspective at the Farallones.

There is thus a great deal of similarity between nesting guilds in Alaska and at the Farallones. With the murre participating in two nesting guilds, four of the six Farallon guilds contain more than one species: talus and burrow (2), flat ground/gentle slope (3), rubble/crevice (3), and cliff/ledge (2). Talus and steep slope each contain only one species. In addition to the murre, however, other species overlap to some degree in guild membership. As populations recover from perturbations, these overlaps underlie the displacements discussed below.

Adjustments to nesting-guild composition

Populations on the Farallones are recovering and, as a consequence, much interspecific encroachment on habitat and apparent guild adjustment is slowly occurring. In the process, nesting niches are becoming more narrowly defined. Until the 1982–83 ENSO, murre numbers were increasing rapidly. As their colonies expanded outward, they took over Brandt's Cormorant nest sites, mainly when male cormorants nesting adjacent to the murres failed to return, leaving sites open. Murres quickly occupied those sites, and although a returning cormorant would viciously oust the murres, a prospecting one would not even consider the site available. Much of the murres' success, of course, stems from their occupying of nesting cliffs four months earlier each season than do Brandt's Cormorants (Figure 12.1). Similarly, should murres ever again reach a population size equivalent to pre-1860 levels, we may see this interaction between them and Pelagic Cormorants as well. The latter, however, also occupy nest sites earlier than do Brandt's Cormorants.

The growth of the Double-crested Cormorants' colony on Maintop indicates that this species, too, can displace Brandt's Cormorants (which have been nesting around the colony's periphery). Displacement occurs as a function of size and, as with the murre, as a function of earlier territory occupation. Moved out by the two other species, Brandt's Cormorants seem to be tending toward occupying slopes intermediate between the steep outcrops of the Double-cresteds' and the gentle slopes of the murres' nesting habitat. Brandt's Cormorants, moved out by murres, easily take over the gulls' nesting space.

As described in Chapter 5, gulls nesting near cormorant colonies often lose their nests to cormorants looking for nest material. Ultimately, with a body smaller than that of the cormorants and nests more dispersed than those of murres or cormorants, gulls lose nesting habitat to both. Farallon Western Gulls begin to occupy nest sites in the fall, well before nesting, but even by doing so they cannot compete with the other species. Among the gulls themselves, however, early occupation likely increases the chances of holding a territory later (Chapter 7). Gulls currently occupy the site of the nineteenth-century colony of Double-crested Cormorants on Southeast Farallon, a site very similar in habitat characteristics to the one the cormorants currently occupy on West End. The Double-crested Cormorant population continues to grow at its ever-increasing rate, and we might soon see gulls displaced even more.

In addition to losing ground to other bird species, Western Gulls, being the only surface nesters using the low, flat marine terraces and rocky shoreline of the island, have been increasingly subjected to disruption by pinnipeds. Before humans appeared and rapidly decimated pinniped populations (Chapter 1), numbers of gulls in particular were likely much, much smaller than they are now, as seals and sea lions probably ruled over much of the marine terrace on Southeast Farallon as well as significant portions of West End. About half the present nesting population of gulls occupies these areas. Now that populations of pinnipeds are rapidly rebounding, the potential impacts of their increase on the seabird populations are becoming clear (Figure 1.4). A similar interaction and impact is becoming ever more apparent on South Georgia, in the South Atlantic (Croxall et al. 1984), and has been a factor in the Benguela Current as well (Shaughnessy 1984). As gulls were pushed out of areas by one species or another, their territories shrank, and they began nesting close to areas of human activities (Chapter 7). There are no areas for further expansion of their numbers, unless space is created by further contraction of territories already small by the standards of large larids (Hunt and Hunt 1975, Pierotti 1976). At present, Western Gulls more than any other Farallon species seem faced with the problem of resorting to suboptimal nesting habitat, as discussed in Chapter 7.

Thus, nesting space, a relatively stable resource, is at a premium on the Farallones, and much evidence for inter- and intraspecific competition exists. To the extent that the birds allow themselves to be crowded, virtually all surfaces touched by air (above and below ground) but not by ocean waves are being used by seabirds for nesting or by pinnipeds for resting. This bears importantly on population regulation among Farallon seabirds, a discussion we will return to below.

Feeding niches during the breeding season

Even with regard to food resources, spatial segregation appeared to be exceedingly important in the ecological structuring of the Farallon seabird community. We can express this segregation by means of three modes, ranked by relative importance. These modes were evident only during warm-water years, when food was limited (Chapters 2 and 3). In order of decreasing importance, these modes are: (1) horizontal (spatial) stratification by water type (estuarine, shelf, slope, oceanic), (2) vertical stratification by depth and a subset of depth, and (3) substrate. Diet composition was a function of this spatial regime, and so was relatively unimportant itself as a structuring factor. In only a few cases did prey size come to bear, and then, when food was limiting, the differences were clearly obvious: the large Pelagic Cormorant vs. the small Pigeon Guillemot in the shelf waters over a rocky bottom (Hutchinsonian ratios of body size [weight] and bill length of 4.4 and 1.3, respectively; Figures 1.1 and 1.2); and the large Brandt's Cormorant vs. the small Common Murre in the middle of the water column over the continental shelf (size ratios 1.4 and 1.2, respectively). The former pair differs widely in body size; the latter pair differs widely in bill size. Spatial stratification was the only factor required to segregate clearly the two groups of congeners, storm-petrels and cormorants. Problematic was the lack of clear segregation between the two puffins, which are the two next most closely related species. Both forage in slope waters, and both dive for their prey. The data, however, are currently insufficient to resolve the problem, particularly in regard to the Rhinoceros Auklet's diet. The Tufted Puffin, however, with twice the body size, appears capable of diving much deeper. Also problematic is the degree to which segregation of feeding by prey size was more a function of habitat than of predator size. Like A. W. Diamond (1983) and Volkman, Presler, and Trivelpiece (1980), we currently favor the habitat factor.

Among the eight species for which we had appreciable data on diet to complement the spatial and behavioral data, the most distinctly segregated in feeding niche was the Double-crested Cormorant (Chapter 3). Without variation, it fed largely on schooling species of fish that occur in estuaries and bays along the mainland coast. Except during 1976 (ENSO), when coastal bays were invaded somewhat by Brandt's Cormorants, the diet and feeding habitat of the Double-crested Cormorant hardly overlapped with those of any other Farallon species. Its foraging behavior was most similar to that of the murre, Rhinoceros Auklet, Tufted Puffin, and, to some extent, Brandt's Cormorant, as all five fed by diving, usually to capture prey (fish, squid) high in the water column.

The species having the next most distinctive feeding ecology was the Cassin's Auklet. Except for the murre and gull, and perhaps the puffin, during its prelaying period (April; PRBO, unpubl. data), the auklet alone fed on zooplankton, especially euphausiids, throughout the breeding season, and, indeed, the entire year. After April, the other species switched to fish. On the other hand, in contrast to its distinctive diet, the auklet's feeding habitat greatly overlapped that of murres and puffins, because all fed mainly in waters over the continental slope and outer shelf north of Southeast Farallon. During warm-water years, murres but not the puffins moved to other habitats. Like the other alcids, the Cassin's Auklet fed by diving, but presumably to shallower depths. Its diet of zooplankton was likely determined in large part by its small size, which may not allow it to capture larger fish. In some years, the auklet did feed to appreciable degrees on the same fish fed upon by other species (*Sebastes*), but it tended to eat smaller individuals. Any apparent segregation on the basis of size, however, was an artifact of the auklet's diet being sampled earlier in the spring than that of the other seabirds, and of the fish available at that time being smaller.

Therefore, ample evidence indicates that two of the eight species with which we are concerned are segregated in feeding ecology. In fact, we also know that the two storm-petrels differ from other species by virtue of their feeding largely in pelagic rather than in shelf and inner-slope waters. We do not know the degree to which the two storm-petrels differ from one another, however. Among the remaining six species, the overlap in feeding ecology during most years was remarkable, especially during 1974–75, 1977, and 1979, when one prey dominated diets: juvenile rockfish. Both diets and feeding habitat overlapped almost completely; in fact, most birds of these species fed in multispecies flocks. To be sure, the gull captured prey at the surface, while the other five species fed by diving. Among species for which data are available, i.e., cormorants, murre, Tufted Puffin, and guillemot, prey of the same size were captured as well. We do not know whether the five diving species in the feeding flocks captured prey at different depths, but given the broad overlap in all other feeding-niche characteristics and the circumstances of superabundant food, we would be surprised if they did. Besides, juvenile rockfish are not necessarily distributed throughout the water column, but fish of different sizes may instead move at different depths.

In great contrast, during ENSO and other warm-water years, diets diversified dramatically and feeding behavior changed. Less niche overlap was evident, and, in fact, niche segregation was at its classical best. The guillemot and Pelagic Cormorant fed where they were "supposed to," i.e., on solitary prey captured on the bottom, on or near a rocky substrate. Diets overlapped somewhat, but the cormorant ate more sculpins and crustaceans and the guillemot ate more sanddabs and octopi. During those years, the Pelagic Cormorant would be more aptly named the "Intertidal Cormorant"; but the guillemot seemed to feed about the same distance from the island as in years of abundant prey. Brandt's Cormorants fed far from the island, mainly in the mouth of San Francisco Bay. They took schooling and nonschooling prey near bottoms having little relief, and they also fed in kelp beds. Murres, too, fed more over the shelf and inshore near the mainland, but also throughout the water column and mainly on schooling prey such as anchovy and smelt. Tufted Puffins fed in deeper water west of the island on schooling prey such as anchovy and saury. Gulls fed much more on garbage and fish offal during the warm-water years. The Double-crested Cormorant continued to feed in estuaries, as it did in other years. Of course, few birds of any species appeared to feed much at all during 1983, and many died.

In addition to feeding on different species, in different areas, and in different habitats, the birds tended to feed on prey of different sizes during the warm-water years. Size, however, was more a function of the type of prey than of selection of different-sized individuals of the same prey species. This was consistent with the findings of A. W. Diamond (1983); in the case of the prey of Farallon seabirds, it is likely that the sizes of the available prey did not cover a continuous range. Segregation of feeding niches on the basis of prey size was approximately consistent with differences in bill and body size, as observed by Ashmole (1968), Bédard (1969a,b), M. P. Harris (1970), and A. W. Diamond (1983).

Fluctuation in feeding-guild composition

In contrast to our exercise of deciding guild composition along the niche dimension of nesting habitat, assigning Farallon species to feeding guilds is a whole different business. We use here a modification of the feeding-guild classification suggested by Croxall and Prince (1980), who categorized seabirds by feeding method and foraging range (or radius; Pennycuick, Croxall, and Prince 1984). Instead of using foraging range we use foraging habitat, because foraging range is an expression of habitat availability. This is particularly evident at a place such as the Farallones, which sit at the outer edge of a broad shelf and, therefore, provide species with a "choice" between feeding either far away or close, in pelagic or neritic waters. Such a situation differs from that at most other seabird nesting communities so far studied, where pelagic waters are necessarily far away and neritic waters are close. Thus, at the Farallones, analysis of "foraging range" provides little insight into habitat preferences. Similarly situated are the Pribilof Islands, where Schneider and Hunt (1984), too, found difficulty with a concept strictly confined to foraging radius.

In our classification, except for the storm-petrels (for which we lack data), Double-crested Cormorant, and Cassin's Auklet, feeding guild was a function of year. In warm-water, food-limited years, we had six guilds: hard-substrate gleaners (Pelagic Cormorant and Pigeon Guillemot), a soft-substrate gleaner (Brandt's Cormorant), neritic/epipelagic pursuit divers (the four remaining alcids), an estuarine pursuit diver (Double-crested Cormorant), epipelagic surface seizers (storm-petrels), and a neritic/intertidal surface seizer (gull). During other years, the substrate gleaners and neritic/epipelagic pursuit divers merged, and the gulls no longer fed intertidally.

Niche overlap and resource stability

The dramatic shift in the composition of feeding guilds from one breeding season to another reveals the instability of the food web in the Gulf of the Farallones, an instability that is characteristic of other eastern boundary-current systems (Chapter 2; Glantz and Thompson 1981, Parrish et al. 1982, MacCall 1986). Figure 2.8, which summarizes the frequency of warm-water events in California during the last century (see also McLain, Brainard, and Norton 1985 for a more detailed look at the past two decades), indicates that the period in which our study took place was not particularly unusual; years or seasons of food stress, as well as of food superabundance, have generally occurred at intervals of three to seven years. In the California Current, especially severe "crunches" (Wiens 1977) probably occurred once every ten to 25 years during this century: 1905–6, 1914–15, 1940–42, 1958–59, and 1982–83 (Figure 2.8). These were episodes two to three years long when mortality of both poorly adapted adults and subadults may have been significant. Episodes of warm water in 1903, 1912, 1919, 1926, 1931, 1952, 1966, 1969, 1972, and 1978 were intense but short. In these short episodes, if 1972 was typical, mainly subadults died (PRBO beached-bird data, 1971–84, Stenzel et al. 1988). Our Farallon data thus support the ideas of Wiens (1977, 1986) by demonstrating that during summer seabirds in this system compete for food only sporadically. Whether Farallon seabirds compete during most winters, however, is an important subject for future research (see also Grant 1986). We feel fortunate indeed that we did not choose the "wrong" one or three years to investigate the summer feeding ecology of Farallon seabirds.

Nesting guilds were far less variable, nesting space being a resource that varies on a scale of decades (or, without modern human interference, centuries) rather than every few years. In virtually every year, we recorded many instances of birds of one burrow-dwelling species or another attempting to nest in a cavity large enough for a larger competitor, and then usually being ejected. Before ENSO 1982–83 and the introduction of gill nets to local fisheries, each year we recorded various colonies of surface nesters encroaching into the domain of others. Unlike competition for food, then, competition for nesting space appeared to be an ongoing, annual process that lasted practically year round for the several species having greatly extended prelaying periods. Unlike the trophic dimension to ecological segregation, in which coexistence could be mediated by frequent periodic resource excess, along the dimension of nesting space, coexistence was a function only of morphological differentiation in body size or of differentiation in behavior (e.g., nesting density or diurnal visitation patterns).

As noted above and in Chapter 1, Roughgarden (1986) analyzed the responses of hypothetical communities to colonization by a species new to the community. He postulated two types of resulting communities, those structured by invasion and those structured by coevolution following invasion. In the first, communities are assembled sequentially, competition between residents and invaders determining the invaders' success. In the second, invaders and residents coevolve if competition is symmetrical (invaders and residents adjust equally), but if it is asymmetrical then one species or the other disappears. He was able to give examples of both types of communities on islands in the Caribbean. Some community histories exhibit structuring by both invasion and coevolution.

It would seem, at the local scale, that the Farallon seabird community is assembled by invasion, much as the Rhinoceros Auklet has been doing in recent years. It is easy to envision that at some time other species arrived and found unoccupied nesting space, particularly the northern species, which may have arrived during glacial periods when the climate at the Farallones was cooler. In the case of the Rhino, it will be interesting to see if it survives competition with Tufted Puffins, if and when the latter recover their former (1800's) numbers. The Rhinoceros Auklet did nest on the islands in the 1800's, but by available accounts it was less abundant than it is now (see Ainley and Lewis 1974; Chapter 11). The species' population appears to be enjoying a period when that of its close relative and competitor is at a low point.

At the scale of the upwelling domain of the eastern Pacific, however, the assembly of the seabird community appears to be structured by coevolution among species. In the context of limited nesting cavities and abundant food, might the enigmatic large size of California and Oregon Pigeon Guillemots (Storer 1952) be a response to competition with smaller alcids that also nest in talus and larger alcids that lack much soil in which to dig their burrows? Perhaps the guillemots' large size is not just a response to the cooler temperatures brought by upwelling, temperatures certainly no cooler than those during the summer in the Aleutian Islands and Bering Sea, where a smaller Pigeon Guillemot nests (all Pigeon Guillemots of western North America likely winter in southeast Alaska and British Columbia). Another example of possible character displacement as a result of coevolution exists among the cormorants. Where it overlaps Brandt's Cormorant, the Double-crested Cormorant is represented by the largest of its subspecies, whereas the Pelagic Cormorant is represented by its smallest (Palmer 1962). As in the burrow-nesting species, the result is a fairly even gradation between a large (Double-crested), medium (Brandt's), and small (Pelagic) cormorant, a type of shift predicted by Roughgarden's analysis. The even distribution in size might well induce greater segregation in the competition for nest space and food, another interesting subject for further study.

BREEDING SUCCESS AND NICHE OVERLAP

Farallon seabirds bred most successfully when their feeding niches overlapped the most, owing to the presence of a superabundant food resource. Especially productive breeding seasons occurred in 1971 (no feeding data), 1974, 1977, and 1979, years when seabirds fed predominantly on juvenile rockfish of the same size, and captured them in multispecies foraging flocks near the island. Conversely, breeding was poor or nonexistent in 1973, 1976, 1978, and 1982 (and 1986), warm-water years when feeding niches diverged in all or most aspects. The situation in 1983 was exceptional in that few seabirds bred and many did not feed. During productive years, not only did parents fledge large broods, but fledglings were heavy. Furthermore, during these years breeding success and chick weights varied little as a function of laying or fledging date; in contrast, during poor years, late nesters fared poorly.

During summers when feeding niches were well defined, and breeding success was poor or zero (1983 not considered), various species' feeding ecologies seemed more typical of what we envision to be their winter situation, or even their "typical" summer situation as defined by observations elsewhere. This is best illustrated by the guillemots, which are among the most widely and intensively studied seabirds with respect to both feeding and nesting ecology. Guillemots are considered to be littoral, benthic predators that feed on nonschooling prey (e.g., Storer 1952, W. Preston 1968, Ainley and Sanger 1979, Cairns 1984); when ice covers the sea, they still seem to prefer feeding in association with a substrate, be it the sea bottom if waters are shallow enough, or the underside of ice floes if waters are deeper (Bradstreet 1980). They also tend to feed within 10 to 15 km of their nests (Kuletz 1983, Cairns 1984). During the nonbreeding season, they do not necessarily remain near nesting areas, but their feeding ecology apparently remains approximately as described above (Preble and McAtee 1923, Storer 1952, Belopol'skii 1961), though their diet may change (Bradstreet and Brown 1985). Interestingly, the classical description of guillemot foraging fits the summer patterns of Farallon Pigeon Guillemots during warm-water years, when their food is limited and breeding success is low, but not during the years when their prey is abundant and breeding success is high!

The Western Gull is another species for which information is available to illustrate the point that feeding ecology during the nonbreeding season is similar to that during summers of poor reproductive success. Large larids are typically considered to be extremely opportunistic omnivores (Ainley and Sanger 1979) that forage in a variety of circumstances, mostly close to or on shore. Not only do they capture live fish and shellfish, but they also scavenge a wide variety of offal and garbage. In areas far from human civilization, for example the Canadian Arctic (Gaston et al. 1985) or the Antarctic (W. Fraser, unpubl. data), gulls forage entirely on marine resources except for the few specialists that frequent seabird rookeries. During most summers, especially when marine prey were abundant, Farallon Western Gulls were actually somewhat exceptional among large larids that have been studied, in the sense that even with garbage available their foraging was confined to offshore marine areas (Chapters 3 and 7; Pierotti 1976, Spear et al. 1987). These were years when they were also reproductively most successful. During warm-water summers, when reproduction was poor, they fed heavily on garbage and a wide variety of other foods, a pattern typical of their nonbreeding (winter) period, as well as of younger, nonbreeding individuals during the summer (Spear 1988a, Spear et al. 1987; Chapters 3 and 7; see also Hunt and Butler 1980).

Less complete information was available for Brandt's Cormorants and murres. When not breeding, as well as during warm-water summers, both tended to feed within 10 km of the mainland coast, the cormorant more so in large, deep estuaries such as Tomales and San Francisco bays. The murre favored schooling prey more than did the cormorant, the cormorant concentrating closer to the bottom, the murre higher in the water column (Chapter 3; PRBO, unpubl.

data 1985–86). In contrast, during summers when reproduction was highly successful, both species fed together near the Farallones on midwater-schooling rockfish.

The point of these examples and discussion is to suggest that feeding-niche segregation in Farallon seabirds is more immediately important to the maintenance and survival of adults than to their production of offspring. During fall, winter, and warm-water periods, when food is scarcer, not only is successful breeding difficult (perhaps even impossible), but mortality is highest (Stenzel et al. 1988). During these periods, the birds' ecological segregation with respect to the food resources appears to be much better developed as well.

That a relationship between resource limitation and ecological segregation exists, of course, is well accepted, although demonstrations of resource limitation or ecological segregation are not (Cody 1974, Wiens 1984). The interesting nuance, at least with regard to the food of long-lived Farallon seabirds, is that the species' ecological (and morphological?) segregation is more a function of surviving difficult times than of maximizing production of offspring. This suggestion is by no means totally novel (for example, see Grant 1986), but it is interesting because it points out a need in the ecological study of higher vertebrates, and especially of seabirds, for a more critical justification of why ecologists and zoologists are so preoccupied with studying breeding adaptations and the breeding period, as if this were necessarily the time of the "ecological crunch" (Wiens 1984). Such a preoccupation is no doubt a function of convenience, but we should realize that, as a result of it, our perceptions of the answers to the "larger" questions of animal ecology, for example, population regulation and overall life-history strategy, might well be distorted.

POPULATION STABILITY AND THE REGULATION OF POPULATION SIZE

Wiens (1977, 1984, 1986) argued that contemporary thinking in ecology assumes that communities are stable or in equilibrium, though not necessarily static. At equilibrium, among other characteristics, species in a community should compete intensely, populations should be resource-limited and density-dependent, niche dimensions should be well defined, and random phenomena should have little impact. The opposite should be true of communities in a nonequilibrium state. If such a continuum in community stability exists, an examination of the Farallon community would be valuable.

It is obvious that most Farallon populations are or have been greatly and often perturbed, and thus are not at equilibrium. Perturbations have resulted from both natural and anthropogenic factors (see Chapter 1; Ainley and Lewis 1974). Anthropogenic factors have decreased population stability dramatically during the last two centuries, but natural factors have been at play as well. For instance, as most chapters detail, populations decreased as a result of increased mortality during intense warm-water/ENSO episodes, particularly those of 1978 and 1983; these events, which might be considered unpredictable, in fact have recurred in the eastern Pacific for as long as records exist (Chapter 2).

Though many seabird researchers follow Lack (1954, 1966) and others by believing that regulation of seabird populations is density-dependent (i.e., that numbers are fine-tuned to resource availability), J. B. Nelson (1980) proposed that the size of seabird populations of the Peru Current, the Southern Hemisphere analog of the California Current, was until recently a function of periodic large die-offs resulting from ENSO. Nelson further proposed that numbers would never reach resource limitation before the next crunch. Of course, this situation does not apply now, because overfishing has destroyed the entire food web, with the result that seabirds are now much fewer and apparently limited by food resources (Idyll 1973, MacCall 1984, Furness and Monaghan 1987). Nelson's hypothesis merits serious consideration, with modification, for populations in upwelling-based ecosystems, including that of California. After all, major ENSO's, such as the 1957–59 and 1982–83 events, occur often enough that most adult seabirds should experience at least one if not two, as well as several less severe but nevertheless significant events, during their long lives—if they survive the events. In the Peru Current, however, because of the long interval between major ENSO's and the incredible richness of the food resources before exploitation, it seems that populations should have been limited by nest space (Murphy 1981, Duffy 1983a,b, and references therein). The same is also likely true of California Current populations (as discussed below).

The preceding chapters present data indicating that nest-site availability directly limits the Farallon breeding populations of at least four species: the Western Gull, Pigeon Guillemot, Cassin's Auklet, and Tufted Puffin. Evidence also exists to suggest competition for nest sites among most cavity-nesting species including the three above, plus the Ashy Storm-Petrel and Rhinoceros Auklet, as well as competition among most surface-nesting species. The evidence consists of (1) the existence of floating populations of several species, (2) many observations of individuals or colonies of one species displacing individuals of other species, (3) the unusually prolonged prebreeding period of territory occupation among many Farallon species, especially the Ashy Storm-Petrel, Pelagic Cormorant, Western Gull, Common Murre, Pigeon Guillemot, and Cassin's Auklet, and (4) the increase in Tufted Puffin and Rhinoceros Auklet numbers following removal of feral rabbits, which had occupied burrows.

The *breeding* portion of many Farallon seabird populations, then, appears to be limited by nest-site availability. These populations would perhaps be at equilibrium were it not for human beings, pinnipeds, and oceanographic events such as ENSO and anti-ENSO. Because of these factors, and the responses to them that we have observed in recent decades, Farallon populations have likely been in a state of flux at least the last two centuries (i.e., the period for which historical records exist), and probably for millennia. Moreover, the Farallones are not alone. Although the Farallones supply only a portion, albeit a significant one, of regional populations, seabird numbers at other sites throughout central and northern California, including the few within the foraging range of Farallon individuals (cf. Osborne and Reynolds 1971, Sowls et al. 1980), also have been changing dramatically during the past two centuries and continue to do so. Virtually all offshore rocks and islands along the coast from central California to Oregon are now occupied by seabirds, except for those disturbed seriously by humans (Sowls et al. 1980). Many geologically stable cliffs, as well as several wharves and the roofs of abandoned waterfront warehouses, are occupied as well. This is not to say that nesting habitat is completely saturated, because at most occupied sites, as at the Farallones, there is still room for the continuing expansion of populations of several species. In some cases this means that some species will displace others to some extent.

While populations breeding at the Farallones and throughout the California Current might ultimately become limited by the availability of nesting space, the existence of floating populations (e.g., J. L. Brown 1969; see Glossary) at least among Farallon Western Gulls (Spear et al. 1987), Pigeon Guillemots (as of 1982; D. A. Nelson 1982), and Cassin's Auklets (Manuwal 1974a) indicates that the limitation to total population size is another story. Thus, although *breeding* numbers reach an asymptote, *total* population can continue to increase, but at a slower overall rate (because floaters do not reproduce). Furness and Monaghan (1987) presented a mathematical model showing how this comes about. Contrasting the Farallon situation, Furness and Birkhead (1984) presented evidence that seabird breeding populations in an area of relatively uniform, low, and predictable ocean productivity in northern Great Britain are limited by the availability of food during the breeding season, as proposed also for tropical, oceanic communities by Ashmole (1963) and A. W. Diamond (1978). These populations should therefore be close to equilibrium. If for these populations food but not nesting space is limited, the occurrence of floating nonbreeders is less likely (but see Schreiber and Chovan 1986).

Eastern boundary currents such as the California and Peru currents are not areas of uniform, predictable, or low productivity, and thus Furness and Birkhead's analysis does not apply. Upwelling, advection, and periodic ENSO see to that (Chapter 2). While Peru and California seabird breeding populations are ultimately limited by nesting space, total populations (i.e., the breeding and nonbreeding portions together) could be limited by food availability during nonbreeding periods, i.e., winter and especially periodic ENSO. This is where a modification of J. B. Nelson's (1980) idea applies, an idea that Murphy (1981) actually proposed 30 years earlier (written in 1954; Glantz and Thompson 1981). This speculative scenario, based on material presented in previous chapters as well as extrapolation to California of historical trends in Peruvian populations (e.g., Murphy 1925, 1936, 1981, Idyll 1973, Paulik 1971, Duffy 1983b, MacCall 1984), is as follows: ENSO disrupts the food web, seabirds do not reproduce and they die in large numbers, and populations decline; when conditions improve, populations begin to rebuild, but because ENSO occurs at relatively frequent intervals, *total* population size conceivably never reaches its ultimate asymptote. *Breeding* portions of populations, however, periodically reach asymptotes, as a function of nesting space, depending on the frequency and intensity of ENSO events. Figure 2.8 shows that ENSO's equivalent to that of 1982–83, when many seabirds died, occurred in the California Current (and off Peru, by association; McLain, Brainard, and Norton 1985) in 1905–6, 1914–15, 1940–42, and 1958–59. The frequency of these events coincides roughly with the expected generation times of many seabirds (i.e., 15–25 years), so certainly most birds that survived to adulthood experienced them. Weaker ENSO's similar to the one in 1972–73, when seabird populations did not decrease significantly, also occurred regularly. Still other ENSO events, plus warm-water intrusions (see Chapter 2)—at least 11 of them—were strong enough to reduce significantly breeding success during this century.

Complicating this scenario is the possibility that food scarcity during winter limits populations before they reach a level at which they would be limited by summer food availability. Lack (1966) and others proposed that winter food supply has ultimate control of seabird population size, although analyses by Gaston and Nettleship (1981), Furness and Birkhead (1984), and Hunt, Eppley, and Schneider (1986) suggest an alternative for populations of northern, less productive waters. Spear et al. (1987) present evidence that winter food availability strongly affects age-specific mortality and dispersal in at least one Farallon species, the Western Gull. This information directly indicates that winter feeding conditions are important in regulation of that species' population and further indicates that the role of winter conditions in regulating populations of other Farallon species should be given serious thought, because like the gull's, many other populations also disperse along the California coast during winter. Furthermore, winter food availability seems to have a great deal to do with feeding adaptations and niche segregation among Farallon seabirds, implying that food may be limiting (see also Grant 1986 and references therein). Therefore, it is possible that winter feeding conditions, along with ENSO and the limitation of nest space, may be important in the regulation of Farallon seabird populations.

In summary, the data presented in Chapters 4 through 11 suggest that Farallon seabird populations are in a state of nonequilibrium brought about by environmental instability, especially ENSO (and also, in historic times, by pinnipeds and humans). The breeding portions of populations reach upper asymptotes through the limitation of nesting space, but major die-offs due to ENSO disrupting food supplies means that stability may be short-lived. When breeding populations reach their space-defined asymptotes, floating portions of populations then develop because food resources are ample, and the total population size continues to increase.

If modern man allows populations to reach their full potential, we might have the chance to see if food ultimately limits total population size at the Farallones. Whether this limitation comes about through periodic collapse of the food web during ENSO's or because of the scarcity of food in winter is another question.

DO FARALLON PATTERNS APPLY TO OTHER SEABIRD COMMUNITIES?

In general, the Farallon seabird community appears to be operating much like the grassland and shrub-steppe communities studied by

Wiens, Rotenberry, and co-workers (summarized by Wiens 1984, 1986), at least with regard to food (the colonial use of nesting space by seabirds is not readily comparable). Seabirds forage opportunistically on a highly variable but nonlimiting resource, and competition is not evident. At the same time, a principle proposed by Schoener (1974) also seems to be operating for Farallon seabirds: species with high niche overlap in one dimension (in this case, food) do not overlap or compete in another important dimension (in this case, colonial use of nesting space). In this regard, it is interesting that at the Farallones the resource of nesting space, for which evidence of competition is strong, is not only highly limiting, because of the paucity of islands within hundreds of kilometers, but is also relatively stable. In great contrast, the resource of food in the breeding season, for which no evidence of direct competition exists, is generally abundant though affected by unpredictable events. On the basis of this contrast, it seems logical that niche segregation should be more difficult to achieve with an abundant, highly variable resource, which is one of the points that Wiens (1984) stressed.

Actually, the importance of nesting space for Farallon seabirds is exacerbated by (1) the average high availability of food during summer, when breeding occurs, which in turn results in overall high nesting success; (2) the long lives of seabirds, which allow them to forgo the increased energetic costs of reproduction in bad years; and (3) seabirds' strong philopatry. The floating populations, which ultimately result from the interplay of these three factors, put nesting space at an even higher premium. One should realize, however, that floating populations, and apparently the conditions that generate them, are not frequently recognized or proven, in many instances because of the effort required to do so (J. L. Brown 1969). The demonstrated existence of floating populations in at least three Farallon species makes this community ostensibly unique among seabirds. But at the same time, other seabird communities have not been as intensively studied, though some individual species have (e.g., Kadlec and Drury 1968).

If Farallon patterns can be applied farther afield, we would look first to other California Current communities from central California (northern Channel Islands) into Oregon—the center of the upwelling system—and then to the communities of other eastern boundary currents, such as those of the Peru Current, where nesting space was once at a premium and where fantastically rich food resources once fostered trophic opportunism (Duffy 1983a,b). The same might apply to the Benguela Current along South Africa, an analog of the California and Peru currents, where the construction of vast platforms to increase the efficiency of the guano harvest may have increased seabird nesting populations (Crawford and Shelton 1978). More recently, however, Duffy and La Cock (1985) contested Crawford and Shelton's conclusion, with evidence to suggest that the platforms increased the local populations only at the expense of others. Otherwise, the presently accepted characterization of the ecology of the Benguela Current, and of the Peru Current for that matter, is that bird populations are now limited by food because of collapse of the fish stocks due to overfishing. Whether space was once a limiting factor in the Benguela system is obscured by historic intensive commercial egging and harvesting of birds, as well as the loss of pinniped populations during the 1800's (Duffy and La Cock 1985).

The other boundary currents are the Canary, off northwest Africa, and the Somali, a unique *western* boundary current, off east Africa (Glantz and Thompson 1981). R. G. B. Brown (1979) commented that the surprisingly low numbers of seabirds in the Canary Current resulted from a paucity of nesting sites (a factor that applies also to the Somali and, as noted in Chapter 1, to all eastern boundary currents), as well as from the "competing attraction" of Benguela Current resources. In comparison to other eastern boundary currents, the immensity of seabird populations formerly nesting in the Peru Current, numbering in the scores of millions, in large measure probably was a function of the number of coastal islands available being larger than in analogous oceanographic areas. Even there, however, artificial nesting areas increased guano harvests (Duffy 1983a). Many fewer islands suitable for nesting occur in the other eastern boundary regions.

Where nesting space is not limited and ocean productivity is less variable and on average much lower, for instance, in northern Great Britain (Furness and Birkhead 1984), the Canadian Arctic (Gaston and Nettleship 1981), the Bering Sea (Hunt, Eppley, and Schneider 1986), and tropical oceanic areas (Ashmole 1963, A. W. Diamond 1978), the patterns with regard to resource limitation likely differ from those of eastern boundary currents. Indirectly adding support to the likelihood of this difference is the comparison of shallow-water and oceanic marine food webs contained in Sherman and Alexander (1988); food-web dynamics in shallow-water systems are affected much more by predator–prey interactions, as opposed to oceanographic factors, than are those in oceanic areas, particularly upwelling systems. Even in the Canadian Arctic and Bering Sea communities, however, the published evidence indicates to us that *interference* competition rather than *exploitative* competition is playing the major role. If so, this brings into question whether space for foraging, instead of food availability, is the resource that is actually in limiting supply. Thus, the situation may be more similar to that of upwelling systems than it at first appears. In any case, investigation of the Farallon patterns has definitely been revealing and, like the studies by Wiens and coworkers, certainly would not have been without several years of data collection.

Appendixes

APPENDIX 2.1

Characterization of upwelling and downwelling in the vicinity of the Farallones, January–June 1970–86

Year	Upwelling/downwelling characterization
1970	Continuous strong DW into mid-March, then continuous UW at ≥ 200 m/sec.
1971	Weak UW and DW pulses into early April, then continuous UW at ≥ 200 m/sec.
1972	Mostly pulsed, weak DW into early April, then continuous UW at ≥ 200 m/sec.
1973	Strong DW through February, then pulsed UW at ≥ 200 m/sec.
1974	Pulsed UW and DW into mid-April, then strong UW at ≥ 200 m/sec.
1975	Pulsed UW and DW into mid-March, then strong UW at ≥ 200 m/sec.
1976	Weak UW through March, then strong UW at ≥ 200 m/sec.
1977	Little UW or DW through February, followed by strong UW and DW pulses in March and April, then continuous UW at ≥ 200 m/sec.
1978	Strong DW through February, weak UW through April, then strong UW at ≥ 200 m/sec.
1979	Pulsed UW and DW through April, then continuous UW at ≥ 200 m/sec.
1980	DW through February, strong UW in March, then weak UW mid-April on at ≈ 200 m/sec.
1981	Pulsed UW and DW through March, then weak UW at ≈ 200 m/sec.
1982	Pulsed UW and strong DW into mid-April, then weak UW at ≤ 200 m/sec.
1983	Strong DW into early April, then weak UW at ≈ 200 m/sec.
1984	Pulsed UW and DW through mid-March, then strong UW at ≥ 200 m/sec.
1985	Pulsed UW and DW through mid-March, then intermittent, moderate UW at ≤ 200 m/sec.
1986	Strong DW through March, then weak UW at ≤ 200 m/sec.

SOURCES: Bakun 1975, Mason & Bakun 1986, NOAA-NMFS unpubl.
NOTE: *DW*, downwelling; *UW*, upwelling.

APPENDIX 3.1

Estimates of reproductive performance by Farallon seabirds, 1984–86

Species	Chicks fledged per pair		
	1984	1985	1986
Ashy Storm-Petrel	0.7	0.7	0.9
Double-crested Cormorant	0.4	NA	NA
Brandt's Cormorant	1.5	2.7	1.3
Pelagic Cormorant	0.7	2.1	0.1
Western Gull	0.8	1.6	1.3
Common Murre	0.7	0.8	0.8
Pigeon Guillemot	1.1	1.2	< 0.1
Cassin's Auklet	0.7	0.7	0.7
Tufted Puffin	0.0	NA	NA

NOTE: *NA*, no data available; sample sizes of nests observed for each species approximately the same as stated in the chapter for that species.

Appendix 3.2

Coefficients of association between species in the Gulf of the Farallones, by census blocks, June 1985 and June 1986

Species and date	1	2	3	4	5	6	7	8	9
June *1985*									
Sooty Shearwater (1)	—								
Ashy Storm-Petrel (2)	1.00								
Brandt's Cormorant (3)	1.00	−1.00							
Pelagic Cormorant (4)	1.00	−1.00	1.00						
Western Gull (5)	−0.26	−1.00	0.60	0.64					
Common Murre (6)	*−0.47*	−1.00	0.94	0.69	0.39				
Pigeon Guillemot (7)	1.00	−1.00	1.00	1.00	0.64	*0.69*			
Cassin's Auklet (8)	0.36	0.02	1.00	1.00	0.43	*−0.01*	1.00		
Rhinoceros Auklet (9)	0.60	0.40	1.00	1.00	*−0.08*	0.43	1.00	0.76	
Tufted Puffin	1.00	−1.00	0.85	0.79	0.49	0.78	0.79	1.00	1.00
June **1986**									
Sooty Shearwater	—								
Ashy Storm-Petrel	−1.00								
Brandt's Cormorant	−0.17	−1.00							
Pelagic Cormorant	0.04	−1.00	0.12						
Western Gull	−1.00	−1.00	0.42	0.58					
Common Murre	*−0.01*	−1.00	0.44	0.58	0.05				
Pigeon Guillemot	0.06	−1.00	0.33	0.85	0.21	*0.76*			
Cassin's Auklet	0.19	−1.00	0.30	0.39	0.20	*0.13*	0.54		
Rhinoceros Auklet	0.26	−1.00	0.02	0.49	*0.42*	0.02	0.61	0.78	
Tufted Puffin	−1.00	−1.00	0.12	−1.00	0.45	0.03	−1.00	0.46	1.00

NOTE: Column numbers correspond to those following species. Italics indicate values statistically greater in 1986. 1 = complete overlap; −1 = no overlap; all SD's = .01.

Appendix 3.3

Between-year overlap in composition of cormorant diets, 1973–77

Species and year	Between-year overlaps			
	1974	*1975*	**1976**	*1977*
Double-crested Cormorant				
1974		*.999*	*.999*	*.999*
1975			*.999*	*.999*
1976				*.999*
Pelagic Cormorant				
1975			.637	*.982*
1976				.521
Brandt's Cormorant				
1973	*.914*	*.882*	.616	.561
1974		*.990*	.405	*.768*
1975			.402	*.853*
1976				.107

NOTE: Numerical data only; extent of overlap expressed by Morisita's index. 1 = complete overlap; 0 = none. Italics indicate similar diets.

Appendix 3.4

Estimated mean standard lengths of fish eaten by cormorants, 1974–77

Species and year	Standard length (mm); n			
	Pelagic (48)[a]	Double-crested (55)[a]	Brandt's (67)[a]	Difference, t test
Sebastes spp.				
1974	69.4 ± 9.6; 48	80.5 ± 26.2; 4	71.0 ± 7.2; 48	ns
1975	56.3 ± 17.3; 48	69.2 ± 34.8; 5	61.2 ± 14.3; 48	ns
1976	70.6 ± 28.8; 48[b]		107.8 ± 14.3; 8	$p < .05$
1977	59.8 ± 7.5; 48	62.0 ± 17.7; 22	65.4 ± 11.6; 48	ns
Citharichthys spp.				
1974		165.4; 1	182.7 ± 12.0; 2	ns
1975	202.8 ± 5.0; 2		134.2 ± 54.0; 48	ns
1976		no otoliths found		
1977		114.6 ± 81.5; 9[b]	188.9 ± 47.5; 6	$p < .05$
Engraulis mordax				
1974		103; 1	98.9 ± 17.4; 3	ns
1975		106.7 ± 13.4; 16	118.2 ± 9.4; 7	ns
1976			114.1 ± 12.9; 24	
Cottids[c]				
1975	60.0 ± 24.0; 48	103.0 ± 15.3; 35[d]	65.0 ± 23.8; 48	$p < .05$
1976	45.6 ± 11.5; 48[b]	97.0 ± 22.4; 6	143.0 ± 29.9; 21[d]	$p < .05$
1977	46.1 ± 9.2; 37	44.0 ± 5.6; 2	49.0 ± 10.5; 14	ns
Cymatogaster aggregata				
1975		74.4 ± 12.4; 48	78.4 ± 8.4; 5	ns
1976		66.4 ± 12.4; 48	70.4 ± 6.4; 48	ns
Microgadus proximus[e]				
1975		4.2 ± 1.1; 48	4.4 ± 0.9; 5	ns
1976		3.8 ± 1.1; 48	4.0 ± 0.8; 48	ns

NOTE: Means are shown ± SD; *ns*, not significant.

[a] Bill size (mm) in parentheses; see Figure 1.1.

[b] Smaller, $p < .05$.

[c] For comparison, the estimated mean total length of cottids eaten by guillemots during 1975, 1976, and 1977 was 50.1 ± 8.1 (n = 12), 68.8 ± 14.6 (n = 88), and 67.0 ± 15.0 (n = 50) mm, respectively.

[d] Larger, $p < .05$.

[e] These are otolith diameters; regression with fish size and weight not available.

Appendix 3.5

Between-year overlaps in composition of diets of cormorants and alcids, 1972–83

Species and years	Between-year and interspecies overlaps									
	Pelagic Cormorant		Brandt's Cormorant		Common Murre			Pigeon Guillemot		
	1975, 1977	**1976**	**1973**–*75, 1977*	**1976**	**1973, 1978, 1983**	**1976**	1974, *1975, 1977, 1979*–**82**	**1972, 1976**	**1978**	**1973**–*75, 1977, 1979*–**82**
Double-crested Cormorant, 1974–*77*										
Numerical	.022	.034	.015	.210	.004	.006	.011	.028	.030	.018
Weight	.003	.008	.016	.331	.000	.000	.000	.004	.004	.001
Pelagic Cormorant, *1975, 1977*										
Numerical		.594	*.892*	.318	.266	.513	*.896*	.517	.337	*.975*
Weight		*.761*	*.736*	.222	.052	.154	*.865*	.252	.159	*.980*
Pelagic Cormorant, **1976**										
Numerical			.254	.318	.070	.130	.246	*.763*	.566	.444
Weight			.021	.183	.029	.085	.475	.434	.307	.505
Brandt's Cormorant, **1973**–*75, 1977*										
Numerical				.453	.290	.540	*.923*	.272	.170	*.933*
Weight				.496	.116	.167	*.762*	.204	.293	*.788*
Brandt's Cormorant, **1976**										
Numerical					.438	.464	.298	.415	.507	.343
Weight					.052	.186	.296	.247	.673	.527
Common Murre, **1973, 1978, 1983**										
Numerical						.662	.386	.089	.086	.282
Weight						.426	.436	.016	.030	.056
Common Murre, **1976**										
Numerical							.560	.161	.093	.525
Weight							.390	.029	.027	.017
Common Murre, 1974, *1975, 1977, 1979*–**82**										
Numerical								.224	.131	*.963*
Weight								.134	.079	*.894*
Pigeon Guillemot, **1972, 1976**										
Numerical									*.738*	.411
Weight									.415	.239
Pigeon Guillemot, **1978**										
Numerical										.284
Weight										.208

NOTE: Extent of overlap expressed by Morisita's index. 1 = complete overlap; 0 = no overlap. Years grouped as in Tables 3.6 and 3.13; similar diets (> .7) indicated by italics.

Appendix 3.6

Between-year overlaps in Common Murre diet, 1973–83

	Between-year overlaps									
Year	**1974**	*1975*	**1976**	*1977*	**1978**	*1979*	1980	1981	**1982**	**1983**
1973	.284	.278	.697	.278	*.912*	.193	.370	.356	.492	*.802*
1974		*.963*	.555	*.962*	.397	*.951*	*.972*	*.957*	*.907*	.308
1975			.558	*.999*	.384	*.990*	*.983*	*.988*	*.930*	.285
1976				.549	*.796*	.483	.633	.607	.681	.544
1977					.381	*.991*	*.980*	*.996*	*.926*	.284
1978						.306	.497	.480	.623	*.893*
1979							*.954*	*.963*	*.885*	.211
1980								*.999*	*.981*	.413
1981									*.975*	.422
1982										.564

NOTE: Numerical data only; extent of overlap expressed by Morisita's index. 1 = complete overlap; 0 = none; particularly extensive overlap indicated by italics.

Appendix 3.7

Estimated mean total lengths of fish eaten by three alcids, 1974–82

Species and year	Total length (mm); *n*		
	Pigeon Guillemot (37)[a]	Common Murre (77)[a]	Tufted Puffin (49)[a]
Sebastes spp.			
1974	74.7 ± 8.1; 96	76.1 ± 0.3; 2,270	72.9 ± 10.0; 25
1975	60.0 ± 9.2; 37[b]	76.7 ± 5.5; 3,532	72.4 ± 6.8; 37
1976	64.1 ± 10.6; 65	66.9 ± 12.8; 706	69.4 ± 4.9; 13
1977	80.9 ± 8.8; 378	80.1 ± 7.9; 2,777	69.0 ± 14.6; 18[b]
1978	64.2 ± 7.7; 20	65.9 ± 5.8; 248	73.8 ± 20.2; 18[c]
1979	73.9 ± 7.7; 127[b]	88.2 ± 5.3; 856	
1980	73.2 ± 11.3; 121	75.2 ± 1.1; 1,187	67.5 ± 7.3; 24
1981	74.4 ± 5.8; 210	77.9 ± 2.1; 1,015	66.0 ± 9.3; 11[d]
1982	74.7 ± 5.8; 91	79.2 ± 1.2; 456	74.3 ± 6.8; 14
Citharichthys spp.			
1974	80.5 ± 20.4; 3	84.1 ± 19.5; 17	
1975		71.0 ± 12.5; 20	
1976	80.5 ± 16.8; 34	79.5 ± 18.8; 23	
1977		102.1 ± 23.1; 11	
1978	83.0 ± 14.6; 13	84.7 ± 5.7; 12	
1979	79.1 ± 12.8; 8	83.5 ± 14.6; 3	
1980	83.8 ± 7.3; 14	82.2 ± 10.9; 11	
1981	86.4 ± 14.3; 16	85.6 ± 10.5; 13	
1982	76.5 ± 9.5; 8	74.1 ± 7.5; 2	
Engraulis mordax			
1974		101.0 ± 17.6; 311[b]	125.6 ± 22.0; 30
1975		102.8 ± 15.2; 288[b]	132.0 ± 12.7; 25
1976		105.2 ± 20.1; 344[b]	140.8 ± 39.1; 31
1977		102.6 ± 19.1; 209[b]	126.6 ± 20.5; 14
1978		104.6 ± 19.8; 578[b]	130.6 ± 18.6; 10
1979		114.1 ± 15.9; 9	
1980		112.4 ± 16.1; 273[b]	122.2 ± 18.6; 10
1981		116.8 ± 11.5; 183	
1982		112.6 ± 11.6; 177[b]	136.9 ± 9.8; 2

NOTE: Means are shown ± SD.

[a] Bill size (mm) in parentheses; see Figure 1.1.

[b] Smaller; t test, $p < .05$.

[c] Larger; t test, $p < .05$.

[d] Smaller than for murre; t test, $p < .05$.

APPENDIX 3.8

Between-year overlaps in diets of Pigeon Guillemots, 1972–82

	Between-year overlaps									
Year	**1973**	1974	*1975*	**1976**	*1977*	**1978**	*1979*	1980	1981	**1982**
1972	.544	.204	.321	*.921*	.203	.696	.184	.383	.412	.395
1973		*.847*	*.943*	*.763*	*.931*	.433	*.848*	*.961*	*.960*	*.970*
1974			*.961*	.481	*.997*	.145	*.995*	*.947*	*.948*	*.937*
1975				.588	*.954*	.286	*.965*	*.989*	*.979*	*.978*
1976					.470	.591	.458	.643	.666	.656
1977						.140	*.993*	*.942*	*.948*	*.936*
1978							.188	.349	.333	.306
1979								*.936*	*.945*	*.936*
1980									*.993*	*.991*
1981										*.994*

NOTE: Numerical data only; extent of overlap expressed by Morisita's index. 1 = complete overlap; 0 = none; particularly extensive overlaps indicated by italics.

APPENDIX 4.1

Ashy Storm-Petrel burrow occupation and nesting success, 1972–83

Year	Sites checked	Site occupied		Eggs laid		Chicks hatched		Chicks fledged			
		n	%[a]	*n*	%[b]	*n*	%[c]	*n*	%[d]	%[c]	Per pair
1972	47	37	78.7%	36	97.3%	30	83.3%	23	63.9%	76.7%	0.64
1973	49	35	71.4	35	100.0	33	94.3	24	68.6	72.7	0.69
1974	47	39	83.0	36	92.3	28	77.8	22	61.1	78.6	0.61
1975	50	36	72.0	34	94.4	26	76.5	22	64.7	84.6	0.65
1976	64	38	59.4	37	97.4	32	86.5	30	81.1	93.8	0.81
1977	70	42	60.0	37	88.1	23	62.2	18	48.6	78.3	0.49
1978	78	36	46.2	30	83.3	24	80.0	20	66.7	83.3	0.67
1979	81	44	54.3	40	90.9	31	77.5	28	70.0	90.3	0.70
1980	87	39	44.8	32	82.1	25	78.1	22	68.8	88.0	0.67
1981	86	32	37.2	30	93.8	25	83.3	23	76.7	92.0	0.77
1982	104	31	29.8	28	90.3	21	75.0	21	75.0	100.0	0.75
1983	96	20	20.8	18	90.0	12	66.7	12	66.7	100.0	0.67
Mean ± SD[e]			54.3 ± 19.6		91.9 ± 5.3		77.9 ± 9.6		69.4 ± 6.0	87.3 ± 9.1	0.69 ± 0.06

[a] Percentage of sites checked.
[b] Percentage of sites occupied; i.e., adult observed in residence at least once.
[c] Percentage of eggs laid.
[d] Percentage of chicks hatched.
[e] Data from 1977 not included (see Methods).

APPENDIX 5.1

Brandt's Cormorant nest-site occupation and nesting success, 1971–83

Year	Sites followed (*n*)		Sites occupied (percent)		Sites with eggs (percent)		Sites with eggs hatched (*n*)		Sites with eggs hatched (percent)		Hatching success (percent)	
	I	II	I	II	I	II	I	II	I	II	I	II
1971	34		100%		77%		16		64%		43%	
1972	28	23	100	100%	89	96%	20	16	80	73%	58	41%
1973	27	23	96	100	70	83	15	9	79	47	61	33
1974	27	26	100	100	96	81	22	10	88	48	66	25
1975	31	21	100	100	94	95	20	14	69	79	51	43
1976	36	24	100	100	83	88	17	13	57	62	30	41
1977	37	26	100	96	81	68	24	10	80	59	57	47
1978	37	29	92	93	3	48	0	5	0	39	0	27
1979	37	30	89	83	76	64	25	11	100	69	71	46
1980	36	30	89	77	69	65	22	12	100	80	89	52
1981	36	32	89	69	78	64	25	10	100	71	81	64
1982	36	33	89	52	69	77	20	12	91	92	58	73
1983	36	33	56	55	15	22	0	0	0	0	0	0
Total/ average	438	330	92%	85%	69%	71%	226	122	70%	60%	51%	41%

NOTE: *I*, study Colony I; *II*, study Colony II (not watched in 1971).

Appendix 5.2

Sizes of clutches initiated by early-, middle-, and late-laying Brandt's Cormorants, Colony I, 1971–83

Year	Clutch size (mean ± SD; n)		
	Early	Middle	Late
1971	3.50 ± 1.05; 6	3.67 ± 0.49; 12	3.50 ± 0.71; 2
1972	3.56 ± 0.53; 9	3.11 ± 0.60; 9	2.83 ± 0.75; 6
1973	3.33 ± 0.82; 6	3.86 ± 0.38; 7	2.83 ± 0.98; 6
1974	2.71 ± 0.49; 7	3.44 ± 0.53; 9	3.29 ± 0.49; 7
1975	3.00 ± 0.63; 11	2.75 ± 0.71; 8	3.20 ± 0.45; 5
1976	3.00 ± 0.50; 9	2.60 ± 0.55; 5	2.30 ± 0.68; 10
1977	2.30 ± 0.68; 10	2.50 ± 0.93; 8	2.64 ± 0.51; 11
1978	No clutches completed		
1979	3.56 ± 0.53; 9	3.10 ± 0.32; 10	2.80 ± 1.17; 6
1980	3.29 ± 0.49; 7	3.13 ± 0.35; 8	3.29 ± 0.49; 7
1981	2.88 ± 0.35; 8	3.17 ± 0.75; 6	3.56 ± 0.53; 9
1982	3.57 ± 0.54; 7	4.00 ± 0.58; 7	3.50 ± 1.05; 6
1983	No clutches completed		
Total/ average[a]	3.12 ± 0.59; 89	3.24 ± 0.55; 89	3.00 ± 0.68; 75

[a] Early vs. middle, $t = 1.40$, DF = 176, $p > .05$; early vs. late, $t = 1.21$, DF = 162, $p > .05$; middle vs. late, $t = 2.49$, DF = 162, $p < .05$.

Appendix 5.3

Reproductive success of early-, middle-, and late-laying Brandt's Cormorants, Colony I, 1971–83

Year	Early			Middle			Late		
	Eggs laid	HS (%)	FS (%)	Eggs laid	HS (%)	FS (%)	Eggs laid	HS (%)	FS (%)
1971	21	14%	67%	46	56%	88%	10	70%	57%
1972	34	82	75	28	46	92	17	41	57
1973	20	50	70	27	81	50	12	47	38
1974	19	58	91	31	74	74	23	65	87
1975	34	68	70	23	35	75	12	35	100
1976	32	22	71	13	77	70	28	21	83
1977	23	39	89	20	60	92	29	66	95
1978	0			1	0		0		
1979	32	69	95	31	74	100	17	65	100
1980	23	96	77	25	92	78	23	78	72
1981	25	76	84	19	89	88	32	81	62
1982	27	63	71	29	55	56	12	62	38
1983	2	0		0			1	0	
Total	292	59%	79%	293	66%	79%	216	58%	72%
Fledglings/ pair ± SD	1.36 ± 1.09			1.63 ± 1.12			1.20 ± 1.07		

NOTE: *HS*, hatching success; *FS*, fledging success.

APPENDIX 5.4

Brandt's Cormorant reproductive success, 1971–83

Year	Sites fledging chicks (%) I	Sites fledging chicks (%) II	FS[a] (%) I	FS[a] (%) II	BS[b] (%) I	BS[b] (%) II	Fledglings per pair (mean ± SD; n) I	Fledglings per pair (mean ± SD; n) II
1971	80%		73%		31%		1.5 ± 1.6; 24	
1972	80	73%	75	86%	43	35%	1.5 ± 1.0; 25	1.4 ± 1.1; 22
1973	68	47	53	62	32	21	1.1 ± 0.9; 19	0.7 ± 0.8; 19
1974	87	43	82	70	53	18	1.7 ± 1.0; 23	0.7 ± 0.9; 21
1975	67	70	77	81	40	34	1.1 ± 0.9; 27	1.0 ± 0.9; 20
1976	57	48	76	39	23	16	0.6 ± 0.6; 30	0.5 ± 0.5; 21
1977	73	59	93	91	53	42	1.4 ± 1.1; 30	1.1 ± 1.1; 17
1978	0	31		60	0	16	0.0; 1	0.5 ± 0.9; 13
1979	100	69	99	92	70	42	2.6 ± 0.6; 25	1.4 ± 1.2; 16
1980	96	73	76	81	68	42	2.2 ± 0.8; 22	1.4 ± 1.1; 15
1981	96	64	72	75	58	48	1.9 ± 0.9; 25	1.5 ± 1.3; 14
1982	91	85	58	77	33	56	1.4 ± 0.7; 22	1.8 ± 1.0; 13
1983	0	0			0	0	0.0; 3	0.0; 4
Total/ average	69%	55%	76%	74%	39%	31%	1.5 ± 0.9; 276	1.0 ± 1.0; 195

NOTE: *I*, study Colony I; *II*, study Colony II (not watched in 1971).
[a] Fledging success: percentage of chicks fledged per number of eggs hatched.
[b] Breeding success: percentage of chicks fledged per number of eggs laid.

APPENDIX 5.5

First clutches and replacement clutch attempts, Brandt's Cormorants, Colonies I and II combined, 1971–83

Year	First clutches: Attempts (n)	First clutches: Eggs laid (n)	First clutches: Chicks fledged Per attempt	First clutches: Chicks fledged Per egg	Replacement attempts: Attempts (n)	Replacement attempts: Eggs laid (n)	Replacement attempts: Chicks fledged Per attempt	Replacement attempts: Chicks fledged Per egg
1971	24	82	1.2	0.4	12	35	0.5	0.2
1972	47	151	1.3	0.4	8	23	0.9	0.3
1973	38	124	0.9	0.3	2	5	0.0	0.0
1974	44	139	1.2	0.4	7	17	0.0	0.0
1975	47	132	1.0	0.4	2	5	1.0	0.4
1976	51	128	0.5	0.2	11	18	0.2	0.1
1977	47	115	1.2	0.5	3	7	1.3	0.6
1978	14	29	0.3	0.1	4	9	0.5	0.2
1979	41	128	1.9	0.6	5	15	1.8	0.6
1980	37	115	1.9	0.6	3	6	0.0	0.0
1981	39	110	1.5	0.5	6	17	1.5	0.5
1982	35	115	1.3	0.4	4	16	1.5	0.4
1983	7	10	0.0	0.0	3	6	0.0	0.0
Total	471	1,378	1.3	0.4	70	179	0.7	0.3

APPENDIX 5.6

Age at which Brandt's Cormorant chicks died, Colonies I and II combined, 1971–83

Year	Chicks hatched	Chicks dead (percent)		
		≤ 10 days	11–20 days	21–30 days
1971	50	8%	8%	6%
1972	86	7	9	2
1973	61	11	16	13
1974	75	11	5	1
1975	67	18	3	0
1976	51	14	20	10
1977	65	5	2	2
1978	10	30	10	0
1979	89	1	1	1
1980	89	4	13	2
1981	95	9	13	2
1982	82	11	16	7
1983	0			
Total	820	10%	10%	4%

Appendix 6.1

Variation in Pelagic Cormorant clutch sizes, 1971–83

Year	Clutches[a]	Clutch size (percent)					Clutch size
		One egg	Two eggs	Three eggs	Four eggs	Five eggs	(mean ± SD)
1971	11			27%	64%	9%	3.8 ± 0.6
1972	35			29	66	6	3.8 ± 0.5
1973	29		3%	48	48		3.4 ± 0.6
1974	24		4	75	21		3.2 ± 0.5
1975	26		4	31	62	4	3.6 ± 0.6
1976	19	5%	32	63			2.6 ± 0.6
1977	23		9	52	39		3.3 ± 0.6
1978	No eggs laid						
1979	27	4	7	39	50		3.4 ± 0.8
1980	22	5	9	64	23		3.0 ± 0.7
1981	20		5	25	70		3.6 ± 0.6
1982	23			39	61		3.6 ± 0.5
1983	No eggs laid						
Total	259	1.2%	6.2%	44.4%	46.7%	1.5%	3.4 ± 0.7

[a] Excludes replacement and incomplete clutches.

Appendix 6.2

Reproductive success of Pelagic Cormorants, 1971–83

Year	A: sites followed (*n*)	B: sites occupied		C: sites with eggs		D: sites with eggs hatched (%C)	E: sites chicks fledged (%C)	Eggs laid (*n*)	HS (%)	FS (%)	BS (%)
		n	%A	*n*	%B						
1971	16	16	100%	13	81%	54%	54%	44	39%	82%	32%
1972	41	40	98	35	88	74	54	132	44	36	16
1973	36	33	92	29	88	66	45	100	49	39	19
1974	28	27	96	24	89	46	29	76	32	42	13
1975	27	27	100	26	96	88	81	95	62	80	49
1976	25	25	100	19	76	No eggs hatched		49			
1977	33	30	91	23	77	83	83	76	74	88	64
1978	29	24	83	0	0	0					
1979	33	29	88	28	97	86	75	93	60	86	52
1980	34	30	88	22	73	32	0	67	19	0	0
1981	33	31	94	21	68	86	86	74	71	94	66
1982	33	27	82	24	89	71	8	85	51	7	4
1983	34	8	24	0	0	0					
Total	402	347	86%	264	76%	65%	48%	891	48%	61%	29%

NOTE: *HS*, hatching success; *FS*, fledging success; *BS*, breeding success.

APPENDIX 6.3

Reproductive success of early-, middle-, and late-breeding Pelagic Cormorants, 1971–83

Year	Early			Middle			Late		
	Eggs laid (n)	HS (%)	FS (%)	Eggs laid (n)	HS (%)	FS (%)	Eggs laid (n)	HS (%)	FS (%)
1971	22	36%	88%	14	57%	75%	8	13%	100%
1972	28	64	33	56	54	37	48	21	40
1973	29	69	50	48	54	31	23	13	33
1974	32	53	53	23	30	14	21	0	
1975	31	65	85	42	60	84	22	64	64
1976	7	0		26	0		16	0	
1977	28	68	89	34	79	81	14	71	100
1978	No eggs laid								
1979	28	64	89	54	63	85	7	57	75
1980	19	42	0	30	17	0	18	0	
1981	18	78	100	46	83	92	10	0	
1982	14	64	0	52	60	11	19	16	
1983	No eggs laid								
Total	256	59%	64%	425	54%	59%	206	22%	62%
Fledglings/pair ± SD		1.3 ± 1.3			1.1 ± 1.3			0.4 ± 0.9	

NOTE: *HS*, hatching success; *FS*, fledging success.

Appendix 7.1

Western Gull egg sizes in relation to laying sequence, 1972–83

	Laying sequence within clutch			
Dimension	First egg	Second egg	Third egg	F[a]
Length (mm)	71.0 ± 2.7	70.4 ± 2.8	68.9 ± 2.6	30.04
Range	58.9–80.1	63.0–76.9	62.4–76.9	
Breadth (mm)	49.2 ± 1.4	49.3 ± 1.5	47.8 ± 1.5	70.37
Range	42.9–53.2	46.1–53.3	44.5–51.1	
Volume (cm^3)[b]	90.2 ± 6.9	89.5 ± 7.5	82.5 ± 7.5	74.52
Range	60.0–104.9	69.0–108.1	65.9–100.7	
Size ranking[c]				
Largest	54%	43%	3%	
Middle	40%	50%	10%	
Smallest	6%	7%	87%	

NOTE: Means shown ± SD.
[a] In all cases, $p < .001$, DF = 2,649.
[b] Calculated from equation of Preston 1974.
[c] Percentage of first, second, and third eggs: $\chi^2 = 438.27$, DF = 4, $p < .001$.

Appendix 7.2

Western Gull egg volumes, 1972–83

Variation among years

Year	All eggs	n
1972	90.3 ± 6.5	45
1974	91.8 ± 7.3	117
1978	84.4 ± 8.0	96
1979	87.2 ± 7.9	273
1983	85.2 ± 7.6	120
F	17.39[a]	

Variation among years, by laying sequence

Year	First egg	Second egg	Third egg
1972	93.0 ± 3.4	92.8 ± 6.4	85.0 ± 6.0
1974	93.6 ± 6.6	94.3 ± 7.0	87.7 ± 6.4
1978	89.0 ± 6.1	86.0 ± 6.5	77.8 ± 6.8
1979	89.2 ± 7.5	89.2 ± 7.5	83.1 ± 6.9
1983	89.3 ± 5.7	87.2 ± 6.2	79.2 ± 7.0
F	4.26[b]	8.76[c]	12.80[c]

NOTE: Volumes expressed in cm^3; means shown ± SD.
[a] $p < .001$, DF = 4,649.
[b] $p < .025$, DF = 4,215.
[c] $p < .001$, DF = 4,215.

Appendix 7.3

Variation in Western Gull egg volumes, by laying sequence and date, 1974–83

Time of laying	Egg volume			Clutches
	First egg	Second egg	Third egg	
1974				
Early	95.1 ± 6.4	96.1 ± 7.2	90.0 ± 6.6	19
Middle	94.8 ± 6.0	94.9 ± 6.3	87.4 ± 5.1	14
Late	87.6 ± 6.0	88.7 ± 5.0	81.4 ± 5.5	6
$F_{2,36}$	3.61	2.87	4.71	
p	.04	.07	.02	
1978				
Early	87.5 ± 5.8	85.1 ± 6.3	76.8 ± 6.7	17
Middle	87.8 ± 7.1	87.8 ± 7.2	77.0 ± 8.8	10
Late	92.0 ± 3.1	87.2 ± 6.7	80.9 ± 3.6	6
$F_{2,30}$	1.39	0.60	0.80	
p	.26	.55	.46	
1979				
Early	92.2 ± 6.1	91.9 ± 6.8	86.3 ± 5.9	35
Middle	88.5 ± 6.4	88.0 ± 6.6	81.8 ± 6.0	20
Late	86.4 ± 8.6	87.2 ± 8.0	80.5 ± 7.2	36
$F_{2,88}$	5.65	3.94	7.41	
p	.01	.02	.001	
1983				
Early	90.9 ± 5.6	84.7 ± 5.1	77.7 ± 2.7	5
Middle	89.5 ± 6.1	89.3 ± 5.8	79.4 ± 6.9	15
Late	88.8 ± 5.7	86.2 ± 6.6	79.5 ± 7.8	20
$F_{2,37}$	0.26	1.54	0.14	
p	.77	.23	.87	

NOTE: Volumes in cm^3; means are shown ± SD.

Appendix 7.4

Western Gull breeding phenology and success, 1971–83

Year	Nests	Mean clutch-initiation date[a]	Mean clutch size[a]	HS (%)	FS (%)	BS (%)	Fledglings per nest[a]
1971	91	5/11	2.8	73%	82%	60%	1.7
1972	91	5/9	2.7	72	80	57	1.6
1973	97	5/8	2.8	83	67	55	1.5
1974	90	5/5	2.8	83	81	67	1.9
1975	93	5/3	2.8	78	78	61	1.7
1976	95	5/10	2.6	74	58	43	1.1
1977	83	5/10	2.7	73	71	52	1.4
1978	74	5/13	2.5	75	55	41	1.0
1979	95	5/14	2.6	69	81	55	1.4
1980	84	5/12	2.6	76	73	56	1.4
1981	45	5/13	2.5	80	52	42	1.1
1982	42	5/8	2.7	88	73	66	1.8
1983	44	5/8	2.6	72	34	25	0.6
Total	1,024	5/9	2.7	76%	71%	54%	1.4

NOTE: *HS*, hatching success; *FS*, fledging success; *BS*, breeding success.
[a] Varies significantly among years; ANOVA, $p < .05$; see Table 7.2.

APPENDIX 7.5

Reproductive success of chicks of early-, middle-, and late-laying Western Gulls, 1971–83

Year	Nests	Hatching success (percent)			Fledging success (percent)			Fledglings per nest[a] (mean)		
		E	M	L	E	M	L	E	M	L
1971	91	86%[b,c]	63%	52%	84%	80%	74%	2.1	1.5	1.0
1972	91	85[b,c]	68	59	79	79	83	1.8	1.5	1.3
1973	97	85	86	74	66	64	74	1.6	1.6	1.3
1974	90	84	84	81	82	74	84	2.0	1.8	1.9
1975	93	79	83	68	82	76	76	1.8	1.8	1.4
1976	95	77	76	64	72[b,c]	52	35	1.6	1.1	0.5
1977	83	82[c]	70	58	65	74	85	1.5	1.5	1.0
1978	74	77	83[d]	61	52	59	55	1.1	1.3	0.7
1979	95	70	67	65	83	81	82	1.6	1.4	1.3
1980	84	84[b]	65	74	84[b]	51	65	2.1	0.9	1.0
1981	45	93[c]	82	58	68[c]	42	33	1.7	0.9	0.4
1982	42	95[c]	79	73	89[b]	68	71	2.9	1.4	1.5
1983	44	86[c]	65	63	42	29	32	1.0	0.5	0.5
Total	1,024	83%[b,c]	75%[d]	67%	75%[b]	66%	70%	1.8[b,c]	1.4[d]	1.1

NOTE: *E*, early; *M*, middle; *L*, late.

[a] Values connected by lines are similar (SNK test, $p > .05$).

[b] Early vs. middle, $\chi^2 \geq 4.12$, $p < .05$.

[c] Early vs. late, $\chi^2 \geq 4.12$, $p < .05$.

[d] Middle vs. late, $\chi^2 \geq 4.12$, $p < .05$.

Appendix 8.1

Reproductive success of Common Murres, 1972–83

Year	First eggs					Replacement eggs					Fledglings per site (mean ± SD)
	n	HS (%)	Lost (%)	Addled (%)	FS (%)	*n*	HS (%)	Lost (%)	Addled (%)	FS (%)	
1972	116	80%	17%	3%	97%	1	0%	100%	0%	0%	0.8 ± 0.4
1973	135	82	18	0	93	9	67	33	0	50	0.8 ± 0.4
1974	173	88	9	3	97	3	100	0	0	100	0.9 ± 0.3
1975	137	93	4	3	98	1	100	0	0	100	0.9 ± 0.3
1976	163	90	9	2	94	3	33	66	0	0	0.8 ± 0.4
1977	123	80	15	4	96	11	64	27	9	57	0.8 ± 0.4
1978	123	81	15	4	85	2	50	50	0	100	0.7 ± 0.5
1979	135	83	8	9	100	5	60	20	20	66	0.8 ± 0.4
1980	144	88	9	3	98	5	60	40	0	100	0.9 ± 0.3
1981	146	89	8	3	94	9	100	0	0	78	0.9 ± 0.3
1982	70	91	7	1	95	2	100	0	0	100	0.9 ± 0.3
1983	41	32	63	5	15	3	0	100	0	0	0.1 ± 0.2
Total	1,506	85%	12.2%	3.2%	95%	54	67%	30%	4%	72%	0.8 ± 0.4

NOTE: *HS*, hatching success; *FS*, fledging success.

APPENDIX 9.1

Nest contents of study and control Pigeon Guillemot sites, 1972–74

Date	Plot,[a] *n*	Nest contents					χ^2; DF = 4	Eggs or chicks per pair[b]
		Two eggs	One egg	Two chicks	One chick	Empty		
1972								
13 July	C, 51	0	2	11	25	13	4.08	0.96
	S, 86	2	1	27	34	22		1.06
1 Aug.[c]	C, 51	0	1	2	25	23	5.42	0.59
	S, 86	0	1	3	26	56		0.38
1973								
24 June	C, 37	3	1	18	7	8	2.13	1.35
	S, 85	11	3	31	15	25		1.20
11 July	C, 51	0	3	23	13	12	1.31	1.22
	S, 85	1	4	35	27	18		1.21
1974								
5 July	C, 50	6	5	16	12	11	12.23	1.22
	S, 85	2	2	41	14	26		1.20
20 July	C, 53	2	2	14	21	14	12.00	1.04
	S, 85	0	2	36	16	31		1.06

NOTE: *C*, control area; *S*, study area.

[a] C vs. S differences, by respective dates, were statistically significant only for the two dates in 1974 ($p = .016$).

[b] None of the differences for chicks or eggs per pair for same-date comparisons were statistically significant ($\chi^2 = .062$, DF = 5).

[c] A number of chicks had fledged by this date.

APPENDIX 9.2

Reproductive success of Pigeon Guillemots, 1971–83

(Percent)

Year	Nests (n)	Nests with eggs	Clutch size[a]		Nests with chicks[b]	Eggs hatched	Fledglings per egg		Fledglings per nest[b]		
			One egg	Two eggs			Laid	Hatched	None	One	Two
1971	61	—	9.8%	90.2%	86.9%	83.6%	76.2%	91.2%	28.2%	21.1%	50.7%
1972	65	100%	26.2	73.8	86.2	81.0	49.1	60.6	23.1	66.2	10.8
1973	90	92	14.5	85.5	80.7	75.8	33.8	44.5	39.8	57.8	2.4
1974	71	82	19.0	81.0	87.9	81.9	66.7	81.4	24.1	31.0	44.8
1975	84	85	15.3	84.7	84.5	80.6	62.8	77.9	25.4	35.2	39.4
1976	104	83	41.9	58.1	82.6	35.7	35.4	44.7	44.2	55.8	0.0
1977	103	80	11.0	89.0	93.9[c]	85.4	65.8	77.0	6.1	26.8	67.1
1978	104	35	82.9	17.1	33.3[d]	30.2[e]	18.6	61.5	77.8	22.2	0.0
1979	101	80	9.9	90.1	96.3[c]	87.0	64.9	74.6	33.3	25.9	40.7
1980	101	80	17.3	82.7	86.4	77.0	42.6	55.3	22.2	74.1	3.7
1981	101	83	14.3	85.7	87.0	81.0	72.0	88.8	19.5	27.3	53.2
1982	107	81	23.0	77.0	83.9	78.6	33.8	43.0	44.8	50.6	4.6
1983	101	0							100.0	0.0	0.0
Total/ average[f]	1,193	80%	23.8%	76.2%	82.5%	73.2%	51.8%	66.7%	30.9%	42.0%	27.3%
± SD		16	20.6	20.6	16.1	19.0	18.6	17.4	19.3	17.8	26.0

NOTE: Includes replacement clutches only if at least one egg hatched; otherwise, data from the initial clutch were used; all table values are percentage except for first two columns.

[a] Frequency of one- vs. two-egg clutches statistically significant ($\chi^2 = 256.2$, DF = 11, $p < .001$), unless 1976 and 1978 (many one-egg clutches) and 1971, 1977, and 1979 (many two-egg clutches) are not included ($\chi^2 = 10.27$, DF = 7, $p > .1$).

[b] Of nests in which eggs were laid.

[c] Significantly higher than any percentage in this column < 85% ($t > 1.96$, $p < .05$).

[d] Significantly lower than all other percentages in this column (arcsin transformation; $t = 6.96$, $p < .01$).

[e] Significantly lower than all other percentages in this column ($t < 5.789$, $p < .01$); no other comparisons in this column are significant except that between 1979 and 1973 ($t = 2.008$, $p < .05$).

[f] Averaged 92 nests per year; means do not include 1983.

APPENDIX 9.3

Intervals between layings and hatchings, Pigeon Guillemot first and second eggs, 1971–79

	Days between layings[a]			Days between hatchings[b]		
Year	< 3	3–5	> 5	< 2	2–4	> 4
1971	25	32	—	17	23	4
1972	19	26	3	10	25	2
1973	31	38	2	4	36	9
1974	19	21	7	4	28	2
1975	23	28	1	6	34	4
1976	25	27	2	2	28	2
1977	33	38	2	17	29	2
1978	3	3	1	—	1	1
1979	26	24	4	12	21	—
Total	204	237	22	73	235	25

NOTE: Body of table shows number of nests in each category.
[a] No significant differences among years; $\chi^2 = 19.577$, DF = 16, $p > .2$.
[b] Significant difference among years (1978 excluded); $\chi^2 = 36.808$, DF = 14, $p < .001$; no significant difference if 1971, 1977, and 1979 are excluded.

APPENDIX 9.4

Growth parameters for single chicks and two-chick broods, Pigeon Guillemots, 1971–77

Parameter and year	Single chicks	*n*	Two-chick broods Chick 1	Two-chick broods Chick 2	*n*
Hatching weight (gm)					
1971	33.7 ± 5.6	8	35.2 ± 6.1	32.7 ± 8.0	12
1972	42.1 ± 4.6	37	42.9 ± 6.4	41.3 ± 6.6	3
1973	35.5 ± 14.3	31	34.2 ± 10.9	24.3 ± 6.8	6
1974	42.3 ± 8.4	11	38.7 ± 6.2	34.0 ± 8.0	20
1975	40.6 ± 7.8	11	42.6 ± 5.1	46.1 ± 8.0	15
1977	44.8 ± 11.4	8	40.8 ± 6.9	32.8 ± 7.5	16
Asymptotic weight (gm)					
1971	430.8 ± 78.5	8	476.8 ± 28.2	431.1 ± 12.1	12
1972	395.9 ± 49.1	37	389.9 ± 8.4	234.9 ± 21.4	3
1973	350.2 ± 73.1	31	343.8 ± 48.0	238.6 ± 15.8	6
1974	453.6 ± 52.8	11	422.8 ± 33.6	393.7 ± 10.3	20
1975	451.5 ± 50.4	11	420.7 ± 28.1	400.2 ± 11.9	15
1976	382.8 ± 88.8	4			
1977	447.8 ± 24.1	8	430.4 ± 31.0	427.1 ± 20.5	16
Growth constant (*K*)					
1971	0.179 ± 0.027	8	0.178 ± 0.016	0.177 ± 0.014	12
1972	0.164 ± 0.034	37	0.171 ± 0.014	0.186 ± 0.014	3
1973	0.193 ± 0.055	31	0.210 ± 0.051	0.278 ± 0.025	6
1974	0.178 ± 0.028	11	0.190 ± 0.018	0.191 ± 0.029	20
1975	0.166 ± 0.020	11	0.177 ± 0.014	0.155 ± 0.029	15
1976	0.172 ± 0.056	4			
1977	0.170 ± 0.016	8	0.185 ± 0.017	0.177 ± 0.031	16

NOTE: Means are shown ± SD. No data on hatching weight, and no two-chick broods at all, in 1976.

Appendix 10.1

Burrow occupation and reproductive success of Cassin's Auklets, first, replacement, and second clutches, 1970–83

		Nests with eggs																
		First clutch					Replacement clutch						Second clutch					
Year	Burrows	*n*	%	HS (%)	FS (%)	C/P	*n*	%[a]	%[b]	HS (%)	FS (%)	C/P	*n*	%[a]	%[c]	HS (%)	FS (%)	C/P
1970[d]	386	362	94%	72%	85%	0.6	35	9%	28%			0.7	2	1%	1%	0		0.0
1971[d]	352	316	90	77	83	0.6	56	15	51	61%	71%	0.4	95	27	42	61%	26%	0.2
1972	66	66	100	80	85	0.7	8	12	62	38	100	0.4	4	6	8	0		0.0
1973	68	65	96	88	91	0.8[e]	7	11	88	43		0.0	2	3	4	0		0.0
1974	66	64	97	80	86	0.7	13	27	100	85	36	0.3	11	17	23	46	20	0.1
1975	72	69	96	78	76	0.6	11	16	73	54	67	0.4	5	7	11	40	0	0.0
1976	75	73	97	77	86	0.7	11	15	65	54	17	0.1	20	27	41	20	0	0.0
1977	79	75	95	75	93	0.7	9	12	47	89	62	0.6	0		0			
1978	84	77	92	86	82	0.7	0						0		0			
1979	83	75	90	92	91	0.8[e]	6	15	100	83	80	0.3	31	41	46		12	0.1
1980	84	78	93	83	75	0.6	5	6	38	60		0.0	0		0			
1981	84	82	98	90	72	0.6	8	11	100	100	38	0.3	7	8	12	43	100	0.4
1982	84	82	98	80	73	0.6	16	30	100	81	77	0.4	10	12	17	0		0.0
1983	84	40	48	30	75	0.2[f]	2	5	7			0.0	0		0			
Total	1,667	1,524	91%	78%	83%	0.6	187	12%	53%	66%	58%	0.3	187	12%	18%	47%	24%	0.1

NOTE: *HS*, hatching success; *FS*, fledging success; *C/P*, fledglings per pair.

[a] Percent of breeding pairs.

[b] Percent of pairs not hatching first egg.

[c] Percent of pairs fledging first chick (first clutch and replacement eggs included).

[d] Data from Manuwal 1972.

[e] Statistically greater than all other values in the column (arcsin transformation, $p < .001$).

[f] Statistically lower than all other values in the column (arcsin transformation, $p < .001$).

Appendix 10.2

Mean fledging weights of Cassin's Auklets on Southeast Farallon Island, 1971–83 (gm)

Year	First brood		Second brood and relays	
	n	*w*	*n*	*w*
1971[a]	15	158.0 ± 13.3	10	115.0 ± 17.6
1972	35	152.6 ± 17.7	2	126.0 ± 43.8
1973	51	157.5 ± 14.1	0	
1974	44	156.8 ± 12.4	5	140.0 ± 18.5
1975	41	154.6 ± 9.4	4	137.2 ± 27.5
1976	48	147.7 ± 10.8[b]	0	
1977	52	152.8 ± 14.8	5	135.4 ± 14.8
1978	54	145.3 ± 15.9[c]	0	
1979	63	150.7 ± 13.9	6	120.5 ± 19.3
1980	49	146.9 ± 19.0[d]	0	
1981	58	154.5 ± 17.8	9	146.6 ± 14.8
1982	57	148.9 ± 15.2[b]	9	144.6 ± 14.3
1983	9	144.6 ± 15.3[e]	0	
Mean total		151.6 ± 14.6[e]		133.2 ± 21.3

NOTE: *n*, number; *w*, mean fledging weight. Means are shown ± SD.

[a] Figures calculated by Manuwal (unpubl. data).

[b] Statistically lower than any mean weight > 156.8 gm (two years); SNK test, $p < .05$.

[c] Statistically lower than any mean weight > 152.8 gm (five years); SNK test, $p < .05$.

[d] Statistically lower than any mean weight > 154.6 gm (three years); SNK test, $p < .05$.

[e] Statistically lower than any mean weight > 152.6 gm (six years); SNK test, $p < .05$.

Appendix 10.3

Cassin's Auklet chicks fledged per breeding pair, 1970–83

Year	Breeding pairs	First and replacement clutches		Second clutches		Total chicks per pair[a]
		Chicks	Chicks per pair[a]	Chicks	Chicks per pair[a,b]	
1970[c]	362	247	0.68	0	0.00	0.68
1971[c]	316	225 (2)	0.71–0.70	15 (9)	0.16–0.06	0.76–0.72
1972	66	48 (5)	0.73–0.65	0	0.00	0.73–0.65
1973	65	52 (2)	0.80–0.77	0	0.00	0.80–0.77
1974	64	48 (5)	0.75–0.67	1 (1)	0.09–0.00	0.76–0.67
1975	69	45 (0)	0.65–0.65	0	0.00	0.65–0.65
1976	73	49 (6)	0.67–0.59	0	0.00	0.67–0.59
1977	75	57 (10)	0.76–0.63	0	0.00	0.76–0.63
1978	77	54 (12)	0.70–0.54	0	0.00	0.70–0.54
1979	75	67 (6)	0.89–0.81	2 (2)	0.06–0.00	0.92–0.81
1980	78	49 (12)	0.63–0.47	0	0.00	0.63–0.47
1981	82	56 (6)	0.68–0.61	3 (0)	0.43–0.43	0.72–0.65
1982	82	58 (8)	0.71–0.61	0	0.00	0.71–0.61
1983	40	9 (2)	0.22–0.18	0	0.00	0.22–0.18
Total[d]	1,162	817 (76)	0.70–0.64	21 (12)	0.14–0.06	0.72–0.64

NOTE: Figures in parentheses are the number of fledglings weighing < 135 gm.

[a] Paired values indicate chicks per pair without and with exclusion of chicks weighing < 135 gm.

[b] On the basis of 94, 11, 33, and 7 pairs laying a second clutch in 1971, 1974, 1979, and 1981, respectively.

[c] Data from Manuwal 1972.

[d] Does not include 1970, for which data were incomplete.

Appendix 11.1

Estimated numbers of Rhinoceros Auklets and Tufted Puffins, and spring arrival of puffins at Southeast Farallon Island, 1971–86

Year	Breeding pairs		First puffin seen
	Auklet	Puffin	
1971			22 Mar.
1972	2	27	2 Apr.
1973	4	30	25 Mar.
1974	6	30	12 Mar.
1975	11	35	14 Mar.
1976	20	35	22 Mar.
1977	40	40	21 Mar.
1978	60	40	29 Mar.
1979	88	45	24 Mar.
1980	105	42	17 Mar.
1981	128	39	30 Mar.
1982	175	50	21 Mar.
1983	< 10	< 5	6 Apr.
1984	75	25	
1985	100	50	
1986	150	40	

NOTE: Blanks indicate lack of data.

Appendix 11.2

Tufted Puffin single-nest laying and hatching dates, and dates first adults were seen carrying fish, 1972–83

Year	Single nest		First adult carrying fish[b]
	Laying	Hatching[a]	
1972	28 Apr.	13 June	
1973	26 Apr.	7 June	
1974	26 Apr.	14 June	
1975	29 Apr.	10 June	
1976	2 May	17 June	
1977	11 May	24 June	
1978	8 May	18 June	
1979	1 May	15 June	15 June
1980	14 June	25 June	1 July
1981	24 May	6 July	25 June
1982	3 May	15 June	11 June
1983			30 June

NOTE: Blanks indicate lack of data.

[a] Estimated hatching dates based on incubation period of 43.9 ± 1.5 days ($n = 9$).

[b] These adults not from the one nest under observation (see text).

[c] In this year, one of the two original adults disappeared; the remaining bird took a new mate (see text).

Reference Matter

Glossary

Certain terms used regularly throughout this book are defined below. Terms are consistent with the definitions of Ainley, LeResche, and Sladen (1983).

Addled: spoiled; said of an egg that fails to hatch although incubated to term; often such eggs are in fact infertile.

Adult: a bird that wears plumage typical of mature individuals.

Breeder: a bird that is mature both physically and behaviorally (i.e., capable of breeding) and that does in fact breed.

Breeding: producing eggs, whether viable or not; a bird is classified as breeding, attempting breeding, or as a breeder only with evidence that it or its mate has laid at least one egg in its nest.

Breeding pair: two birds of opposite sex that mate, defend a territory, and attempt breeding.

Breeding season: the period extending from the time of initial territory occupation, through the egg stage (the period during which the majority of breeders are incubating eggs) and the chick stage (when the majority of breeders are tending chicks), until all breeders and chicks of that season have departed. In some species, such as the Ashy Storm-Petrel, Common Murre, or Cassin's Auklet, territory occupation may begin in the calendar year preceding the one in which eggs are laid. See also *Nonbreeding season*.

Breeding success: the number of chicks that fledged divided by the number of nests in which eggs were laid ($\times 100$).

Chick: a bird from the time of its complete emergence from the egg until it becomes independent of its parents. A chick is often defined by its plumage characteristics, i.e., it is a chick up to the time it acquires its complete juvenal plumage, at which point it is known as a dependent juvenile. For simplicity we use the terms "chick" and "nestling" synonymously; the bird then leaves the nest as a fledgling.

Clutch: the eggs laid in one nesting attempt. The original clutch is the first one laid by a breeding pair within a breeding season; the *relay* or *replacement clutch* is laid by a pair following an unsuccessful breeding attempt that season; a *second clutch* is sometimes laid after chicks from the first clutch are successfully fledged. See also *Incomplete clutch*.

Clutch size: the number of eggs laid in one nesting attempt (see *Incomplete clutch*).

Colony: a discrete and contiguous group of nesting birds of one species. The number of nests in a colony at the Farallones can vary from a few to many thousands. Since nest-to-nest or next-neighbor distances vary among species, the word "contiguous" has a different spatial connotation for each.

Community: the assemblage of interacting, ecologically interrelated species within a habitat; for the Farallones, the assemblage of all seabird populations considered together, including breeding and nonbreeding segments (see *Population*).

Crêche: two or more chicks from different nests that group closely together for protection from predators or for warmth.

Date of hatching: the first day that a chick is observed completely free of its egg.

Date of laying: the first day that an egg is observed at a nest, plus or minus the interval between nest checks.

Failed breeder: an adult that has occupied a nest site and helped in the production of eggs but then loses the eggs or chicks prematurely; i.e., eggs are laid but no fledglings are produced.

Fledging: the stage in nestling development when the bird departs the nest and becomes independent of its parents. From a practical standpoint in field research, fledging or fledgling is defined differently than this for species such as the gull, murre, or cormorants. Chicks of these species leave the nest at a relatively young age but are still dependent on their parents for food. At some point, we can no longer follow these chicks, and we consider them to have fledged.

Fledging period: the number of elapsed days between hatching and fledging. See also *Nestling period*.

Fledging success: the number of chicks that fledged divided by the number of eggs that hatched (×100).

Fledgling: a chick that leaves the nest and swims or flies off to sea, not to return until at least the following year.

Floater: an adult that does not attempt breeding because not enough nesting sites are available. The *floating population* is a special segment of the nonbreeding population (see *Population*).

Food web: the system of interrelated predator–prey relationships in an ecosystem.

Hatching: the complete emergence of the chick from its egg.

Hatching success: the number of eggs hatched divided by the number of eggs laid (×100).

Incomplete clutch: a clutch that has lost one or more of its complement of eggs. On occasion, eggs disappear from nest sites of species that lay multiple-egg clutches within a few days of clutch initiation but within the range of time that more eggs could be added if the female has not finished laying. These clutches are considered incomplete, and are not included in clutch-size calculations.

Incubation period: the number of elapsed days from the date an egg is laid, day zero, to the day on which the chick is free of the egg.

Juvenile: subadult; a post-fledging individual that has not matured physiologically to become an adult.

Neritic waters: waters overlying the continental shelf.

Nestling: see *Chick*.

Nestling period: the number of elapsed days from the date a chick has emerged from its egg, day zero, to the day on which it is observed absent (and is therefore a fledgling), given that it was last observed healthy, there was no evidence of predation, and it was at a morphological stage of development where fledging was expected.

Niche: the ecological role of an organism in an ecosystem.

Niche dimension: refers to the conceptual model of a niche as a "volume," with each axis or dimension being a single characteristic (e.g., foraging niche is composed of the habitat, depth, temporal, and prey-size dimensions).

Niche segregation: the condition said to occur when characteristics of a species separate it ecologically from other species along a specific niche dimension.

Nonbreeder: an adult that has not been observed to breed during the current breeding season, and may or may not have occupied a nest site. This does not include failed breeders. See also *Floater*.

Nonbreeding season: the period during which individuals of a species do not defend territories on the nesting grounds; at the Farallones the nonbreeding season generally occurs during fall and winter (but see *Breeding season*).

Population: all the interacting individuals of a single species that are associated with one breeding locality, e.g., the Farallones; a population is composed of both breeding and nonbreeding segments.

Season: see *Breeding season* and *Nonbreeding season*.

Territory: a defended space with geographic reference.

Trophic: pertaining to food or the food web.

Literature Cited

Throughout the text, the form "et al." has been used only in references to works of four or more co-authors. Such works appear below listed under the name of the principal author, secondarily alphabetized by co-authors' names, and not necessarily in chronological order.

Abbott, D. P., and R. Albee. 1967. Summary of thermal conditions and phytoplankton volumes measured in Monterey Bay, California 1961–1966. Calif. Coop. Ocean. Fish. Invest., Rept. 11: 155–56.

Ainley, D. G. 1975. The development of reproductive maturity in Adélie Penguins, pp. 139–57. *In* B. Stonehouse (ed.), *The Biology of Penguins*. Macmillan, London.

———. 1976. The occurrence of seabirds in the coastal region of California. West. Birds 7: 33–68.

———. 1977. Feeding methods in seabirds: A comparison of polar and tropical nesting communities in the eastern Pacific Ocean, pp. 669–85. *In* G. A. Llano (ed.), *Adaptations within Antarctic Ecosystems*. Gulf Publ., Houston.

———. 1980. Geographic variation in Leach's Storm-Petrel. Auk 97: 837–53.

———. 1983. Further notes on variation in Leach's Storm-Petrel. Auk 100: 230–33.

Ainley, D. G., D. W. Anderson, and P. R. Kelly. 1981. Feeding ecology of marine cormorants in southwestern North America. Condor 83: 120–31.

Ainley, D. G., and R. J. Boekelheide. 1983. An ecological comparison of oceanic seabird communities of the South Pacific Ocean. Studies Avian Biol. 8: 2–23.

Ainley, D. G., H. R. Carter, D. W. Anderson, K. T. Briggs, M. C. Coulter, F. & J. B. Cruz, C. A. Valle, G. Merlen, S. I. Fefer, S. A. Hatch, E. A. & R. W. Schreiber, and N. G. Smith. 1986. "El Niño"–Southern Oscillation 1982–83 and effects on Pacific Ocean marine bird populations, pp. 1747–58. *In* H. Ouellet (ed.), Proc. XIX Int. Ornithol. Congr., Natl. Mus. Nat. Sci., Ottawa.

Ainley, D. G., C. R. Grau, T. E. Roudybush, S. H. Morrell, and J. M. Utts. 1981. Petroleum ingestion reduces reproduction in Cassin's Auklets. Mar. Pollut. Bull. 12: 314–17.

Ainley, D. G., R. E. LeResche, and W. J. L. Sladen. 1983. *Breeding Biology of the Adélie Penguin*. Univ. Calif. Press, Berkeley.

Ainley, D. G., and T. J. Lewis. 1974. The history of Farallon Island marine bird populations 1843–1972. Condor 76: 432–46.

Ainley, D. G., T. J. Lewis, and S. Morrell. 1976. Molt in Leach's and Ashy storm-petrels. Wilson Bull. 88: 76–95.

Ainley, D. G., S. Morrell, and T. J. Lewis. 1974. Patterns in the life histories of storm-petrels on the Farallon Islands. Living Bird 13: 295–312.

Ainley, D. G., E. F. O'Connor, and R. J. Boekelheide. 1984. The marine ecology of birds in the Ross Sea, Antarctica. Am. Ornithol. Union, Monogr. 32.

Ainley, D. G., and G. A. Sanger. 1979. Trophic relations of seabirds in the northeastern Pacific Ocean and Bering Sea, pp. 95–122. *In* J. C. Bartonek and D. N. Nettleship (eds.), *Conservation of Marine Birds in Northern North America*. U.S. Dept. Inter., Wildl. Res. Rept. 11.

Allan, R. G. 1962. The Madeiran Storm-Petrel *Oceanodroma castro*. Ibis 103b: 274–95.

American Ornithologists' Union. 1957. *Checklist of North American Birds*. 5th Ed. Lawrence, Kansas.

———. 1983. *Checklist of North American Birds*. 6th Ed. Lawrence, Kansas.

Anderson, D. W., and I. T. Anderson. 1976. Distribution and status of Brown Pelicans in the California Current. Am. Birds 30: 3–12.

Anderson, D. W., F. Gress, and K. Mais. 1982. Brown Pelicans: Influence of food supply on reproduction. Oikos 39: 23–31.

Asbirk, S. 1979. The adaptive significance of the reproductive pattern in the Black Guillemot, *Cepphus grylle*. Vidensk. Medd. Naturh. Foren. 141: 29–80.

Ashmole, N. P. 1963. The regulation of numbers of tropical oceanic birds. Ibis 103b: 458–73.

———. 1968. Body size, prey size, and ecological segregation in five sympatric tropical terns (Aves: Laridae). Syst. Zool. 17: 292–304.

———. 1971. Seabird ecology and the marine environment, pp. 223–86. *In* D. S. Farner, J. R. King, and K. C. Parkes (eds.), *Avian Biology*, Vol. I. Academic Press, New York.

Ashmole, N. P., and M. J. Ashmole. 1967. Comparative feeding ecology of seabirds of a tropical oceanic island. Peabody Mus. Nat. Hist., Yale Univ., Bull. 24: 1–131.

Baird, P. A., S. A. Hatch, R. D. Jones, D. R. Nysewander, and M. R. Petersen. 1983. The breeding biology and feeding ecology of marine birds in the Gulf of Alaska. U.S. Fish Wildl. Serv. (Anchorage), Outer Continental Shelf Environmental Assessment Program, Final Rept.

Bakun, A. 1973. Coastal upwelling indices, West Coast of North America, 1946–71. NOAA Tech. Rep. NMFS SSRF-671.

———. 1975. Daily and weekly upwelling indices, West Coast of North America, 1967–73. NOAA Tech. Rep. NMFS SSRF-693.

Bakun, A., and R. H. Parrish. 1982. Turbulence, transport and pelagic fish in the California and Peru current systems. Calif. Coop. Ocean. Fish. Invest., Rept. 23: 99–112.

Balfour, E., A. Anderson, and G. M. Dunnet. 1967. Orkney cormorants—Their breeding distribution and dispersal. Scott. Birds 4: 481–93.

Baltz, D. M., and G. V. Morejohn. 1977. Food habits and niche overlap of seabirds wintering on Monterey Bay, California. Auk 94: 526–43.

Baptista, L. F. 1966. Albinistic feathers in storm-petrels (Hydrobatidae). Condor 68: 512–14.

Bartholomew, G. A., Jr. 1942. The fishing activities of Double-crested Cormorants on San Francisco Bay. Condor 44: 13–21.

———. 1943. The daily movements of cormorants on San Francisco Bay. Condor 45: 3–18.

Batchelor, G. R. 1978. Population structure and apparent predation on the Red-breasted Hurper, *Tilapia rendalli*, in a northern Transvaal impoundment. S. Afr. J. Wildl. Res. 8: 39–41.

Beck, J. R., and D. W. Brown. 1971. The breeding biology of the Black-backed Storm-Petrel *Fregetta tropica*. Ibis 113: 73–90.

———. 1972. The biology of Wilson's Storm-Petrel, *Oceanites oceanicus* (Kuhl), at Signy Island, South Orkney Islands. Brit. Antarc. Surv., Sci. Rept. No. 69.

Bédard, J. 1967. *Ecological Segregation among Plankton-Feeding Alcidae (Aethia and Cyclorrhynchus)*. Unpubl. Ph.D. thesis., Univ. Brit. Columbia, Victoria.

———. 1969a. Feeding of the Least, Crested, and Parakeet auklets around St. Lawrence Island, Alaska. Can. J. Zool. 47: 1025–50.

———. 1969b. Adaptive radiation in Alcidae. Ibis 111: 189–98.

———. 1969c. The nesting of the Crested, Least, and Parakeet auklets on St. Lawrence Island, Alaska. Condor 71: 386–98.

———. 1976. Coexistence, coevolution, and convergent evolution in seabirds: A comment. Ecology 57: 177–84.

———. 1985. Evolution and characteristics of the Atlantic Alcidae, pp. 1–53. *In* D. N. Nettleship and T. R. Birkhead (eds.), *The Atlantic Alcidae*. Academic Press, Orlando.

Beer, C. G. 1963. Incubation and nest-building behavior of the Black-headed Gulls. IV. Nest-building in the laying and incubation periods. Behaviour 21: 155–76.

Bellrose, C. A. 1983. *The Breeding Biology and Ecology of a Small Mainland Colony of Western Gulls*. Unpubl. Master's thesis, Calif. State Univ., San Jose.

Belopol'skii, L. O. 1961. *The Ecology of Sea Colony Birds of the Barents Sea*. Israel Prog. Sci. Transl., Jerusalem.

Bent, A. C. 1922. Life histories of North American petrels and pelicans and their allies. U.S. Natl. Mus., Bull. 121.

Benz, C., and R. Garrett. 1978. Colony development and nesting behavior of Double-crested and Pelagic cormorants. Pacific Seabird Group, Bull. 5: 82. (Abstract.)

Bergman, G. 1971. Gryllteisten *Cepphus grylle* in einem Randgebiet: Nahrung, Brutresultat, Tagesrhythmus und Ansiedlung. Commentat. Biol. Soc. Sci. Fenn. 42: 1–26.

———. 1978. Av näringsbrist förorsakade stöningar i tejstens (tobisgrisslans) *Cepphus grylle* häckning. Mem. Soc. Fauna Flora Fenn. 54: 31–32.
Bernal, P. A. 1981. A review of low-frequency response of the pelagic ecosystem in the California Current. Calif. Coop. Ocean. Fish. Invest., Rept. 22: 49–62.
Berry, H. H. 1976. Physiological and behavioral ecology of the Cape Cormorant, *Phalacrocorax capensis*. Madoqua 9: 5–55.
Berry, H. H., R. P. Millar, and G. W. Louw. 1979. Environmental cues influencing the breeding biology and circulating levels of various hormones and triglycerides in the Cape Cormorant. Comp. Biochem. Physiol. 62: 879–84.
Bierman, W. H., and K. H. Voous. 1950. Birds observed and collected during the whaling expeditions of the "William Barendsz" in the Antarctic, 1946–47 and 1947–48. Ardea 37: 1–123.
Birkhead, T. R. 1974. The movements and survival of British guillemots. Bird Study 21: 241–54.
———. 1976. *Breeding Biology and Survival of Guillemots (Uria aalge)*. Unpubl. Ph.D. thesis, Oxford Univ., Oxford.
———. 1977a. Adaptive significance of the nestling period of guillemots (*Uria aalge*). Ibis 119: 544–49.
———. 1977b. The effect of habitat and density on breeding success in the Common Guillemot (*Uria aalge*). J. Anim. Ecol. 46: 751–64.
———. 1978a. Attendance patterns of guillemots (*Uria aalge*) at breeding colonies at Skomer Island. Ibis 120: 219–29.
———. 1978b. Behavioral adaptations to high density nesting in the Common Guillemot (*Uria aalge*). Anim. Behav. 26: 321–31.
———. 1985. Coloniality and social behaviour in the Atlantic Alcidae, pp. 353–83. *In* D. N. Nettleship and T. R. Birkhead (eds.), *The Atlantic Alcidae*. Academic Press, Orlando.
Birkhead, T. R., and R. W. Furness. 1985. The regulation of seabird populations, pp. 145–67. *In* R. M. Sibly and R. H. Smith (eds.), *Behavioural Ecology*. Blackwell, Oxford.
Birkhead, T. R., and P. J. Hudson. 1977. Population parameters for the Common Guillemot (*Uria aalge*). Ornis Scand. 8: 145–54.
Birkhead, T. R., and D. N. Nettleship. 1980. Census methods for murres, *Uria* species, a unified approach. Can. Wildl. Serv., Occ. Pap. No. 43.
———. 1981. Reproductive biology of Thick-billed Murres (*Uria lomvia*): An intercolony comparison. Auk 98: 258–69.
Birkhead, T. R., and A. M. Taylor. 1977. Moult of the guillemot *Uria aalge*. Ibis 119: 80–85.
Birt, V. L., T. P. Birt, D. Goulet, D. K. Cairns, and W. A. Montevecchi. 1987. Ashmole's halo: Direct evidence for prey depletion by a seabird. Mar. Ecol. Prog. Ser. 40: 205–8.
Boekelheide, R. J., and D. G. Ainley. 1989. Age, resource availability, and breeding effort in Brandt's Cormorant. Auk 106: 389–401.
Boersma, P. D. 1978. Breeding patterns of Galápagos Penguins as an indicator of oceanographic conditions. Science 200: 1481–83.
Boersma, P. D., M. K. Nerini, and E. S. Wheelwright. 1980. The breed-

ing biology of the Fork-tailed Storm-Petrel (*Oceanodroma furcata*). Auk 97: 268–82.

Boersma, P. D., and N. T. Wheelwright. 1979. The costs of egg neglect in the Procellariiformes: Reproductive adaptations in the Fork-tailed Storm-Petrel. Condor 81: 157–65.

Bolin, R. L., and D. P. Abbott. 1963. Studies on the marine climate and phytoplankton of the central coastal area of California, 1954–60. Calif. Coop. Ocean. Fish. Invest., Rept. 9: 23–45.

Bowmaker, A. P. 1963. Cormorant predation on two central African lakes. Ostrich 34: 2–26.

Bowman, R. I. 1961. Late spring observations on birds of South Farallon Island, California. Condor 63: 410–16.

Boyce, M. S., and C. M. Perrins. 1987. Optimizing Great Tit clutch size in a fluctuating environment. Ecology 68: 142–53.

Bradstreet, M. S. W. 1979. Thick-billed Murres and Black Guillemots in the Barrow Strait area, N.W.T., during spring: Distribution and habitat use. Can. J. Zool. 57: 1789–1802.

———. 1980. Thick-billed Murres and Black Guillemots in the Barrow Strait area, N.W.T., during spring: Diets and food availability along ice edges. Can. J. Zool. 58: 2120–40.

———. 1982. Pelagic feeding ecology of Dovekies *Alle alle* in Lancaster Sound and western Baffin Bay, Canada. Arctic 35: 126–40.

Bradstreet, M. S. W., and R. G. B. Brown. 1985. Feeding ecology of the Atlantic Alcidae, pp. 264–318. *In* D. N. Nettleship and T. R. Birkhead (eds.), *The Atlantic Alcidae*. Academic Press, Orlando.

Briggs, K. T. 1977. *Social Dominance in Young Western Gulls: Its Importance in Survival and Dispersal*. Unpubl. Ph.D. thesis, Univ. California, Santa Cruz.

Briggs, K. T., and E. W. Chu. 1986. Sooty Shearwaters off California: Distribution, abundance, and habitat use. Condor 88: 355–64.

Briggs, K. T., W. B. Tyler, D. B. Lewis, and D. R. Carlson. 1987. Bird communities at sea off California: 1975–1983. Studies Avian Biol. 11.

Brinton, E. 1962. The distribution of Pacific euphausiids. Bull. Scripps Inst. Oceanogr. 8(2): 51–270.

———. 1981. Euphausiid distributions in the California Current during the warm-water spring of 1977–78, in the context of a 1949–66 time series. Calif. Coop. Ocean. Fish. Invest., Rept. 22: 135–54.

Brown, C. R. 1986. Cliff Swallow colonies as information centers. Science 234: 83–85.

Brown, J. H., and M. A. Bowers. 1984. Patterns and processes in three guilds of terrestrial vertebrates, pp. 282–96. *In* D. R. Strong, D. Simberloff, L. G. Abele, and A. B. Thistle (eds.), *Ecological Communities: Conceptual Issues and the Evidence*. Princeton Univ. Press, Princeton.

Brown, J. L. 1969. Territorial behavior and population regulation in birds. Wilson Bull. 81: 293–329.

Brown, R. G. B. 1979. Seabirds of the Senegal upwelling and adjacent waters. Ibis 121: 283–92.

———. 1985. The Atlantic Alcidae at sea, pp. 384–427. *In* D. N. Nettleship and T. R. Birkhead (eds.), *The Atlantic Alcidae*. Academic Press, Orlando.

Brown, R. G. B., S. P. Barker, D. E. Gaskin, and M. R. Sandeman. 1981. The foods of Great and Sooty shearwaters *Puffinus gravis* and *P. griseus* in eastern Canadian waters. Ibis 123: 19–30.

Bryant, W. E. 1888. Birds and eggs from the Farallon Islands. Proc. Calif. Acad. Sci., Ser. 2, 1: 25–50.

Burger, J. 1977. Role of visibility in nesting behavior of *Larus* gulls. J. Comp. Physiol. Psychol. 91: 1347–58.

Cairns, D. K. 1979. Censusing hole-nesting auks by visual counts. Bird-banding 50: 358–64.

———. 1980. Nesting density, habitat structure and human disturbance as factors in Black Guillemot reproduction. Wilson Bull. 92: 352–61.

———. 1981. Breeding, feeding and chick growth of the Black Guillemot *Cepphus grylle* in southern Quebec, Canada. Can. Field-Nat. 95: 312–18.

———. 1984. *The Foraging Ecology of the Black Guillemot (Cepphus grylle)*. Unpubl. Ph.D. thesis, Carleton Univ., Ottawa.

California Dept. Fish and Game. 1982. Review of some California fisheries for 1980 and 1981. Calif. Coop. Ocean. Fish. Invest., Rept. 23: 8–14.

———. 1983. Review of some California fisheries for 1982. Calif. Coop. Ocean. Fish. Invest., Rept. 24: 6–10.

———. 1984. Review of some California fisheries for 1983. Calif. Coop. Ocean. Fish. Invest., Rept. 25: 7–15.

———. 1985. Review of some California fisheries for 1984. Calif. Coop. Ocean. Fish. Invest., Rept. 26: 9–16.

———. 1986. Review of some California fisheries for 1985. Calif. Coop. Ocean. Fish. Invest., Rept. 27: 7–15.

Carter, H. R., K. A. Hobson, and S. G. Sealy. 1984. Colony-site selection by Pelagic Cormorants (*Phalacrocorax pelagicus*) in Barkley Sound, British Columbia. Colonial Waterbirds 7: 25–34.

Chelton, D. B. 1981. Interannual variability in the California Current—Physical factors. Calif. Coop. Ocean. Fish. Invest., Rept. 22: 34–48.

———. 1982. Large-scale response of the California Current to forcing by the wind stress curl. Calif. Coop. Ocean. Fish. Invest., Rept. 23: 130–48.

Chelton, D. B., P. A. Bernal, and J. A. McGowan. 1982. Large-scale interannual physical and biological interaction in the California Current. J. Mar. Res. 40: 1095–1125.

Chesson, P. L. 1986. Environmental variation and the coexistence of species, pp. 240–56. *In* J. Diamond and T. R. Case (eds.), *Community Ecology*. Harper and Row, New York.

Chu, E. W. 1984. Sooty Shearwaters off California: Diet and energy gain, pp. 64–71. *In* D. N. Nettleship, G. A. Sanger, and P. F. Springer (eds.), *Marine Birds: Their Feeding Ecology and Commercial Fisheries Relationships*. Minister Supply Serv. Can., Ottawa.

Clark, G. A., Jr. 1969. Spread-wing postures in Pelecaniformes, Ciconiiformes, and Falconiformes. Auk 86: 136–39.

Cline, D. R., and E. Dornfeld. 1968. The Agassiz Refuge cormorant colony. Loon 40: 68–72.

Cody, M. L. 1968. On the methods of resource division in grassland bird communities. Am. Nat. 102: 107–47.

———. 1973. Coexistence, coevolution and convergent evolution in seabird communities. Ecology 54: 31–44.

———. 1974. *Competition and the Structure of Bird Communities*. Monogr. Popul. Biol. No. 7. Princeton Univ. Press, Princeton.

Cole, L. C. 1949. The measurement of interspecific association. Ecology 30: 411–24.

Connor, E. F., and D. Simberloff. 1984. Neutral models of species' co-occurrence patterns, pp. 282–96. *In* D. R. Strong, D. Simberloff, L. G. Abele, and A. B. Thistle (eds.), *Ecological Communities: Conceptual Issues and the Evidence*. Princeton Univ. Press, Princeton.

Cooper, J. 1986. Diving patterns of cormorants Phalacrocoracidae. Ibis 128: 562–70.

Coulson, J. C. 1961. Movements and seasonal variation in mortality of shags and cormorants ringed on the Farne Islands, Northumberland. Brit. Birds 54: 225–35.

———. 1966. The influence of the pair-bond of the Kittiwake Gull, *Rissa tridactyla*. J. Anim. Ecol. 35: 269–79.

———. 1968. Differences in the quality of birds nesting in the centre and on the edges of a colony. Nature 217: 478–79.

———. 1973. Comparison of the breeding biology of the shag and kittiwake. Ibis 115: 474–75.

Coulson, J. C., and M. G. Brazendale. 1968. Movements of cormorants ringed in the British Isles and evidence of colony-specific dispersal. Brit. Birds 61: 1–21.

Coulson, J. C., N. Duncan, and C. Thomas. 1982. Changes in the breeding biology of the Herring Gull (*Larus argentatus*) induced by reduction in the size and density of the colony. J. Anim. Ecol. 51: 739–56.

Coulson, J. C., and J. M. Porter. 1985. Reproductive success of the kittiwake *Rissa tridactyla*: The roles of clutch size, chick growth and parental quality. Ibis 127: 450–66.

Coulson, J. C., G. R. Potts, and J. Horobin. 1969. Variation in the eggs of the shag (*Phalacrocorax aristotelis*). Auk 86: 232–45.

Coulson, J. C., and C. S. Thomas. 1985. Changes in the biology of the kittiwake *Rissa tridactyla*: A 31-year study of a breeding colony. J. Anim. Behav. 54: 9–26.

Coulson, J. C., and E. White. 1957. Mortality rates of the shag estimated by two independent methods. Bird Study 4: 166–71.

Coulter, M. C. 1973. *Breeding Biology of the Western Gull Larus occidentalis*. Unpubl. Master's thesis, Oxford Univ., Oxford.

———. 1975. Post-breeding movements and mortality in the Western Gull *Larus occidentalis*. Condor 77: 243–49.

———. 1977. *Growth, Mortality and Third Chick Disadvantage in the Western Gull, Larus occidentalis*. Unpubl. Ph.D. thesis, Univ. Pennsylvania, Philadelphia.

Coulter, M. C., and J. Higbee. Unpubl. ms. Food budget of chicks of Cassin's Auklet, *Ptychoramphus aleuticus*, on the Farallon Islands, California. PRBO, Stinson Beach, California.

Crawford, R. J. M., and P. A. Shelton. 1978. Pelagic fish and seabird interrelationships off the coast of Southwest and South Africa. Biol. Conserv. 14: 85–109.

Crossin, R. S. 1974. The storm-petrels (Hydrobatidae), pp. 154–205. *In* W. B. King (ed.), *Pelagic Studies of Sea Birds in the Central and Eastern Pacific Ocean*. Smithson. Contrib. Zool. 158.

Croxall, J. P. 1984. Seabirds, pp. 533–620. *In* R. M. Laws (ed.), *Antarctic Ecology*, Vol. 2. Academic Press, New York.

Croxall, J. P., and P. A. Prince. 1980. Food, feeding ecology and ecological segregation of seabirds at South Georgia. Biol. J. Linnean Soc. 14: 103–31.

Croxall, J. P., P. A. Prince, J. Hunter, S. J. McInnes, and P. G. Copstake. 1984. The seabirds of the Antarctic Peninsula, islands of the Scotia Sea, and Antarctic continent between 80° and 20° W: Their status and conservation, pp. 637–66. *In* J. P. Croxall, P. G. A. Evans, and R. W. Schreiber (eds.), *Status and Conservation of the World's Seabirds*. Int. Coun. Bird Preserv., Tech. Publ. No. 2, Cambridge, England.

Curry-Lindahl, K. 1970. Spread-wing postures in Pelecaniformes and Ciconiiformes. Auk 87: 371–72.

Daan, S., and J. Tinbergen. 1979. Young guillemots (*Uria lomvia*) leaving their arctic breeding cliffs: A daily rhythm in numbers and risk. Ardea 67: 97–100.

Darling, F. F. 1938. *Bird Flocks and the Breeding Cycle*. Cambridge Univ. Press, Cambridge, England.

Davis, P. 1957. The breeding of the storm petrel. Brit. Birds 50: 86–101, 371–84.

Dawson, W. L. 1911. Another fortnight on the Farallones. Condor 13: 171–83.

Deans, I. R. 1972. Shags laying two clutches. Brit. Birds 65: 166–67.

DeLong, R. L., and R. S. Crossin. Unpubl. ms. Status of seabirds on Islas Guadalupe, Natividad, Cedros, San Benito, and Los Coronados. Pacific Ocean Biol. Surv. Progr., U.S. Natl. Mus. Nat. Hist., Washington, D.C.

Derenne, P., G. Mary, and J.-L. Mougin. 1976. Le cormoran à ventre blanc *Phalacrocorax albiventer melanogenis* (Blyth) de l'Archipel Crozet. Com. Natl. Fr. Rech. Antarc. 40: 191–219.

DeSante, D. S., and D. G. Ainley. 1980. The avifauna of the South Farallon Islands, California. Studies Avian Biol. 4.

Dewar, J. M. 1924. *The Bird as Diver*. Witherby, London.

Diamond, A. W. 1978. Feeding strategies and population size in tropical seabirds. Am. Nat. 112: 215–23.

———. 1983. Feeding overlap in some tropical and temperate seabird communities. Studies Avian Biol. 8: 24–46.

Diamond, J. 1986. Evolution of ecological segregation in the New Guinea montane avifauna, pp. 98–125. *In* J. Diamond and T. J. Case (eds.), *Community Ecology*. Harper and Row, New York.

Diamond, J., and T. J. Case (eds.). 1986. *Community Ecology*. Harper and Row, New York.

Divoky, G. J., G. E. Watson, and J. C. Bartonek. 1974. Breeding of the Black Guillemot in northern Alaska. Condor 76: 339–43.

Doughty, R. W. 1971. San Francisco's nineteenth-century egg basket: The Farallons. Geogr. Rev. 61: 554–72.

———. 1974. The Farallones and the Boston men. Calif. Hist. Quar. 53: 309–16.

Dow, D. D. 1964. Diving times of wintering water birds. Auk 81: 556–58.

Drent, R. H. 1965. Breeding biology of the Pigeon Guillemot, *Cepphus columba*. Ardea 53: 99–160.

———. 1973. The natural history of incubation, pp. 262–311. *In* D. S. Farner (ed.), *Breeding Biology of Birds*. Natl. Acad. Sci., Washington, D.C.

Drent, R. H., and C. J. Guiguet. 1961. A catalogue of British Columbia seabird colonies. British Columbia Prov. Mus., Occ. Pap. No. 12.

Drent, R. H., G. F. van Tets, F. Tompa, and K. Vermeer. 1964. The breeding birds of Mandarte Island, British Columbia. Can. Field-Nat. 78: 208–63.

Drury, W. H. 1973. Population changes in New England seabirds. Bird-banding 44: 267–313.

Duffy, D. C. 1983a. Competition for nesting space among Peruvian guano birds. Auk 100: 680–88.

———. 1983b. Environmental uncertainty and commercial fishing: Effects on Peruvian guano birds. Biol. Conserv. 26: 227–38.

———. 1983c. The foraging ecology of Peruvian seabirds. Auk 100: 800–810.

Duffy, D. C., W. E. Arntz, H. T. Serpa, P. D. Boersma, and R. L. Norton. 1986. A comparison of the effects of El Niño and the Southern Oscillation on birds in Peru and the Atlantic Ocean, pp. 1740–46. *In* H. Ouellet (ed.), Proc. XIX Int. Ornithol. Congr., Natl. Mus. Nat. Sci., Ottawa.

Duffy, D. C., and S. Jackson. 1986. Diet studies of seabirds: A review of methods. Colonial Waterbirds 9: 1–17.

Duffy, D. C., and G. D. La Cock. 1985. Partitioning of nesting space among seabirds of the Benguela upwelling region. Ostrich 56: 186–201.

Duncan, K. W. 1968. The food of the Black Shag (*Phalacrocorax carbo novaehollandiae*) in Otago inland waters. Trans. R. Soc. N.Z., Biol. Sci., 11: 9–23.

Dunn, E. H. 1975. The role of environmental factors in the growth of tern chicks. J. Anim. Ecol. 44: 743–54.

———. 1976. Development of endothermy and existence energy expenditure of nesting Double-crested Cormorants. Condor 78: 350–56.

Dunning, J. B., Jr. 1984. Body weights of 686 species of North American birds. West. Bird Banding Assoc., Monogr. No. 1.

du Plessis, S. S. 1957. Growth and daily food intake of the White-breasted Cormorant in captivity. Ostrich 28: 197–201.

Elkins, N., and M. R. Williams. 1974. Shag movements in northeast Scotland. Bird Study 21: 149–51.

Ellison, L. N., and L. Cleary. 1978. Effects of human disturbance on breeding of Double-crested Cormorants. Auk 95: 510–17.

Elowson, A. M. 1984. Spread-wing postures and the water repellancy of feathers: A test of Rijke's hypothesis. Auk 101: 371–83.

Emerson, W. O. 1904. The Farallone Islands revisited, 1887–1903. Condor 6: 661–68.

Enfield, D. B. 1981. El Niño: Pacific eastern boundary response to interannual forcing, pp. 213–54. *In* M. H. Glantz and J. D. Thompson (eds.), *Resource Management and Environmental Uncertainty*. Wiley, New York.

Erskine, A. J. 1972. The Great Cormorants of eastern Canada. Occ. Pap. Can. Wildl. Serv. 14: 13–21.

Ewald, P. W., G. L. Hunt, Jr., and M. W. Hunt. 1980. Territory size in Western Gulls: Importance of intrusion pressure, defense investments and vegetation structure. Ecology 61: 80–87.

Farner, D. S. 1965. Circadian systems in the photoperiodic responses of vertebrates, pp. 357–69. *In* J. Aschoff (ed.), *Circadian Clocks*. North-Holland, Amsterdam.

Favorite, F., A. J. Dodimead, and K. Nasu. 1976. Oceanography of the subarctic Pacific region, 1960–71. Int. N. Pacific Fish. Comm., Bull. 33.

Fiedler, P. C. 1984. Some effects of El Niño 1983 on the Northern Anchovy. Calif. Coop. Ocean. Fish. Invest., Rept. 25: 53–58.

Fisher, H. J., and M. L. Fisher. 1969. The visits of Laysan Albatrosses to the breeding colony. Micronesica 5: 173–221.

Follett, W. I., and D. G. Ainley. 1976. Fishes collected by Pigeon Guillemots, *Cepphus columba* (Pallas), nesting on Southeast Farallon Island, California. Calif. Fish and Game 62: 28–31.

Frey, H. W. 1971. *California's Living Marine Resources and Their Utilization*. Calif. Res. Agency, Dept. Fish and Game, Sacramento.

Furness, R. W., and S. R. Baillie. 1981. Factors affecting capture rate and biometrics of storm petrels on St. Kilda. Ringing Migr. 3: 137–48.

Furness, R. W., and R. T. Barrett. 1985. The food requirements and ecological relationships of a seabird community in north Norway. Ornis Scand. 16: 305–13.

Furness, R. W., and T. R. Birkhead. 1984. Seabird colony distributions suggest competition for food supplies during the breeding season. Nature 311: 655–56.

Furness, R. W., and P. Monaghan. 1987. *Seabird Ecology*. Blackie, Glasgow.

Gaston, A. J. 1985. Development of the young in the Atlantic Alcidae, pp. 319–54. *In* D. N. Nettleship and T. R. Birkhead (eds.), *The Atlantic Alcidae*. Academic Press, Orlando.

Gaston, A. J., D. K. Cairns, R. D. Elliott, and D. G. Noble. 1985. *A Natural History of Digges Sound*. Can. Wildl. Serv., Rept. Ser. No. 46.

Gaston, A. J., G. Chapdelaine, and D. G. Noble. 1983. The growth of Thick-billed Murre chicks at colonies in Hudson Strait: Inter- and intra-colony variation. Can. J. Zool. 61: 2465–75.

Gaston, A. J., and D. N. Nettleship. 1981. *The Thick-billed Murres of Prince Leopold Island*. Can. Wildl. Serv., Monogr. Ser. No. 6.

Gaston, A. J., D. G. Noble, and M. A. Purdy. 1983. Monitoring breeding biology parameters for murres *Uria* spp.: Levels of accuracy and sources of bias. J. Field Ornithol. 54: 275–82.

Gill, A. E., and A. J. Clarke. 1974. Wind-induced upwelling, coastal currents and sea level changes. Deep-Sea Res. 21: 325–45.

Glantz, M. H., and J. D. Thompson (eds.). 1981. *Resource Management and Environmental Uncertainty*. Wiley, New York.

Gochfeld, M. 1980. Mechanisms and adaptive value of reproductive synchrony in colonial seabirds, pp. 207–70. *In* J. Burger, B. L. Olla, and H. E. Winn (eds.), *Behavior of Marine Animals*, Vol. 4: *Marine Birds*. Plenum, New York.

Grant, P. R. 1986. Interspecific competition in fluctuating environments, pp. 173–91. *In* J. Diamond and T. J. Case (eds.), *Community Ecology*. Harper and Row, New York.

Greenwood, J. 1964. The fledging of the guillemot *Uria aalge*, with notes on the Razorbill *Alca torda*. Ibis 106: 469–81.

———. 1972. The attendance of guillemots and razorbills at a Scottish colony. Proc. XV Int. Ornithol. Congr., p. 648.

Greichus, A., and Y. A. Greichus. 1973. Some factors affecting the nesting success of pelicans and cormorants in South Dakota. Proc. S. Dak. Acad. Sci. 52: 241–47.

Gress, F., R. W. Risebrough, D. W. Anderson, L. F. Kiff, and J. R. Jehl, Jr. 1973. Reproductive failures of Double-crested Cormorants in southern California and Baja California. Wilson Bull. 85: 197–208.

Gress, F., R. W. Risebrough, and F. C. Sibley. 1971. Shell thinning in eggs of the Common Murre *Uria aalge* from the Farallon Islands. Condor 73: 368–69.

Gross, W. A. O. 1935. The life history cycle of Leach's Petrel (*Oceanodroma leucorhoa*) on the outer sea islands in the Bay of Fundy. Auk 52: 382–99.

Grover, J. J., and B. L. Olla. 1983. The role of the Rhinoceros Auklet *Cerorhinca monocerata* in mixed species feeding assemblages of sea birds in the Strait of Juan de Fuca, Washington. Auk 100: 979–82.

Hahn, D. C. 1981. Asynchronous hatching in the Laughing Gull: Cutting losses and reducing rivalry. Anim. Behav. 29: 421–27.

Haley, D. 1984. *Seabirds of Eastern North Pacific and Arctic Waters*. Pacific Search Press, Seattle.

Hanson, W. C. 1968. Recent history of Double-crested Cormorant colonies in southeastern Washington. Murrelet 49: 25–27.

Harper, C. A. 1971. Breeding biology of a small colony of Western Gulls, *Larus occidentalis wymani*, in California. Condor 73: 337–41.

Harper, P. C., J. P. Croxall, and J. Cooper. 1985. *A Guide to Foraging Methods Used by Marine Birds in Antarctic and Subantarctic Seas*. BIOMASS Handbook No. 24, Sci. Com. Antarc. Res., Cambridge, England.

Harris, M. P. 1966. Age of return to the colony, age of breeding and adult survival of Manx Shearwaters. Bird Study 13: 84–95.

———. 1969a. The biology of storm petrels in the Galápagos Islands. Proc. Calif. Acad. Sci., 4th Ser., 37: 95–166.

———. 1969b. Food as a factor controlling the breeding of *Puffinus lherminieri*. Ibis 111: 139–56.

———. 1970. Differences in the diet of British auks. Ibis 112: 540–41.

———. 1984. *The Puffin*. Poyser, Calton, England.

Harris, M. P., and T. R. Birkhead. 1985. Breeding ecology of the Atlantic Alcidae, pp. 155–204. *In* D. N. Nettleship and T. R. Birkhead (eds.), *The Atlantic Alcidae*. Academic Press, Orlando.

Harris, M. P., and P. Rothery. 1985. The post-fledging survival of young puffins *Fratercula arctica* in relation to hatching date and growth. Ibis 127: 243–50.

Harris, S. W. 1974. Status, chronology and ecology of nesting storm petrels in northwestern California. Condor 76: 249–61.

Harrison, C. S., T. Hida, and M. Seki. 1983. The feeding ecology of seabirds in the northwestern Hawaiian Islands. Wildl. Soc., Wildl. Monogr. No. 85.

Harvey, J. T. Unpubl. ms. Daytime surface swarming of euphausiids in Monterey Bay, California. Hatfield Mar. Sci. Center, Newport, Oregon.

Harvey, T. 1982. *The Breeding Biology of Three Species of Larids in Elkhorn Slough*. Unpubl. Master's thesis, Calif. State Univ., San Francisco.

Hatler, D. F., R. W. Campbell, and A. Dorst. 1978. Birds of Pacific Rim National Park. Brit. Columbia Prov. Mus., Occ. Pap. No. 20.

Heath, C. B., and J. M. Francis. 1983. Population dynamics and feeding ecology of the California sea lion with applications for management. Natl. Mar. Fish. Serv., Southwest Fish. Center Admin. Rept. LJ-83-04C.

Hedgren, S. 1979. Seasonal variation in fledging weight of guillemots *Uria aalge*. Ibis 121: 356–61.

———. 1980. Reproductive success of guillemots *Uria aalge* on the island of Stora Karlsö. Ornis Fenn. 57: 49–57.

Hedgren, S., and A. Linnman. 1979. Growth of guillemot *Uria aalge* chicks in relation to time of hatching. Ornis Scand. 10: 29–36.

Heermann, A. L. 1859. Report upon birds collected on survey. U.S. Pacific Railroad Surv., Rept. 10: 29–80.

Henry, F. D. 1986. Status of the coastwide Chilipepper (*Sebastes goodei*) fishery. In *Status of the Pacific Coast Groundfish Fishery Through 1986 and Recommended Acceptable Biological Catches for 1987*. Pacific Fish. Manage. Council, Portland, Oregon.

Hewitt, R. P. 1985. The spawning biomass of the Northern Anchovy. Calif. Coop. Ocean. Fish. Invest., Rept. 26: 17–25.

Hickey, B. M. 1979. The California Current system—Hypothesis and facts. Prog. Oceanogr. 8: 191–279.

Hightower, J. E., and W. H. Lenarz. 1986. Status of the Widow Rockfish fishery. In *Status of the Pacific Coast Groundfish Fishery Through 1986 and Recommended Acceptable Biological Catches for 1987*. Pacific Fish. Manage. Council, Portland, Oregon.

Hobson, E., P. Adams, J. Chess, D. Howard, and W. Samiere. 1986. Temporal and spatial variations in the numbers of first-year juvenile rockfishes off northern California. Natl. Mar. Fish. Serv., Southwest Fish. Center Admin. Rept. T-86-02.

Hobson, K. A., and S. G. Sealy. 1985. Diving rhythms and diurnal roosting times of Pelagic Cormorants. Wilson Bull. 97: 116–19.
Hodder, J., and M. R. Graybill. 1985. Reproduction and survival of seabirds in Oregon during the 1982–83 El Niño. Condor 87: 535–41.
Hoffman, W., D. Heinemann, and J. A. Wiens. 1981. The ecology of seabird feeding flocks in Alaska. Auk 98: 437–56.
Hoffman, W., J. A. Wiens, and J. M. Scott. 1978. Hybridization between gulls (*Larus glaucescens* and *L. occidentalis*) in the Pacific Northwest. Auk 95: 441–58.
Houston, C. S. 1971. Cormorant ringing in Saskatchewan, Canada. Ring 67: 125–28.
Hubbs, C. L. 1960. The marine vertebrates of the outer coast. Syst. Zool. 9: 134–47.
Huber, H. R., C. Beckham, J. Nisbet, A. Rovetta, and J. Nusbaum. 1985. Studies of marine mammals at the Farallon Islands, 1982–1983. Natl. Mar. Fish. Serv., Southwest Fish. Center Admin. Rept. LJ-85-01C.
Hunt, G. L., Jr. 1972. Influence of food distribution and human disturbance on the reproductive success of Herring Gulls. Ecology 53: 1051–61.
———. 1980. Mate selection and mating systems in seabirds, pp. 113–52. *In* J. Burger, B. L. Olla, and H. E. Winn (eds.), *Behavior of Marine Animals*, Vol. 4: *Marine Birds*. Plenum, New York.
Hunt, G. L., Jr., B. Burgeson, and G. A. Sanger. 1981. Feeding ecology of seabirds of the eastern Bering Sea, pp. 629–48. *In* D. W. Hood and J. A. Calder (eds.), *The Eastern Bering Sea Shelf: Oceanography and Resources*. Univ. Wash. Press, Seattle.
Hunt, G. L., Jr., and J. Butler. 1980. Reproductive ecology of Western Gulls and Xantus' Murrelets with respect to food resources in the Southern California Bight. Calif. Coop. Ocean. Fish. Invest., Rept. 21: 62–67.
Hunt, G. L., Jr., Z. Eppley, and W. H. Drury. 1981. Breeding distribution and reproductive biology of marine birds in the eastern Bering Sea, pp. 649–87. *In* D. W. Hood and J. A. Calder (eds.), *The Eastern Bering Sea Shelf: Oceanography and Resources*. Univ. Wash. Press, Seattle.
Hunt, G. L., Jr., Z. A. Eppley, and D. C. Schneider. 1986. Reproductive performance of seabirds: The importance of population and colony size. Auk 103: 306–17.
Hunt, G. L., Jr., and M. W. Hunt. 1973. Clutch size, hatching success, and eggshell thinning in Western Gulls. Condor 75: 483–86.
———. 1975. Reproductive ecology of the Western Gull: The importance of nest spacing. Auk 92: 270–79.
———. 1976a. Exploitation of fluctuating food resources by Western Gulls. Auk 93: 301–37.
———. 1976b. Gull chick survival: The significance of growth rates, timing of breeding and territory size. Ecology 57: 62–75.
———. 1977. Female–female pairing in Western Gulls (*Larus occidentalis*) in southern California. Science 196: 1466–67.
Hunt, G. L., Jr., A. L. Newman, M. H. Warner, J. C. Wingfield, and J. Kaiw.

1984. Comparative behavior of male–female and female–female pairs among Western Gulls prior to egg laying. Condor 86: 157–62.

Hunt, G. L., Jr., R. L. Pitman, M. Naughton, K. Winnett, A. Newman, P. R. Kelly, and K. T. Briggs. 1981. Summary of marine mammal and seabird surveys of the Southern California Bight area 1975–1978. Final Rept., U.S. Dept. Interior, Bur. Land Manage.

Huntington, C. E. 1963. Population dynamics of Leach's Petrels, *Oceanodroma leucorhoa*. Proc. XIII Int. Ornithol. Congr., pp. 701–5.

Hurtubia, J. 1973. Trophic diversity measurement in sympatric predatory species. Ecology 54: 885–90.

Husby, D. M., and C. S. Nelson. 1982. Turbulence and vertical stability in the California Current. Calif. Coop. Ocean. Fish. Invest., Rept. 23: 113–29.

Hutchinson, G. E. 1959. Homage to Santa Rosalia; or, Why are there so many kinds of animals? Am. Nat. 93: 145–49.

Idyll, C. P. 1973. The anchovy crisis. Sci. Amer. 228: 22–29.

Isleib, M. E., and B. Kessel. 1973. Birds of the north Gulf coast–Prince William Sound region, Alaska. Univ. Alaska Biol. Pap. No. 14.

James-Veitch, E. 1970. *The Ashy Petrel, Oceanodroma homochroa, at Its Breeding Grounds on the Farallon Islands, California*. Unpubl. Ph.D. thesis, Loma Linda Univ.

Jarvis, M. J. F. 1974. The ecological significance of clutch size in the South African Gannet (*Sula capensis* Lichtenstein). J. Anim. Ecol. 43: 1–17.

Johnson, R. A. 1941. Nesting behavior of the Atlantic murre. Auk 58: 153–63.

Johnson, S. R., and G. C. West. 1975. Growth and development of heat regulation in nestlings and metabolism of adult Common and Thick-billed murres. Ornis Scand. 6: 109–15.

Jones, R. E. 1981. Food habits of smaller marine mammals from northern California. Proc. Calif. Acad. Sci. 42(16): 409–33.

Jordán, R. 1959. El fenómeno de las regurgitaciones en el Guanay (*Phalacrocorax bougainvillii* L.) y un método para estimar la ingestión diaria. Bol. Cia. Admin. Guano 35(4): 23–40.

———. 1967. The predation of guano birds on the Peruvian Anchovy (*Engraulis ringens* Jenyns). Calif. Coop. Ocean. Fish. Invest., Rept. 11: 105–9.

Junor, F. J. R. 1972. Estimation of the daily food intake of piscivorous birds. Ostrich 43: 193–205.

Kadlec, J. A., and W. H. Drury. 1968. Structure of the New England Herring Gull population. Ecology 49: 644–76.

King, W. B. 1970. The trade wind zone oceanography pilot study. Part VII: Observations of sea birds March 1964 to June 1965. U.S. Fish Wildl. Serv., Spec. Sci. Rept. Fish. No. 586.

Knudsen, R. L., Jr. 1976. *Sexing and the Structure of the Beak in Cassin's Auklet (Ptychoramphus aleuticus [Pallas])*. Unpubl. Master's thesis, Calif. State Univ., Sacramento.

Koelink, A. F. 1972. *Bioenergetics of Growth in the Pigeon Guillemot, Cepphus columba*. Unpubl. Ph.D. thesis, Univ. Brit. Columbia, Victoria.

Kortlandt, A. 1942. Levensloop, samenstelling en structuur der Nederlandse aalscholverbevolking. Ardea 31: 175–280.

Kuletz, K. J. 1983. *Mechanisms and Consequences of Foraging Behavior in a Population of Breeding Pigeon Guillemots*. Unpubl. Master's thesis, Univ. Calif., Irvine.

Lack, D. 1945. The ecology of closely related species with special reference to the cormorant (*Phalacrocorax carbo*) and shag (*P. aristotelis*). J. Anim. Ecol. 14: 12–16.

———. 1946. Competition for food by birds of prey. J. Anim. Ecol. 15: 123–29.

———. 1954. *The Natural Regulation of Animal Numbers*. Clarendon Press, Oxford.

———. 1966. *Population Studies of Birds*. Clarendon Press, Oxford.

———. 1968. *Ecological Adaptations for Breeding in Birds*. Methuen, London.

———. 1971. *Ecological Isolation in Birds*. Harvard Univ. Press, Cambridge, Massachusetts.

Lenarz, W. H. 1980. Shortbelly Rockfish, *Sebastes jordani*: A large unfished resource in waters off California. Mar. Fish. Rev. 42: 34–40.

Lenarz, W. H., and D. R. Gunderson (eds.). 1987. Widow Rockfish: Proceedings of a workshop, Tiburon, California, December 11–12, 1980. Natl. Mar. Fish. Serv., NOAA Tech. Rept. NMFS 48.

Lenarz, W. H., and S. Moreland. 1985. Progress report on rockfish recruitment studies at the Tiburon Laboratory. Natl. Mar. Fish. Serv., Southwest Fish. Center Admin. Rept. T-85-02.

Lewis, H. F. 1929. *The Natural History of the Double-crested Cormorant (Phalacrocorax auritus auritus L.)*. Ru-Mi-Lou Books, Ottawa.

Lloyd, C. S. 1975. Timing and frequency of census counts on cliff-nesting auks. Brit. Birds 68: 507–13.

———. 1979. Factors affecting breeding of Razorbills *Alca torda* on Skokholm. Ibis 121: 165–75.

Lloyd, C. S., and C. M. Perrins. 1977. Survival and age of first breeding in the Razorbill (*Alca torda*). Bird-banding 48: 239–52.

Lock, A. R., and R. K. Ross. 1973. The nesting of the Great Cormorant (*Phalacrocorax carbo*) and the Double-crested Cormorant (*P. auritus*) in Nova Scotia in 1971. Can. Field-Nat. 87: 43–49.

Loomis, L. M. 1896. California water birds. No. III. South Farallon Island in July. Proc. Calif. Acad. Sci., 2nd Ser. 6: 353–66.

Lynn, R. J. 1983. The 1982–83 warm episode in the California Current. Geophys. Res. Lett. 10: 1093–95.

MacCall, A. D. 1984. Seabird–fishery trophic interactions in eastern Pacific boundary currents: California and Peru, pp. 136–49. *In* D. N. Nettleship, G. A. Sanger, and P. F. Springer (eds.), *Marine Birds: Their Feeding Ecology and Commercial Fisheries Relationships*. Minister Supply Serv. Can., Ottawa.

———. 1986. Changes in the biomass of the California Current ecosystem, pp. 33–54. *In* K. Sherman and L. M. Alexander (eds.), *Variability and Management of Large Marine Ecosystems*. Westview Press, Boulder, Colorado.

MacGregor, J. S. 1986. Relative abundance of four species of *Sebastes* off California and Baja California. Calif. Coop. Ocean. Fish. Invest., Rept. 27: 121–35.

Mahoney, S. P., and W. Threlfall. 1982. Notes on the agonistic behavior of Common Murres. Wilson Bull. 94: 595–98.

Mais, K. F. 1974. Pelagic fish surveys in the California Current. Calif. Dept. Fish and Game, Fish Bull. 162: 1–79.

———. 1981. Age-composition changes in the anchovy, *Engraulis mordax*, central population. Calif. Coop. Ocean. Fish. Invest., Rept. 22: 82–87.

Manuwal, D. A. 1972. *The Population Ecology of the Cassin's Auklet on Southeast Farallon Island, California*. Unpubl. Ph.D. thesis, Univ. Calif., Los Angeles.

———. 1974a. Effects of territoriality on breeding in a population of Cassin's Auklet. Ecology 55: 1399–1406.

———. 1974b. The natural history of Cassin's Auklet. Condor 76: 421–31.

———. 1974c. The incubation patches of Cassin's Auklet. Condor 76: 481–84.

———. 1979. Reproductive commitment and success of Cassin's Auklet. Condor 81: 111–21.

———. 1984. Alcids—Dovekie, murres, guillemots, murrelets, auklets, and puffins, pp. 168–87. *In* D. Haley (ed.), *Seabirds of Eastern North Pacific and Arctic Waters*. Pacific Search Press, Seattle.

Manuwal, D. A., and R. W. Campbell. 1979. Status and distribution of breeding seabirds of southeastern Alaska, British Columbia, and Washington. U.S. Dept. Interior, Fish Wildl. Serv., Wildl. Rept. 11: 73–91.

Manuwal, D. A., T. R. Wahl, and S. M. Speich. 1979. *The Seasonal Distribution and Abundance of Marine Bird Populations in the Strait of Juan de Fuca and Northern Puget Sound in 1978*. Mar. Ecol. Anal. Progr., NOAA, Boulder, Colorado.

Mason, J. E., and A. Bakun. 1986. Upwelling index update, U.S. west coast, 33N–48N latitude. NOAA Tech. Mem. NOAA-TM-NMFS-SWFC-67.

McLagan, R. L., and J. F. Piatt. Unpubl. ms. Breeding success and phenology of Common Murres (*Uria aalge*) at Cape St. Mary's, Newfoundland, 1980–1984. Newfoundland Inst. Cold Ocean Sci., Memorial Univ., St. John's.

McLain, D. R., R. E. Brainard, and J. G. Norton. 1985. Anomalous warm events in eastern boundary current systems. Calif. Coop. Ocean. Fish. Invest., Rept. 26: 51–64.

McLain, D. R., and D. H. Thomas. 1983. Year-to-year fluctuations of the California Countercurrent and effects on marine organisms. Calif. Coop. Ocean. Fish. Invest., Rept. 24: 165–80.

McLeod, J. A., and G. F. Bondar. 1953. A brief study of the Double-crested Cormorant on Lake Winnepegosis. Can. Field-Nat. 67: 1–11.

Mead, C. J. 1974. The results of ringing auks in Britain and Ireland. Ireland Bird Study 21: 45–86.

Mendall, H. L. 1936. The home-life and economic status of the Double-crested Cormorant, *Phalacrocorax auritus auritus*. Univ. Maine Studies No. 38.

Michael, C. W. 1935. Nesting habits of cormorants. Condor 37: 36–37.

Michaelson, J. C. 1977. *North Pacific Sea Surface Temperatures and California Precipitation*. Unpubl. Master's thesis, Univ. Calif., Berkeley.

Mitchell, R. M. 1977. Breeding biology of the Double-crested Cormorant on Utah Lake. Great Basin Nat. 37: 1–23.

Montevecchi, W. A., I. R. Kirkham, D. D. Roby, and K. L. Brink. 1983. Size, organic composition, and energy content of Leach's Storm-Petrel (*Oceanodroma leucorhoa*) eggs with reference to position in the precocial spectrum and breeding biology. Can. J. Zool. 61: 1456–63.

Morisita, M. 1959. Measuring of interspecific association and similarity between communities. Mem. Fac. Sci., Kyushu Univ., Series E, 3: 65–80.

Morrison, M. L., E. Shanley, Jr., and R. D. Slack. 1979. Breeding biology and age-specific mortality of Olivaceous Cormorants. Southwest. Nat. 24: 259–66.

Mougin, J.-L. 1968. Etude écologique de quatre espèces de pétrels antarctiques. L'Ois. et Rev. Fr. Ornithol., 38 (no. spéc.): 2–52.

Munro, J. A., and I. M. Cowan. 1947. *A Review of the Bird Fauna of British Columbia*. Brit. Columbia Prov. Mus., Spec. Publ. No. 2.

Murphy, R. C. 1925. *The Bird Islands of Peru*. Putnam, New York.

———. 1936. *Oceanic Birds of South America*. Macmillan, New York.

———. 1981. The guano and anchoveta fishery, pp. 81–106. *In* M. H. Glantz and J. D. Thompson (eds.), *Resource Management and Environmental Uncertainty*. Wiley, New York.

Namais, J. 1976. Some statistical and synoptic characteristics associated with El Niño. J. Phys. Oceanogr. 6: 130–38.

Nelson, A. 1971. King Shags in the Marlborough Sound. Notornis 18: 30–37.

Nelson, C. S. 1977. Wind stress and wind curl over the California Current. NOAA Tech. Rept. NMFS SSRF-714.

Nelson, D. A. 1981. Sexing Cassin's Auklets by bill measurements. J. Field Ornithol. 52: 233–34.

———. 1982. *The Communication Behavior of the Pigeon Guillemot Cepphus columba*. Unpubl. Ph.D. thesis, Univ. Michigan, Ann Arbor.

———. 1987. Factors influencing colony attendance by Pigeon Guillemots on Southeast Farallon Island, California. Condor 89: 340–48.

Nelson, J. B. 1978. *The Sulidae: Gannets and Boobies*. Oxford Univ. Press, Oxford.

———. 1980. *Seabirds: Their Biology and Ecology*. Hamlyn, London.

Nettleship, D. N. 1972. Breeding success of the Common Puffin (*Fratercula arctica*) on different habitats at Great Island, Newfoundland. Ecol. Monogr. 42: 239–68.

Nettleship, D. N., and P. G. H. Evans. 1985. Distribution and status of the Atlantic Alcidae, pp. 54–155. *In* D. N. Nettleship and T. R. Birkhead (eds.), *The Atlantic Alcidae*. Academic Press, Orlando.

Nordhoff, C. 1874. The Farallon Islands. Harper's New Monthly Mag. 48: 617–25.

Norman, F. I. 1974. Notes on the breeding of the Pied Cormorant near Weiribee, Victoria, in 1971, 1972, and 1973. Emu 74: 223–27.

Nysewander, D., and D. B. Barbour. 1979. The breeding biology of marine birds associated with Chumah Bay, Kodiak Island, 1975–1978, pp. 21–106.

In *Environmental Assessment of the Alaskan Continental Shelf*, Ann. Rept. Principal Invest., Vol. 2. U.S. Dept. Commerce, NOAA, Boulder, Colorado.

O'Brien, J. J., A. Busalacchi, and J. Kindle. 1981. Ocean models of El Niño, pp. 159–212. *In* M. H. Glantz and J. D. Thompson (eds.), *Resource Management and Environmental Uncertainty*. Wiley, New York.

O'Connor, R. J. 1976. Weight and body composition in nesting Blue Tits *Parus caeruleus*. Ibis 118: 108–12.

———. 1978. Brood reduction in birds: Selection for fratricide, infanticide, and suicide. Anim. Behav. 26: 79–96.

Oliver, M. D., and M. A. Kuyper. 1978. Breeding biology of the White-breasted Cormorant in Natal. Ostrich 49: 25–30.

Osborne, T. O. 1972. Survey of seabird use of the coastal rocks of northern California from Cape Mendocino to the Oregon line. Calif. Dept. Fish and Game, Wildl. Manage. Branch Admin. Rept. No. 71–4.

Osborne, T. O., and J. G. Reynolds. 1971. California seabird breeding ground survey 1969–70. Calif. Dept. Fish and Game, Wildl. Manage. Admin. Rept. No. 71–3.

Owre, O. T. 1967. Adaptations for locomotion and feeding in the Anhinga and the Double-crested Cormorant. Am. Ornithol. Union, Monogr. No. 6.

Palmer, R. S. (ed.). 1962. *Handbook of North American Birds*, Vol. 1. Yale Univ. Press, New Haven.

Paludan, K. 1951. Contributions to the breeding biology of *Larus argentatus* and *L. fuscus*. Vidensk. Medd. Dansk Naturh. Foren. 114: 1–128.

Parrish, R. H., A. Bakun, D. M. Husby, and C. S. Nelson. 1982. Comparative climatology of selected environmental processes in relation to eastern boundary current pelagic fish reproduction. Food Agric. Org., Fish. Rept. 291: 731–77.

Parrish, R. H., D. L. Mallicoate, and K. F. Mais. 1985. Regional variations in the growth and age composition of Northern Anchovy, *Engraulis mordax*. Fish. Bull. 83: 483–96.

Parrish, R. H., C. S. Nelson, and A. Bakun. 1981. Transport mechanisms and reproductive success of fishes in the California Current. Biol. Oceanogr. 1: 175–203.

Parsons, J. 1970. Relationship between egg size and post-hatching mortality in the Herring Gull (*Larus argentatus*). Nature 228: 1221–22.

———. 1975. Asynchronous hatching and chick mortality in the Herring Gull *Larus argentatus*. Ibis 117: 517–20.

Paulik, G. J. 1971. Anchovies, birds, and fishermen in the Peru Current. *In* W. W. Murdoch (ed.), *Environment: Resources, Pollution, and Society*. Sinauer Assoc., Stamford, Connecticut.

Payne, R. B. 1965. The molt of breeding Cassin's Auklets. Condor 67: 220–28.

Pearson, T. H. 1968. The feeding biology of sea-bird species breeding on the Farne Islands, Northumberland. J. Anim. Ecol. 37: 521–52.

Penhale, L. B. 1972. Reproductive failure of Pelagic Cormorants, San Luis Obispo Co., California, 1970. Calif. Fish and Game 58: 238.

Pennycuick, C. J. 1956. Observations on a colony of Brunnich's Guillemots (*Uria lomvia*) in Spitzbergen. Ibis 98: 80–99.

Pennycuick, C. J., J. P. Croxall, and P. A. Prince. 1984. Scaling of foraging radius and growth rate in petrels and albatrosses (Procellariiformes). Ornis Scand. 15: 145–54.

Perrins, C. M. 1965. Population fluctuations and clutch-size in the Great Tit *Parus major* L. J. Anim. Ecol. 34: 601–47.

———. 1966. Survival of young Manx Shearwaters *Puffinus puffinus* in relation to their presumed date of hatching. Ibis 108: 132–35.

———. 1979. *British Tits*. Collins, London.

Perrins, C. M., M. P. Harris, and C. K. Britton. 1973. Survival of Manx Shearwaters *Puffinus puffinus*. Ibis 115: 535–48.

Piatt, J. F., and R. L. McLagan. 1987. Attendance patterns of Common Murres (*Uria aalge*) at Cape St. Mary's, Newfoundland, 1980–1984. Can. J. Zool. 65: 1530–34.

Piatt, J. F., and D. N. Nettleship. 1985. Diving depths of four alcids. Auk 102: 293–97.

Pierotti, R. J. 1976. *Sex Roles, Social Structures, and the Role of the Environment in the Western Gull (Larus occidentalis Audubon)*. Unpubl. Master's thesis, Calif. State Univ., Sacramento.

———. 1981. Male and female parental roles in the Western Gull under different environmental conditions. Auk 81: 532–49.

———. 1982. Habitat selection and its effect on reproductive output in the Herring Gull *Larus argentatus* in Newfoundland, Canada. Ecology 63: 854–68.

———. 1983. Gull *Larus argentatus* and puffin *Fratercula arctica* interactions on Great Island, Newfoundland, Canada. Biol. Conserv. 26: 1–14.

———. 1987. Behavioral consequences of habitat selection in the Herring Gull. Studies Avian Biol. 10: 119–28.

Pierotti, R. J., and C. A. Annett. 1986. Reproductive consequences of dietary specialization and switching in an ecological generalist, pp. 417–42. *In* A. C. Kamil, J. R. Krebs, and H. R. Pulliam (eds.), *Foraging Behavior*. Plenum, New York.

Pierotti, R. J., and C. A. Bellrose. 1986. Proximate and ultimate causation of egg size and the "third-chick disadvantage" in the Western Gull. Auk 103: 401–7.

Pitman, R. L., M. R. Graybill, J. Hodder, and D. H. Varoujean. Ms. *The Catalog of Oregon Seabird Colonies*. U.S. Dept. Interior, Fish Wildl. Serv. FWS/OBS. (In press.)

Porter, J. M., and J. C. Coulson. 1987. Long-term changes in recruitment to the breeding group and the quality of recruits at a Kittiwake *Rissa tridactyla* colony. J. Anim. Ecol. 56: 675–89.

Porter, J. M., and S. G. Sealy. 1982. Dynamics of seabird multispecies feeding flocks: Age-related feeding behavior. Behaviour 81-2/4: 91–109.

Potts, G. R. 1969. The influence of eruptive movements, age, population size and other factors on the survival of the Shag (*Phalacrocorax aristotelis* L.). J. Anim. Ecol. 38: 53–102.

Preble, E. A., and W. L. McAtee. 1923. A biological survey of the Pribilof Islands, Alaska. U.S. Bur. Biol. Surv., N. Am. Fauna No. 46.

Preston, F. W. 1974. The volume of an egg. Auk 91: 132–38.
Preston, W. 1968. *Breeding Ecology and Social Behavior of the Black Guillemot, Cepphus grylle*. Unpubl. Ph.D. thesis, Univ. Michigan, Ann Arbor.
Pugesek, B. H. 1981. Increased reproductive effort with age in the California Gull (*Larus californicus*). Science 212: 822–23.
Pugesek, B. H., and K. L. Diem. 1983. A multivariate study of the relationship of parental age to reproductive success in California Gulls. Ecology 64: 829–39.
Quinn, W. H., D. O. Zopf, K. S. Short, and R. T. W. K. Yang. 1978. Historical trends and statistics of the Southern Oscillation, El Niño, and Indonesian droughts. Fish. Bull. 76: 663–78.
Rand, R. W. 1960. The biology of guano-producing seabirds: 3. The distribution, abundance, and feeding habits of the cormorants Phalacrocoracidae of the southwest coast of Cape Province. Invest. Rept. Div. Fish., Union S. Afr. 42: 1–32.
———. 1963. The biology of guano-producing seabirds: 4. Composition of colonies on the Cape Islands. Invest. Rept. Div. Fish., Union S. Afr. 43: 1–32.
Recksiek, C. W., and H. W. Frey. 1978. Background of Market Squid research program, basic life history, and the California fishery. Calif. Dept. Fish and Game, Fish. Bull. 169: 7–10.
Reid, J. L., Jr., G. I. Roden, and J. G. Wyllie. 1958. Studies of the California Current system. Calif. Coop. Ocean. Fish. Invest., Rept. 4: 27–56.
Reid, W. V. 1987. Constraints on clutch size in the Glaucous-winged Gull. Studies Avian Biol. 10: 8–25.
Richardson, F. 1961. Breeding biology of the Rhinoceros Auklet on Protection Island, Washington. Condor 63: 456–73.
Richdale, L. E. 1949. A study of a group of penguins of known age. Biol. Monogr. 1, Dunedin.
———. 1950. The pre-egg stage in the albatross family. Biol. Monogr. 3, Dunedin.
———. 1963. Biology of the Sooty Shearwater *Puffinus griseus*. Proc. Zool. Soc. London 141: 1–117.
———. 1965. Biology of the birds of Whero Island, New Zealand, with special reference to the Diving Petrel and the White-faced Storm Petrel. Trans. Zool. Soc. Lond. 31: 1–86.
Ricklefs, R. E. 1968. Patterns of growth in birds. Ibis 110: 419–51.
———. 1983. Avian postnatal development, pp. 1–83. *In* D. S. Farner, J. R. King, K. E. Parkes (eds.), *Avian Biology*, Vol. 7. Academic Press, New York.
Ricklefs, R. E., C. H. Day, E. E. Huntington, and J. B. Williams. 1985. Variability in feeding rate and meal size of Leach's Storm-Petrel at Kent Island, New Brunswick. J. Anim. Ecol. 54: 883–98.
Ricklefs, R. E., D. Duffy, and M. Coulter. 1984. Weight gain of Blue-footed Booby chicks: An indicator of marine resources. Ornis Scand. 15: 162–66.
Ricklefs, R. E., S. White, and J. Cullen. 1980. Postnatal development of Leach's Storm-Petrel. Auk 97: 768–81.

Ridgway, R. 1919. The birds of North and Middle America. U.S. Natl. Mus., Bull. 50, Part 8.
Rijke, A. M. 1967. The water repellancy and feather structure of cormorants, Phalacrocoracidae. J. Exp. Biol. 48: 185–89.
Robert, H. C., and C. J. Ralph. 1975. Effects of human disturbance on the breeding suceess of gulls. Condor 77: 495–99.
Roberts, B. 1940. The life cycle of Wilson's Petrel *Oceanites oceanicus* (Kuhl). Brit. Graham Land Exped. 1934–37, Sci. Rept. 1: 141–94.
Robertson, I. 1974. The food of nesting Double-crested and Pelagic cormorants at Mandarte Island, British Columbia, with notes on feeding ecology. Condor 76: 346–48.
Roney, K. N. 1979. Preliminary results on the food consumed by nesting Double-crested Cormorants at Cypress Lake, Saskatchewan. Proc. Colonial Waterbird Group 3: 257–58.
Root, R. B. 1967. The niche exploitation pattern of the Blue-gray Gnatcatcher. Ecol. Monogr. 37: 317–50.
Ross, R. K. 1977. A comparison of the feeding and nesting requirements of the Great Cormorant (*Phalacrocorax carbo* L.) and Double-crested Cormorant (*P. auritus* Lesson) in Nova Scotia. Proc. Nova Scotian Inst. Sci. 27: 114–32.
Roudybush, T. E., C. R. Grau, M. R. Petersen, D. G. Ainley, K. V. Hirsch, A. P. Gilman, and S. M. Patten. 1979. Yolk formation in some charadriiform birds. Condor 81: 293–98.
Roughgarden, J. 1986. A comparison of food-limited and space-limited animal competition communities, pp. 492–516. *In* J. Diamond and T. J. Case (eds.), *Community Ecology*. Harper and Row, New York.
Ryder, J. 1980. The influence of age on the breeding biology of colonial nesting seabirds, pp. 153–68. *In* J. Burger, B. L. Olla, and H. E. Winn (eds.), *Behavior of Marine Animals*, Vol. 4: *Marine Birds*. Plenum, New York.
Sander, T. G. 1986. Aspects of the breeding biology and recolonization history of Rhinoceros Auklets (*Cerorhinca monocerata*) on Southeast Farallon Island, California. Unpubl. Rept., PRBO/Evergreen State College, Olympia, Washington.
Sanger, G. A. 1972. Preliminary standing stock and biomass estimates of seabirds in the subarctic Pacific region, pp. 581–611. *In* A. Yositada Takenouti et al. (eds.), *Biological Oceanography of the Northern North Pacific Ocean*. Idemitsu Shoten, Tokyo.
———. 1982. The feeding habits and food web relationships of seabirds in the Gulf of Alaska and adjacent marine regions. Environmental Assessment of the Alaskan Continental Shelf, Final Rept. 01-5-022-2538. U.S. Dept. Commerce, NOAA, Boulder, Colorado.
Sayce, J. R., and G. L. Hunt, Jr. 1987. Sex ratios of prefledging Western Gulls. Auk 104: 33–37.
Schaffner, F. 1982. *Aspects of the Reproductive Ecology of the Elegant Tern (Sterna elegans) at San Diego Bay*. Unpubl. Master's thesis, San Diego State Univ.
Schlatter, R. P., and C. A. Moreno. 1976. Habitos alimentarios del Cormoran

Antartico, *Phalacrocorax atriceps bransfieldensis* (Murphy) en Isla Green, Antarctica. Ser. Cient. Inst. Antart. Chileno 4: 69–88.

Schneider, D., and G. L. Hunt, Jr. 1984. A comparison of seabird diets and foraging distribution around the Pribilof Islands, Alaska, pp. 86–95. *In* D. N. Nettleship, G. A. Sanger, and P. F. Springer (eds.), *Marine Birds: Their Feeding Ecology and Commercial Fisheries Relationships*. Minister Supply Serv. Can., Ottawa.

Schoener, T. W. 1974. Resource partitioning in ecological communities. Science 185: 27–39.

———. 1986. Patterns in terrestrial versus arthropod communities: Do systematic differences in regularity exist? pp. 556–86. *In* J. Diamond and T. J. Case (eds.), *Community Ecology*. Harper and Row, New York.

Schreiber, R. 1970. Breeding biology of Western Gulls on San Nicolas Island, California, 1968. Condor 72: 133–40.

Schreiber, R., and J. L. Chovan. 1986. Roosting of pelagic seabirds: Energetic, populational, and social considerations. Condor 88: 487–92.

Schreiber, R., and E. A. Schreiber. 1984. Central Pacific seabirds and the El Niño–Southern Oscillation: 1982–1983 perspectives. Science 225: 713–16.

Scott, D. A. 1970. *The Breeding Biology of the Storm Petrel Hydrobates pelagicus*. Unpubl. Ph.D. thesis, Oxford Univ., Oxford.

Scott, J. M. 1971. Inter-breeding of the Glaucous-winged Gull and Western Gull in the Pacific Northwest. Calif. Birds 2: 129–33.

———. 1973. *Resource Allocation in Four Syntopic Species of Marine Diving Birds*. Unpubl. Ph.D. thesis, Oregon State Univ., Corvallis.

Scott, J. M., W. Hoffman, D. Ainley, and C. F. Zeillemaker. 1974. Range expansion and activity patterns in Rhinoceros Auklets. West. Birds 5: 13–20.

Sealy, S. G. 1968. *A Comparative Study of Breeding Ecology and Timing in Plankton-feeding Alcids (Cyclorrhynchus and Aethia spp.) on St. Lawrence Island, Alaska*. Unpubl. Master's thesis, Univ. Brit. Columbia, Victoria.

———. 1972. *Adaptive Differences in Breeding Biology in the Marine Family Alcidae*. Unpubl. Ph.D. thesis, Univ. Michigan, Ann Arbor.

———. 1973a. Adaptive significance of post-hatching development patterns and growth rates in the Alcidae. Ornis Scand. 4: 113–21.

———. 1973b. Interspecific feeding assemblages of marine birds off British Columbia. Auk 90: 796–802.

———. 1981. Variation in fledging weight of Least Auklets *Aethia pusilla*. Ibis 123: 230–33.

Sergeant, D. E. 1951. Ecological relationships of the guillemots *Uria aalge* and *Uria lomvia*. Proc. X Int. Ornithol. Congr., pp. 578–87.

Serventy, D. L. 1938. The feeding habits of cormorants in south-western Australia. Emu 38: 293–316.

Shaughnessy, P. D. 1984. Historical population levels of seals and seabirds on islands off southern Africa, with special reference to Seal Island, False Bay. S. Afr. Sea Fish. Inst., Invest. Rept. 127: 1–61.

Sherman, K., and L. M. Alexander (eds.). 1988. *Biomass and Geography of Large Marine Ecosystems*. Westview Press, Boulder, Colorado.

Siegel-Causey, D., and G. L. Hunt, Jr. 1981. Colonial defense behavior in Double-crested and Pelagic cormorants. Auk 98: 522–31.
———. 1986. Breeding-site selection and colony formation in Double-crested and Pelagic cormorants. Auk 103: 230–34.
Siegfried, W. R., A. J. Williams, P. G. H. Frost, and J. B. Kinahan. 1975. Plumage and ecology of cormorants. Zool. Afr. 10: 183–92.
Simmons, K. E. L. 1972. Some adaptive features of seabirds' plumage types. Brit. Birds 65: 465–79, 510–21.
Skogsberg, T. 1936. Hydrography of Monterey Bay, California. Thermal conditions, 1929–1933. Trans. Am. Philos. Soc., New Ser., 29: 1–152.
Slagsvold, T., J. Sandvik, G. Rofstad, D. Lorentsen, and M. Husby. 1984. On the adaptive value of intraclutch egg-size variation in birds. Auk 101: 685–97.
Slater, P. J. B., and E. P. Slater. 1972. Behavior of the Tystie during feeding of the young. Bird Study 19: 105–13.
Smith, S. E., and P. B. Adams. 1988. Daytime surface swarms of *Thysanoessa spinifera* (Euphausiacea) in the Gulf of the Farallones, California. Bull. Mar. Sci. 42: 76–84.
Snow, B. K. 1960. The breeding biology of the Shag *Phalacrocorax aristotelis* on the island of Lundy, Bristol Channel. Ibis 102: 554–75.
Sokal, R. R., and F. J. Rohlf. 1969. *Biometry*. Freeman, San Francisco.
Southern, W. E. 1987. Gull research in the 1980s: Symposium overview. Studies Avian Biol. 10: 1–7.
Sowls, A. L., A. R. DeGange, J. W. Nelson, and G. S. Lester. 1980. *Catalog of California Seabird Colonies*. U.S. Dept. Interior, Fish Wildl. Serv. FWS/OBS-80/37.
Sowls, A. L., S. A. Hatch, and C. J. Lensink. 1978. *Catalog of Alaskan Seabird Colonies*. U.S. Dept. Interior, Fish Wildl. Serv. FWS/OBS-78/78.
Spear, L. B. 1988a. Dispersal patterns of Western Gulls from Southeast Farallon Island. Auk 105: 128–41.
———. 1988b. The Halloween mask episode. Nat. Hist. Mag. 97: 4–8.
Spear, L. B., D. G. Ainley, and R. P. Henderson. 1986. Post-fledging parental care in the Western Gull. Condor 88: 194–99.
Spear, L. B., T. M. Penniman, J. F. Penniman, H. R. Carter, and D. G. Ainley. 1987. Survivorship and mortality factors in a population of Western Gulls. Studies Avian Biol. 10: 26–43.
Speich, S., and D. A. Manuwal. 1974. Gular pouch development and population structure of Cassin's Auklet. Auk 91: 291–306.
Speich, S., and T. R. Wahl. 1989. *Catalog of Washington Seabird Colonies*. U.S. Dept. Interior, Fish Wildl. Serv. FWS/OBS-88(6).
Spratt, J. D. 1981a. Status of the Pacific Herring (*Clupea harengus pallasi*) resource in California 1972 to 1980. Calif. Dept. Fish and Game, Fish Bull. 171.
———. 1981b. Age and growth of the Market Squid, *Loligo opalescens* Berry, in Monterey Bay from statoliths. Calif. Dept. Fish and Game, Fish Bull. 169: 35–44.
Spring, L. 1971. A comparison of functional and morphological adaptations

in the Common Murre (*Uria aalge*) and Thick-billed Murre (*Uria lomvia*). Condor 73: 1–27.

Squibb, R., and G. L. Hunt, Jr. 1983. A comparison of nesting-ledges used by seabirds on St. George Island. Ecology 64: 727–34.

Steel, R. G. D., and J. H. Torrie. 1960. *Principles and Procedures of Statistics*. McGraw-Hill, New York.

Stenzel, L. E., G. W. Page, H. R. Carter, and D. G. Ainley. 1988. Seabird mortality in California as witnessed through 14 years of beached bird censuses. Unpubl. Rept., PRBO, Stinson Beach, California.

Stirling, D., and F. Buffam. 1966. The first breeding record of Brandt's Cormorant in Canada. Can. Field-Nat. 80: 117–18.

Stonehouse, B. 1967. Feeding behavior and diving rhythms of some New Zealand shags, Phalacrocoracidae. Ibis 109: 600–605.

Storer, R. W. 1945. Structural modifications in the hind limb of the Alcidae. Ibis 87: 433–56.

———. 1952. A comparison of variation, behavior, and evolution in the seabird genera *Uria* and *Cepphus*. Univ. Calif. Publ. Zool. 52: 121–22.

———. 1960. Evolution in the diving birds. Proc. XII Int. Ornithol. Congr., pp. 694–707.

Summers, K. R. 1970. *Growth and Survival of the Rhinoceros Auklet on Cleland Island*. Unpubl. Bachelor's thesis, Univ. Brit. Columbia, Victoria.

Summers, K. R., and R. H. Drent. 1979. Breeding biology and twinning experiments of Rhinoceros Auklets on Cleland Island, British Columbia. Murrelet 60: 16–22.

Sverdrup, H. U., R. H. Fleming, and M. W. Johnson. 1942. *The Oceans: Their Physics, Chemistry, and General Biology*. Prentice-Hall, Englewood Cliffs, New Jersey.

Swann, R. L., and A. D. K. Ramsey. 1979. An analysis of shag recoveries from northwest Scotland. Ringing Migr. 2: 137–43.

Swartz, L. G. 1966. Sea-cliff birds, pp. 611–78. *In* N. J. Wilimovsky and J. N. Wolfe (eds.), *Environment of the Cape Thompson Region, Alaska*. U.S. Atomic Energy Comm., Oak Ridge, Tennessee.

Swennen, C., and P. Duiven. 1977. Size of food objects of three fish-eating seabird species: *Uria aalge, Alca torda*, and *Fratercula arctica* (Aves, Alcidae). Neth. J. Sea. Res. 11: 92–98.

Takekawa, J., H. C. Carter, and T. E. Harvey. Ms. Decline of the Common Murre in central California, 1980–1986. Studies Avian Biol. (In press.)

Thomas, D. H. 1986. Status of California's Bocaccio stock. Calif. Dept. Fish and Game, Menlo Park. (Mimeo.)

Thompson, J. D. 1981. Climate, upwelling, and biological productivity: Some primary relationships, pp. 13–34. *In* M. H. Glantz and J. D. Thompson (eds.), *Resource Management and Environmental Uncertainty*. Wiley, New York.

Thoresen, A. C. 1960. Notes on winter and early spring bird activity on the Farallon Islands, California. Condor 62: 408.

———. 1964. The breeding behavior of the Cassin's Auklet. Condor 66: 456–76.

———. 1969. Observations on the breeding behavior of the Diving Petrel *Pelecanoides u. urinatrix* (Gmelin). Notornis 16: 241–60.

———. 1983. Diurnal activity and social displays of Rhinoceros Auklets *Cerorhinca monocerata* on Teuri Island, Japan. Condor 85: 373–75.

Thoresen, A. C., and E. S. Booth. 1958. Breeding activities of the Pigeon Guillemot, *Cepphus columba columba* (Pallas). Walla Walla College, Publ. Dept. Biol. Sci. and Biol. Sta., No. 23.

Tiburon Laboratory. 1987. Progress in rockfish recruitment studies. Natl. Mar. Fish. Serv., Admin. Rept. T-87-01.

Tovar S., H. 1978. Las poblaciones de aves guaneras en los ciclos reproductivos de 1969/70 a 1973/74. Inf. Inst. Mar. Peru–Callao 45: 1–13.

Tuck, L. M. 1961. *The Murres*. Can. Wildl. Serv., Monogr. Ser. No. 1.

Udvardy, M. D. F. 1963. Zoogeographical study of the Pacific Alcidae, pp. 85–111. *In* J. L. Gressitt (ed.), *Pacific Basin Biogeography*. Tenth Pacific Sci. Conf., Bishop Mus., Honolulu.

———. 1978. *World Biogeographical Provinces*. CoEvolution Quart., Sausalito, California.

Uspenski, S. M. 1958. *The Bird Bazaars of Novaya Zemlya*. [Transl.] Russian Game Rept., Vol. 4. Queen's Printer, Ottawa.

Van Dobben, W. H. 1952. The food of the cormorant in the Netherlands. Ardea 40: 1–63.

Varoujean, D. H. 1979. *Seabird Colony Catalog: Washington, Oregon, and California*. U.S. Dept. Interior, Fish Wildl. Serv., Portland, Oregon.

Vermeer, K. 1963. The breeding ecology of the Glaucous-winged Gull on Mandarte Island, B.C. Occ. Pap. Brit. Columbia Prov. Mus., No. 13.

———. 1969a. Colonies of Double-crested Cormorants and White Pelicans in Alberta. Can. Field-Nat. 83: 36–39.

———. 1969b. Some aspects of the breeding chronology of Double-crested Cormorants at Lake Newell, Alberta, in 1968. Murrelet 50: 19–20.

———. 1970. Arrival and clutch initiation of Double-crested Cormorants at Lake Newell, Alberta. Blue Jay 28: 124–25.

———. 1973. Great Blue Heron and Double-crested Cormorant colonies in the prairie provinces. Can. Field-Nat. 87: 427–32.

———. 1977. Some observations of Arctic Loons, Brandt's Cormorants, and Bonaparte's Gulls at Active Pass, B.C. Murrelet 58: 45–47.

———. 1978. Extensive reproductive failure of Rhinoceros Auklets and Tufted Puffins. Ibis 120: 112.

———. 1979. Nesting requirements, food and breeding distribution of Rhinoceros Auklets, *Cerorhinca monocerata*, and Tufted Puffins, *Lunda cirrhata*. Ardea 67: 101–10.

———. 1980. The importance of timing and type of prey to reproductive success of Rhinoceros Auklets *Cerorhinca monocerata*. Ibis 122: 343–50.

———. 1981. The importance of plankton to Cassin's Auklets during breeding. J. Plankton Res. 3: 315–30.

Vermeer, K., and L. Cullen. 1982. Growth comparison of a plankton- and a fish-eating alcid. Murrelet 63: 34–39.

Vermeer, K., L. Cullen, and M. Porter. 1979. A provisional explanation of the reproductive failure of Tufted Puffins *Lunda cirrhata* on Triangle Island, British Columbia. Ibis 121: 348–54.
Vermeer, K., K. Devito, and L. Rankin. 1988. Comparison of nesting biology of Fork-tailed and Leach's storm-petrels in the Queen Charlotte Islands, British Columbia. Colonial Waterbirds 11: 46–57.
Vermeer, K., and M. Lemon. 1986. Nesting habits and habitats of Ancient Murrelets and Cassin's Auklets in the Queen Charlotte Islands, British Columbia. Murrelet 67: 33–45.
Vermeer, K., and L. Rankin. 1984. Population trends in nesting Double-crested and Pelagic cormorants in Canada. Murrelet 65: 1–9.
Vermeer, K., J. Robertson, R. W. Campbell, G. Kaiser, and M. Lemon. 1983. *Distribution and Densities of Marine Birds on the Canadian West Coast*. Can. Wildl. Serv., Ottawa.
Vermeer, K., R. A. Vermeer, K. R. Summers, and R. R. Billings. 1979. Numbers and habitat selection of Cassin's Auklets breeding on Triangle Island, British Columbia. Auk 96: 143–51.
Volkman, N. J., P. Presler, and W. Trivelpiece. 1980. Diets of pygoscelid penguins at King George Island, Antarctica. Condor 82: 373–78.
Vrooman, A. M., P. A. Palona, and J. R. Zweifel. 1981. Electrophoretic, morphometric, and meristic studies of subpopulations of Northern Anchovy, *Engraulis mordax*. Calif. Fish and Game 67: 39–51.
Wanless, S., D. D. French, M. P. Harris, and D. R. Langslow. 1982. Detection of annual changes in the numbers of cliff-nesting seabirds in Orkney, 1976–80. J. Anim. Ecol. 51: 785–95.
Wanless, S., and M. P. Harris. 1986. Time spent at the colony by male and female guillemots *Uria aalge* and Razorbills *Alca torda*. Bird Study 33: 169–76.
Warheit, K. I., D. R. Lindberg, and R. J. Boekelheide. 1984. Pinniped disturbance lowers reproductive success of Black Oystercatcher *Haematopus bachmani* (Aves). Mar. Ecol. Prog. Ser. 17: 101–4.
Waters, W. E. 1964. Arrival times and measurements of small petrels on St. Kilda. Brit. Birds 57: 309–15.
Watson, G. E. 1968. Synchronous wing and tail molt in diving petrels. Condor 70: 182–83.
Wetmore, A. 1927. The amount of food consumed by cormorants. Condor 29: 273.
Whittam, T. S., and D. Siegel-Causey. 1981. Species interactions and community structure in Alaskan seabird colonies. Ecology 62: 1515–24.
Wickett, W. P. 1967. Ekman transport and zooplankton concentration in the North Pacific Ocean. J. Fish. Res. Board Can. 24: 581–94.
Wiens, J. 1969. An approach to the study of ecological relationships among grassland birds. Am. Ornithol. Union, Monogr. 8.
———. 1977. On competition and variable environments. Am. Sci. 65: 590–97.
———. 1984. On understanding a non-equilibrium world: Myth and reality

in community patterns and processes, pp. 439–57. *In* D. R. Strong, D. Simberloff, L. G. Abele, and A. B. Thistle (eds.), *Ecological Communities: Conceptual Issues and the Evidence*. Princeton Univ. Press, Princeton.

———. 1986. Spatial scale and temporal variation in studies of shrubsteppe birds, pp. 154–72. *In* J. Diamond and T. J. Case (eds.), *Community Ecology*. Harper and Row, New York.

Wiens, J., and J. M. Scott. 1975. Model estimation of energy flow in Oregon coastal seabird populations. Condor 77: 439–52.

Wilbur, H. M. 1969. The breeding biology of Leach's Petrel *Oceanodroma leucorhoa*. Auk 86: 433–42.

Williams, A. J. 1974. Site preferences and interspecific competition among guillemots *Uria aalge* and *Uria lomvia* on Bear Island, Norway. Ornis Scand. 5: 113–22.

Williams, A. J., and A. E. Burger. 1979. Aspects of the breeding biology of the Imperial Cormorant, *Phalacrocorax atriceps*, at Marion Island. Gerfaut 69: 407–23.

Williams, L. 1942. Display and sexual behavior of the Brandt's Cormorant. Condor 44: 85–104.

Winkler, D. W., and J. R. Walters. 1983. The determination of clutch size in precocial birds, pp. 33–68. *In* R. F. Johnston (ed.), *Current Ornithology*, Vol. 1. Plenum, New York.

Winnett, K. A. 1979. *The Influence of Cover on the Breeding Biology of Western Gulls on Santa Barbara Island, California*. Unpubl. Master's thesis, Calif. State Univ., Northridge.

Wolfson, A. 1965. Circadian rhythm and the photoperiod regulation of the annual reproductive cycle in birds, pp. 370–78. *In* A. J. Aschoff (ed.), *Circadian Clocks*. North-Holland, Amsterdam.

Wyllie Echeverria, T. W. 1987. Thirty-four species of California rockfishes: Maturity and seasonality of reproduction. Fish. Bull. 85: 229–50.

Wyrtki, K. 1975. El Niño—The dynamic response of the equatorial Pacific Ocean to atmospheric forcing. J. Phys. Oceanogr. 5: 572–84.

Index

In this index, "f" means a second mention on the next page; "ff" means separate references on the next two pages; two numbers with a dash between them mark a discussion spanning two or more pages; and *passim* denotes separate mentions on three or more pages in close but not necessarily consecutive sequence.

Library of Congress Cataloging-in-Publication Data

Seabirds of the Farallon Islands: ecology, dynamics, and structure of an upwelling-system community / edited by David G. Ainley and Robert J. Boekelheide; contributors, David G. Ainley . . . [et al.].

p. cm.

Includes bibliographical references.

ISBN 0-8047-1530-0 (alk. paper):

1. Sea birds—California—Farallon Islands. I. Ainley, David G. II. Boekelheide, Robert J.

QL684.C2S44 1990

598.29'24'097946—dc20 89-19728

CIP